Federal Ecosystem Management

Federal Ecosystem Management

Its Rise, Fall, and Afterlife

James R. Skillen

University Press of Kansas

Published by the University Press of Kansas (Lawrence, Kansas 66045), which was organized by the Kansas Board of Regents and is operated and funded by Emporia State University, Fort Hays State University, Kansas State University, Pittsburg State University, the University of Kansas, and Wichita State University

Library of Congress Cataloging-in-Publication Data

Skillen, James.
Federal ecosystem management : its rise, fall, and afterlife / James R. Skillen.
pages cm
Includes bibliographical references and index.
ISBN 978-0-7006-2127-9 (cloth : alk. paper)
ISBN 978-0-7006-2164-4 (ebook)
1. Ecosystem management—West (U.S.) 2. Ecosystem management—Yellowstone National Park Region. I. Title.
QH76.5.W34S55 2015
333.720978—dc23

2015020376

British Library Cataloguing-in-Publication Data is available.

Printed in the United States of America

10 9 8 7 6 5 4 3 2 1

The paper used in this publication is recycled and contains 30 percent postconsumer waste. It is acid free and meets the minimum requirements of the American National Standard for Permanence of Paper for Printed Library Materials Z39.48-1992.

For Sam & Ben

Contents

Acknowledgments

This project started as an essay on the integration of scientific and political authority in federal ecosystem management. As the project expanded, so did the list of people that deserve my gratitude. Thank you first to my family: Beth, Sam, and Ben. They do not necessarily share my interest in appropriations riders, judicial rulings, or the vast array of federal acronyms, but they certainly share my interest in public lands throughout the American West. Thank you for all of the time we have spent exploring.

This project took shape with assistance from numerous scholars, public servants, and research assistants. I am particularly grateful to John Freemuth, Bill Rowley, Paul Culhane, and John Nagle for long conversations and for reviewing portions of this manuscript. I am also grateful to many, many current and retired employees from the Forest Service and the BLM, including folks at the Forest Service's Pacific Northwest Research Station and the Public Lands Foundation. For the NWFP chapter, thank you especially to Elaine Zielinski, and for the ICBEMP chapter, thank you especially to Geoff Middaugh. I am grateful for general research assistance from Rachel Stehouwer, Bethany Van Kooten, and Seth Kenbeek, and for general editing assistance from Katie Sauer. Finally, I am grateful to my colleague Melissa Sexton for her extensive editorial comments. All of these people contributed to the project, but none are responsible for any errors or misinterpretations.

This project was made possible by financial support from a number of sources. I am grateful to Calvin College for several research fellowships that provided time for research and writing and to the Mellema Program in Western American Studies for funding travel to various archives. I am also grateful to the Oregon Extension for generous support that provided time for me to finish the manuscript.

Finally, thank you to my children, Sam (11) and Ben (9), who volunteered to make pancakes for dinner one night during the late stages of writing so that I could stay hunched over my keyboard. You may not want to read the whole book, but please at least read the dedication.

Abbreviations

ACS	Aquatic Conservation Strategy
AMA	adaptive management area
APHIS	USDA Animal and Plant Health Inspection Service
BLM	Bureau of Land Management
CCA	Candidate Conservation Agreement
CEQ	Council on Environmental Quality
DSEIS	Draft SEIS
DFWP	Department of Fish, Wildlife, and Parks
EIS	environmental impact statement
EPA	Environmental Protection Agency
ESA	Endangered Species Act of 1973
ESC	executive steering committee
FEMAT	Forest Ecosystem Management Assessment Team
FLPMA	Federal Land Policy and Management Act of 1976
FSEIS	Final EIS
FWS	US Fish and Wildlife Service
GAO	Government Accountability Office
GYC	Greater Yellowstone Coalition
GYCC	Greater Yellowstone Coordinating Committee
GYE	Greater Yellowstone Ecosystem
HCA	Habitat Conservation Area
HCP	habitat conservation plan
IBMP	Interagency Bison Management Plan
ICBEMP	Interior Columbia Basin Ecosystem Management Project
ICRB	Interior Columbia River Basin
IGBST	Interagency Grizzly Bear Study Team
INFISH	Inland Native Fish Strategy
LCAT	Labor and Community Assistance Team
MOU	Memorandum of Understanding
MUSY	multiple use and sustained yield
NAS	National Academy of Sciences
NBS	National Biological Survey
NEPA	National Environmental Policy Act of 1969

NFMA	National Forest Management Act of 1976
NMFS	National Marine Fisheries Service
NPS	National Park Service
NWFP	Northwest Forest Plan
PACFISH	Interim Strategies for Managing Anadromous Fish-Producing Watersheds
PARC	Division of Predator and Rodent Control (US Fish and Wildlife Service)
RCA	riparian conservation area
RIEC	Regional Interagency Executive Committee
REO	Regional Ecosystem Office
ROD	Record of Decision
S&M	Survey and Manage
SBO	supplemental biological opinion
SDEIS	Supplemental Draft EIS
SIT	Science Integration Team
SEIS	Supplemental EIS
USGS	US Geological Survey

Introduction

"It has become apparent that we are witnessing one of the most interesting periods in natural resource management since the heady days of Progressivism."
—John Freemuth, "The Emergence of Ecosystem Management: Reinterpreting the Gospel?" (1996)

"Some have suggested that these differences amount to no less than a radical revision of professional perspectives, values, and management practices—in other words, a paradigm shift."
— Hanna Cortner and Margaret Moote, *The Politics of Ecosystem Management* (1999)

"Is a new era dawning on the western public domain?"
— Robert Keiter, *Keeping Faith with Nature* (2003)

Background

In the 1990s, scholars, policymakers, land managers, and many others were talking about a "new paradigm" for federal lands and resources: ecosystem management. Advocates of this new approach promised that it would yield a host of benefits: integrating federal land and resource management across jurisdictional boundaries; protecting biodiversity *and* economic development; and making federal management more collaborative and less hierarchical. And it could do all of this, if necessary, without any major legislative changes by using the vast discretionary authority that Congress had delegated to the executive branch.

Why did federal agencies need a new paradigm? In brief, ecosystem management proponents argued, the existing paradigms—preservation and multiple-use conservation—could no longer meet the ecological, legal, or political realities the agencies faced. They were neither comprehensive enough nor coherent enough to deal with the complex challenges of managing the federal estate: national park managers could not mitigate external threats on their own; the US Fish and Wildlife Service (FWS) could not keep up with its Endangered Species Act obligations, finding the act's species-based approach expensive and time-consuming; and the Bureau of Land Management (BLM) and the Forest Service could not sustain existing levels of resource development—logging, grazing, and mining—without endless rounds of environmental litigation. As one political scientist had been arguing since the 1960s, "Our national tendency is to deal with environmental problems segmentally [which] has again and again produced some very impractical results."[1]

Proponents offered ecosystem management as a new paradigm that was, above all else, a comprehensive paradigm sufficient to transcend and integrate preservation and conservation. Ecosystem management did not reject preservation and conservation; it simply reframed them in more comprehensive ecological and political terms. Indeed, it gained broad support in the 1990s in part because it was broad enough to afford wide-ranging definitions that appealed to preservationists and conservationists alike. As environmentalist Andy Kerr wrote skeptically, "*ecosystem management* was born of crisis—an ecological crisis that expanded into a political crisis. It was born to appeal to (and to appease) all factions. Environmentalists hear 'ecosystem,' while industrialists hear 'management.'"[2]

In the 1990s support emerged for different models of ecosystem management. Some embraced *ecosystem* management as a new, substantive paradigm that required federal agencies to prioritize biodiversity and ecosystem integrity above other land and resource values, within the limits of existing law. They saw ecosystem management as an ecological reinterpretation of the preservation paradigm. The FWS, for example, embraced ecosystem management as a broader approach to its strict preservation requirements under the Endangered Species Act (ESA) of 1973. Others saw ecosystem *management* as a procedural paradigm that simply enlarged the emphases of multiple-use management to include a wider range of ecological, economic, and social goals. The Forest Service, for example, explained that ecosystem management was simply a more holistic form of multiple-use decision making that emphasized the insights of ecology and accountability from extensive public participation to ensure a sustainable yield of diverse resources.

Academic scholarship in the 1990s also advanced wide-ranging definitions and emphases. Some scholars described ecosystem management as a substantive paradigm that prioritized ecological protection; some described it as a procedural paradigm that resolved tensions between ecological protection and resource development through inclusive deliberation. In 1994, biologist R. Edward Grumbine published one of the most widely cited essays on the topic, entitled "What Is Ecosystem Management?" Surveying the existing literature, primarily in the biological sciences, Grumbine identified ten common themes suggesting that ecosystem management was a substantive paradigm focused on environmental protection: "Ecosystem management integrates scientific knowledge of ecological relationships within a complex sociopolitical and values framework toward the general goal of protecting native ecosystem integrity over the long term."[3] Indeed, Grumbine argued, it served as "an early stage in a fundamental reframing of how humans value nature . . . adjusting management to stave off mass extinctions and habitat destruction will not only help to reduce our negative impact on the biosphere but will also give us the opportunity to reinterpret our place on the planet as one species among many."[4]

In 1999, social scientists Hanna Cortner and Margaret Moote likewise explained in *The Politics of Ecosystem Management* that "the overall goal of ecosystem management is sustaining ecological attributes and functions into perpetuity." However, they emphasized institutional and procedural elements of the approach, arguing that ecosystem management rested on four main principles: "(1) socially defined goals and objectives; (2) holistic, integrated science; (3) adaptable institutions; and (4) collaborative decision making."[5] For scholars, then, interpretations of ecosystem management differed in the relative weight placed on substantive vs. procedural goals. Some advocated for *ecosystem* management; some advocated for ecosystem *management.*

The Clinton administration, which championed ecosystem management in the 1990s, was careful to accommodate these diverse interpretations. For example, the Interagency Ecosystem Management Task Force (IEMTF), which Vice President Gore established in 1993, described ecosystem management variously as a "method for sustaining or restoring natural systems and their functions and values," "a common sense way for public and private managers to carry out their mandates with greater efficiency," "a mechanism for resolving conflicts that protects our national economy and the resources on which it is based," and an approach that "helps build local and regional consensus."[6] One did not need to choose between environmental protection and resource development, the task force suggested; in ecosystem management, one could enjoy both. Furthermore one did not need to choose between substantive and procedural goals; in ecosystem management one could have substantive goals of ecosystem protection and resource development as well as procedural goals of scientific assessment and open, democratic deliberation: "The goal of the ecosystem approach is to restore and sustain the health, productivity, and biological diversity of ecosystems . . . based on a collaboratively developed vision of desired future conditions."[7]

Ecosystem management's broad, or ambiguous, definition was key to its political appeal and enabled eighteen federal agencies with diverse missions, cultures, mandates, and clientele to adopt ecosystem management as official policy in the 1990s. The agencies, ecosystem management proponents argued, had simply focused too narrowly on jurisdictional boundaries, specific resource flows, and expert judgment. If agencies would reframe their work as adaptive and collaborative management of large-scale ecosystems, they could resolve the paradoxes that plagued federal land and resource management. Mike Dombeck, an acting BLM director and later Forest Service chief in the 1990s, wrote with two co-authors, "Sometimes the wrong solutions are applied because the overall goal of land management was mistaken in the first place. If the goal is systemwide—such as restoring the overall health of the watershed—then the proper solution becomes obvious."[8] The larger planning scales would make it easier to find room for both protection and use, and collaboration would help agencies and the public form shared goals and resolve conflicts outside the courts. In short, ecosystem management gained popularity because it promised to turn adversarial, zero-sum conflicts into win-win management opportunities.

Said another way, the Clinton administration relied on the broad and ambiguous meaning of ecosystem management in its promise to resolve two basic tensions or paradoxes embedded in federal land and resource management. First, it promised to balance environmental protection and use, arguing that one did not need to choose between protecting ecosystems and developing their resources. Careful planning and management could produce significant timber, forage, minerals, and recreation while simultaneously protecting ecosystem integrity. Second, it promised that scientific expertise and democratic deliberation could work harmoniously in federal land and resource management. These two sets of commitments—protection and use, and what here will be referred to as scientific authority and democratic authority—certainly do not always conflict. Some management practices, such as a timber harvest designed to enhance wildlife habitat, can advance ecological protection while producing resources. And sometimes expert scientific judgment about how best to protect ecosystems aligns with the majority outcome of public deliberation. What the Clinton administration avoided addressing in any final sense was how it would deal with unavoidable trade-offs between environmental protection and use, and whether it would privilege expert scientific judgments or public opinion when the two conflicted. In other words, the administration avoided the choice between a substantive model of ecosystem management as science-driven ecological protection and a procedural model of ecosystem management as a collaborative and participatory process, informed by scientific information, in which the substantive goals were truly open to debate.

The Clinton administration touted significant successes in ecosystem management, from the Chesapeake Bay to the northwest forests, arguing that the new paradigm had broken through gridlock of the past and opened a new era in federal land and resource management. But critics on both sides expressed disappointment with the trade-offs that emerged in ecosystem management projects. Those who supported *ecosystem* management expressed frustration at compromises between environmental protection and resource production. Those who supported ecosystem *management* complained that instead of providing a genuinely balanced approach, the administration had simply replaced the conservation paradigm with an ecologically oriented preservation paradigm. Ecosystem management, they argued, subordinated human needs to those of other species, and instead of creating win-win solutions, it simply ensured that environmental interests won more often.[9]

When President George W. Bush took office in 2001, those favoring ecosystem *management* found a more sympathetic administration. The new administration rejected the substantive principle of protecting ecological processes and functions over resource outputs and expressed concern about the principle of managing along ecological rather than political boundaries, but it celebrated the principle of adaptive, collaborative management. Interior Secretary Gale Norton, for example, promised over and over again to use "Consultation, Communication, and Cooperation, all in the service of Conservation." The Bush administration essentially worked to reframe ecosystem management in the rhetoric of conservation as efficient use, guided not by scientific or administrative expertise but by cooperation with those who lived and worked on the public lands. The shift prompted a number of scholars to publish essays during the Bush administration with titles such as "Whatever Happened to Ecosystem Management?" and "The U.S. Forest Service: Whither the New Resource Management Paradigm?"[10]

Such is the risk of administrative initiatives, which can shift substantially from one president to the next. In the 1990s, with ecosystem management on the ascendency, legal scholars such as Robert Keiter pointed to this vulnerability. In 1996, Keiter wrote, "Within our democratic system, Congress is the ultimate policy-making institution. . . . Until Congress speaks, ecosystem management can only claim a tenuous legitimacy, which also leaves the concept undefined for legal purposes."[11] Congress refused to pass ecosystem management legislation, partly because Republicans who controlled Congress beginning in 1995 generally opposed the Clinton administration's public land reform initiatives and partly because Congress could not agree on a legal definition of ecosystem management. Senator Mark Hatfield (R-OR) encountered this in 1994 when he introduced an unsuccessful bill that would have given the BLM a procedural ecosystem management mandate.[12] As a result, the paradigm that took shape in the 1990s never solidified as a controlling framework for the federal estate.

Argument

An enormous amount of literature has addressed ecosystem management over the last twenty years. Most of that literature has been in the natural and social sciences and has been preoccupied with questions about its definition, predictive models of its success or failure, and empirical evaluations of its impacts. Far less has been written in the humanities, though many of the exceptions are notable.[13] This project takes the humanistic road less traveled, providing a selective history of federal ecosystem management. It does not answer the question "What is ecosystem management?" in a categorical or generalized sense, nor does it evaluate the merits of federal ecosystem management efforts in the 1990s over and against other approaches. Rather, it explains why ecosystem management emerged in the 1990s as official federal policy, how federal environmental law shaped two very different models of ecosystem management, and what remained of ecosystem management during the first decade of the twenty-first century; and it does so through two interrelated arguments.

The first argument draws on the intellectual history of ecology and political science. Both fields went through profound shifts in the middle of the twentieth century, from models that emphasized stability and predictability to models that emphasized dynamism and contingency. The earlier models had promised that agencies could achieve a steady balance between environmental protection and use; the newer models suggested that a fixed balance was illusory. The earlier models had promised managerial control based on scientific and administrative expertise; the later models suggested a far more modest project of managerial coping, informed by scientific expertise but guided by direct public participation. Ecosystem management gained popularity because it reinterpreted the project of federal land and resource management in light of these new ecological and administrative models.

The second, and closely related, argument focuses on the multiple interpretations of ecosystem management—substantive and procedural—and their emphases on scientific and democratic authority. Both science-driven ecosystem protection and open public deliberation were emphasized by the environmental movement of the 1960s and 1970s, and both were written into various environmental laws. When adequate consensus emerged on specific environmental protection goals, Congress passed statutes such as the ESA that established clear environmental protection goals and established science as the primary source of decision-making authority. When adequate consensus was lacking or the issues were too complex to establish clear goals, Congress passed statutes such as the Federal Land Policy and Management Act (FLPMA) of 1976 and the National Forest Management Act (NFMA) of 1976 that required agencies to gather scientific information but to use a participatory process for prioritizing substantive goals. Ecosystem management also emphasized both science-driven ecosystem protection and open public deliberation, promising to integrate them more harmoniously and consistently. But without significant legislative changes, two very different models of ecosystem management emerged under federal law: one substantive, the other procedural.

Substantive ecosystem management was associated with and driven by the ESA and, for the Forest Service, select provisions of NFMA. These laws gave the agencies predetermined goals and priorities, whether it was the requirement that all federal agencies protect listed threatened and endangered species and their habitat or the requirement that the Forest Service maintain viable populations of all vertebrate species. In the implementation of these laws, scientists became key policymakers, and given the scientific emphasis on ecological complexity, dynamism, and uncertainty in ecological systems, scientists placed enormous constraints on resource development programs. Ecosystem management implemented through the ESA certainly included public participation and deliberation, but primarily as a means of gathering and disseminating information and trying to convince constituencies to support agency actions. This is the model that conservation biologists and environmentalists insisted was the real ecosystem management, namely a model in which ecosystem protection is the prerequisite for all other management actions.

Procedural ecosystem management was associated with and driven by the National Environmental Policy Act of 1969 (NEPA). The stated purpose of the act is to "encourage productive and enjoyable harmony between man and his environment," but the judicially enforceable requirements are procedural. As the courts have interpreted NEPA, its binding requirement is that agencies, using "a systematic, interdisciplinary approach," prepare a comprehensive environmental impact statement before undertaking any "major Federal action significantly affecting the quality of the human environment."[14] Ecosystem management implemented under the NEPA model required agencies to gather all of the relevant scientific information, consider all of the potential environmental impacts of their proposed action, and consider all of the management alternatives available. The agencies were then to use public participation to help them make management decisions that best met public needs, whether or not they privileged what scientists defined as ecosystem integrity. This is essentially the model that the multiple-use agencies and resource extraction industries said was the real ecosystem management. It was an open process of rational decision making in which trade-offs were genuine political calculations.

By the end of the Clinton administration, both models of ecosystem management had been tested. In some cases, the ESA had driven substantive ecosystem management; in other cases, NEPA had supported procedural ecosystem management. The former generated enormous opposition, with critics complaining that the ecosystem management paradigm would invariably expand federal land and resource preservation at the expense of resource use. Thus, while the Bush administration never renounced ecosystem management as federal policy, it emphasized procedural ecosystem management and worked effectively to block substantive ecosystem management applications.

Despite ecosystem management's decline during the Bush administration and now the Obama administration, it has not disappeared entirely. The basic challenges that prompted ecosystem management and the basic principles that ecosystem management proponents advanced have enduring importance for federal land and resource management. The fact that Congress has not provided any significant new guidance for federal land and resource management means that federal agencies are still caught by two dominant paradoxes: preservation and use; scientific authority and democratic authority. The fact that scholarship and public opinion continue to highlight the inadequacy of the older preservation and conservation paradigms means that federal agencies cannot simply fall back on their earlier sense of mission. Ecology and conservation biology continue to emphasize the complexities and uncertainties of ecological systems, particularly in light of climate change. Political science and public opinion continue to emphasize the need for democratic accountability and for agencies to cooperate across political boundaries. Indeed, as recent developments in collaborative management and partnerships suggest, federal agencies are actively looking for more coherent and integrated ways to address the paradoxes of federal land and resource management. And many of the principles advanced under the banner of ecosystem management in the 1990s continue to influence federal land and resource management in the twenty-first century, albeit prioritized in different ways. The need continues for management that follows ecological boundaries, that protects large-scale ecological processes and functions, and that encourages adaptation and collaboration. What remains unclear is whether or not public land law and politics can address this need in a consistent and effective way.

The chapters that follow are divided into two relatively discrete sections. The first three chapters explain the interconnected roots of ecosystem management: intellectual, political/legal, and practical. Federal ecosystem management was not a developed paradigm, theory, or framework that federal agencies took out of a box and implemented systematically across the federal estate. Rather, ecosystem management was a banner under which federal agencies gathered a variety of emergent concepts and tools, and its application was shaped directly by specific federal laws and specific management problems.

Chapter 1 explains the intellectual roots of ecosystem management in the fields of ecology and public administration/public policy. The four federal land agencies were all created in the first half of the twentieth century, when the models of ecological and political systems emphasized comparatively static or predictable dynamics. The concepts of ecosystem management—adaptive management, range of variability, population viability analysis, collaborative management, public participation, etc.—reflected a dramatic shift that took place in both ecology and public administration/public policy in the mid- to late twentieth century. The new ecology and the new public administration eschewed the earlier emphasis on stability and detailed predictability, assuming instead that ecological and political systems are complex, dynamic, and fundamentally unpredictable. These assumptions called for dramatic reform of federal land-use planning and management. In particular, scholars called for a shift away from fixed resource production goals and static administrative practices. Federal agencies adopted ecosystem management in part because it promised a complex and flexible approach that better matched ecological and political systems. This is not to say that federal agencies adopted ecosystem management simply because the *ideas* it offered were compelling; these ideas were but one of the roots of federal ecosystem management.

Chapter 2 explains the political and legal roots of ecosystem management, largely in the 1960s and 1970s. These roots are certainly related closely to the intellectual roots, since the American environmental movement that emerged in the 1960s and shaped the political and legal landscape reflected an ecological perspective on the natural world, understanding it not as a collection of resources or objects but as complex webs of interactions and relationships. But this environmental movement was a social and political movement rather than a scientific movement, so these social and political realities require a separate chapter of discussion. As environmental historian Thomas Dunlap explains, the environmental movement was Newton's disciple and Emerson's child.[15] In other words, the environmental movement merged a scientific description of nature with a Romantic understanding of its meaning and value. The laws that Congress passed in response to the environmental movement utterly transformed the federal role in environmental management. Environmentalists demanded both specific protection measures and a comprehensive environmental policy, and Congress responded with wide-ranging statutes such as the Wilderness Act of 1964, NEPA (1969), the Clean Air Act amendments of 1970, the Clean Water Act amendments of 1972, the ESA (1973), etc. Congress also updated federal land and resource policy as well in statutes such as NFMA (1976) and FLPMA (1976).

Congress could not and did not appreciate the full implications of these new, legally binding commitments. And as federal agencies implemented them in the context of federal land and resource management, they discovered just how demanding meeting these requirements was. Thus, chapter 2 explains the growing tension for federal land and resource agencies between their original mandates to produce and protect resources—economic and recreational—with the new, ecologically informed requirements for environmental protection. They adopted ecosystem management in part because it promised that these were not, in fact, conflicting demands and that they did not need to choose between them.

By focusing on Yellowstone National Park and the surrounding landscape, chapter 3 shows how new scientific ideas and new environmental laws worked together to drive a more ecologically oriented approach to federal land and resource management. As the National Park Service shifted from its initial zoological model of wildlife management, in which it worked to ensure that visitors could enjoy wildlife exhibits, to a more ecologically informed model of protecting both wildlife and the natural processes that sustain it, it increasingly argued that park boundaries were insufficient for the task. Passage of the ESA in 1973 and listing of the grizzly bear drew new planning and management boundaries around all of the federal lands in what the agencies called the Greater Yellowstone Area or Greater Yellowstone Ecosystem. The Park Service, the Forest Service, and the FWS were forced, with mixed reactions, to collaborate on endangered/threatened species protection across their various jurisdictional boundaries. This led in 1990 to what could be considered the first large-scale federal ecosystem management plan, which the Bush administration rejected as a potential extension of the ESA preservation model to the entire Yellowstone ecosystem. The Republican-controlled Congress in the 1990s reiterated this fear and fought the Clinton administration's efforts to make ecosystem management the overarching paradigm for the federal estate.

The second section of the book, chapters 4 through 7, provides a political history of federal ecosystem management in the late 1980s and 1990s, as it grew out of the roots described in chapters 1 through 3. As with previous administrative reforms, ecosystem management was an effort to renegotiate the complex terms of federal land and resource management within the limits of existing statutes and political will. The thousands of statutes governing the federal estate do not add up to a clear formula for prioritizing management goals, and they lead to various contradictions and tensions for each agency. Ecosystem management was a framework for addressing some of the tensions most pressing on federal agencies at the end of the twentieth century.

Chapter 4 explains how the stage was set for ecosystem management during the George H. W. Bush administration, 1989–1993. Vice President Bush campaigned in 1988 with the promise that he would be "the environmental president," and he took office with the explicit promises to stop wetlands destruction, address climate change, and reduce acid rain. He also took office with staunch commitments to the Reagan administration's war on federal regulation and to economic growth as well as with a promise not to let environmental protection slow economic productivity. The administration demonstrated that it was committed to environmental protection, but it disagreed with scientists, the courts, and environmentalists about what this required. The president wanted environmental policy that was predictable, was contained in discrete initiatives, and, above all, did not restrict private property rights and economic development. Ecology and conservation biology both indicated that this desire was unrealistic when dealing with complex ecological systems. The standoff led both to the northern spotted owl crisis and to the surprise twist in which the Forest Service officially adopted ecosystem management in 1992.

Chapter 5 explains the broad adoption of and implementation of ecosystem management during the Clinton administration. In contrast to the Bush administration, the Clinton administration embraced the expansive perspective of ecological science, arguing that it was less of a threat to economic interests and more of a tool that would balance environmental protection and resource development. The administration embraced the broad, somewhat ambiguous banner of ecosystem management, offering it both as a means of science-driven ecosystem protection and as a collaborative, consensus-driven process to establish management goals and carry them out. This pragmatic view of ecosystem management led to widespread experimentation.

Chapters 6 and 7 explain two of the largest and most ambitious ecosystem management projects undertaken during the 1990s: the Northwest Forest Plan (NWFP) and the Interior Columbia Basin Ecosystem Management Project (ICBEMP). These two projects are important because they show the differences between a substantive ecosystem management project driven by the strict requirements of the ESA and NFMA and a procedural ecosystem management project driven by the procedural requirements of NEPA. The NWFP was created quickly with limited political deliberation, since it was created to address a court injunction that had shut down the federal timber program in western Washington, Oregon, and northwestern California. Drafted by scientists, the challenge was not creating the plan but implementing it over the next decade. ICBEMP, by contrast, was initiated partly to avoid the kind of crisis that had necessitated the NWFP. It did not begin with clear substantive goals but with a commitment to develop them in a true collaboration between scientific research and political deliberation. What followed was essentially a decade of deliberation in which the parties could not reach consensus, and ICBEMP ended without an ecosystem management plan.

At the end of the Clinton administration, support for and opposition to ecosystem management was high. Some projects had been consensus-driven environmental victories, some had been contentious efforts driven by the courts, and some had ended without decisive action. The George W. Bush administration did not attack ecosystem management as such. Rather it attacked what it argued was anti-democratic rule by value-laden science, and it promised to return federal land and resource management to a more collaborative political posture that favored the interests of those whose livelihood depended on resource development. Thus, ecosystem management remained official federal policy for the agencies, but the substantive/procedural ambiguity was resolved. The Bush administration reduced ecosystem management to a NEPA-like process without enforceable substantive goals of environmental protection.

While ecosystem management never replaced preservation and conservation as a single, integrated paradigm for all of the federal estate, some of its central tenets still influence how agencies carry out their preservation and conservation responsibilities. The sciences of ecology and conservation biology still emphasize the complexities and irreducible uncertainties involved in ecological systems and still push federal agencies toward adaptive management that seeks to cope with ecological change rather than control ecological systems in any final sense. While resistance remains, these scientific insights have grown more rather than less persuasive, particularly in light of the uncertainties surrounding global climate change; federal agencies continue to find that individual success requires collaboration across jurisdictional boundaries, both interagency collaboration and cooperation with private and state landowners; and negative forces such as budget cuts and high costs of wildland fire suppression continue to force agencies to look for holistic management approaches that can achieve multiple goals at lower costs. In some ways, then, the banner of ecosystem management has fallen, but the concepts and tools of ecosystem management still influence federal land and resource management today. Undoubtedly, some future administration or Congress will attempt to repackage these tools under another banner and provide the federal estate with more integrated, holistic, and coherent management direction.

I | The Roots of Federal Ecosystem Management

1 | The Intellectual Context of Ecosystem Management

Introduction

Ecosystem management emerged in the 1990s as a framework for integrating the existing preservation and multiple-use conservation paradigms, which critics said were outdated scientifically and politically. Though ecosystem management lacked a precise definition, many proponents emphasized common themes: managing along ecological rather than political boundaries, maintaining ecological processes and functions rather than resource outputs, and building adaptive and collaborative management processes. These themes reflected new scholarly and popular interpretations of land and resource management and posed direct challenges to the earlier paradigms. Indeed, they reinterpreted sources of scientific and political authority, as distinct from legal authority, to which federal agencies appealed in their planning and management activities.

To understand the political history of ecosystem management, then, it is essential first to consider reinterpretations of the scientific and political models to which federal agencies appealed. Rather than offering an encyclopedia of definitions, this chapter traces developments in the academic fields of ecology and political science during the twentieth century. Changes in both fields were critically important for the rise of ecosystem management, though they were important in different ways. Ecology shaped federal management directly, both because many land and resource managers had formal training in the natural sciences and because Congress established explicit requirements in statutes such as the Endangered Species Act (ESA) and the National Environmental Policy Act (NEPA) for agencies to incorporate the best available science in their decision making. Thus, federal agencies appealed directly to ecology and related scientific fields as sources of authority to legitimize their decisions. Political science, particularly public administration and public policy, did not have the same direct and controlling impact on federal land and resource management. It more often reflected than led substantial shifts in the public perceptions of, and confidence in, federal land and resource agencies. Federal agencies appealed to political processes, such as public participation in land and resource planning, as sources of authority to legitimize their decisions.

Ecology and political science serve as sources of authority for federal land agencies in two important respects. First, they provide theoretical and empirical explanations of how ecological and political systems work. In particular, they identify the determinative factors that make these systems potentially predictable for the express purpose of control. Second, they provide prescriptive principles for ecological and political systems, explaining how these systems ought to be organized and managed.

So it is significant that both ecology and political science shifted dramatically in the second half of the twentieth century in ways that made prediction and prescription more difficult. The ecological and political sciences that supported preservation and conservation in the first half of the twentieth century promised to provide federal agencies with clear predictive and prescriptive knowledge, justifying substantial administrative autonomy. But ecologists and political scientists in the second half of the twentieth century argued that their fields had promised more than they could deliver. Scholars in both fields increasingly rejected the idea that communities/institutions are closed and stable systems (descriptive) that should be maintained in some static form (prescriptive). Instead, they described communities/institutions as open, dynamic systems that shape and are shaped by a wide range of external environmental factors (descriptive), and they advocated managing them so as to maintain their dynamism, variability, and resilience (prescriptive). The new paradigms in ecology and political science pushed the rhetoric and, to some degree, the practices of federal agencies from an emphasis on autonomous, professional bureaucracies producing optimal flows of particular natural resources to an emphasis on participatory and collaborative administration that maintained things like the "natural range of variability" within ecological and political systems.

In the 1990s, federal agencies adopted ecosystem management precisely because it acknowledged scientific and political uncertainty and promised to integrate scientific and political authority in one comprehensive framework.[1] The rhetoric of ecosystem management, in other words, redefined the very project of land and resource management in ways that were consonant with prevailing ecological and political ideas explained in this chapter.

Ecology

Early Twentieth Century

American ecology emerged at the end of the nineteenth century as a way to explain changes in plant communities over time, particularly after a disturbance. It was dominated by Henry Cowles and Frederic Clements, who developed a process-oriented science referred to as "dynamic ecology." Both botanists theorized that plant communities changed predictably through succession as a response to soil, climate, and competition, and both botanists viewed plant communities as fundamental units of study. It was within this framework that they reconciled the observed complexity of plant communities with their stability and predictability.[2]

Of the two men, Cowles emphasized the complexity of plant communities to a greater extent, writing in 1901 that "succession is not a straight-line process. Its stages may be slow or rapid, direct or tortuous and often they are retrogressive." Succession is, he wrote, "a variable approaching a variable."[3] Clements, by contrast, viewed plant communities as complex organisms that go through predictable, developmental changes that lead inexorably to a stable climax community.[4] He theorized that after a disturbance eliminated the climax community —what he called nudation or retrogression—a successional process would ultimately return the landscape to its predetermined climax community. As historian William Rowley puts it, "Few boast to have a key to the future, but Clementsian plant science could clearly make bold claims to it. A successional process once set in motion would predictably lead to an envisioned result."[5]

It was Clements's teleological theory that came to dominate ecology during the first half of the twentieth century. Clements published his basic succession/monoclimax theory in 1905, and in 1913 he began applying it to questions of federal range management.[6] Later, against the devastating backdrop of the Dust Bowl, Clements argued that ecology could provide "a point of view and a plan of attack" for achieving optimal forage and crop outputs within the land's fixed ecological limits.[7]

The value of Clementsian theory for range management was clear, and it was nothing less than the promise of control over nature. Within this theory of succession, each area of rangeland had a single climax plant community, and succession proceeded at a relatively steady pace toward that community. Grazing was viewed as a counter-successional pressure that could "be made equal and opposite to the successional tendency, producing an equilibrium in the vegetation at a set stocking rate. A sustainable yield of livestock products [could] be harvested from such an equilibrium."[8]

Clementsian ecology was firmly embedded in range science by the 1930s and persisted largely unchallenged for several decades.[9] Indeed, by 1950 the standard view of range health among federal range managers was that it could be measured as the degree of variation from the climax plant community. And since the climax community was the most stable but not the most productive for grazing, the goal of management was to maintain the western range in a subclimax state that produced the most valuable forage for livestock. After all, allowing range plant communities to move to the climax would be a waste of resource potential, particularly with fire and disease constantly threatening the range.[10]

Clements focused on range science, but silviculture also developed under a Clementsian influence in the early twentieth century. Silviculturalists accepted the basic idea that plant communities developed through a series of successional stages, leading ultimately to a climax forest. As with range science, silviculturalists determined that maximizing timber production required keeping forests from reaching the climax stage and maintaining them in the subclimax stage in which trees grew most rapidly.[11] This thinking contributed by the 1960s to shifts in the US Forest Service's timber program from selective harvesting to clearcutting, since it kept the forest in a vigorous early stage of succession and allowed the agency to select a monoculture of the most desirable species.

For both range science and silviculture, then, Clementsian ecological theory supported a simple model of sustained yield. By focusing primarily on two variables—vegetative growth rates and vegetation removal through logging or livestock grazing—agencies could promise to deliver the maximum, perpetual supply of resources. The model of ecological balance or equilibrium also supported a policy of forest protection that included little to no room for natural disturbances. Fires, insects, diseases, etc., were all exogenous threats to the forest that would waste timber resources prior to harvest, and the Forest Service in particular set out to eliminate these forces to the extent practicable.

This is not to say that Clementsian ecology went unchallenged. The British botanist Arthur Tansley supported Clements's holistic perspective, but he rejected the teleological assumptions embedded in his organism metaphor. In particular, he rejected Clements's assertion that each climatic region had a single climax plant community or association (monoclimax theory), and he rejected the notion that succession could only occur in a single direction toward that climax. However, Tansley shared a broader conviction with Clements, namely that a climax community was "a permanent or apparently permanent condition reached when the vegetation is in equilibrium with all the incident factors."[12] Tansley offered his own alternative concept for understanding succession in 1935: an "ecosystem." This concept was, for Tansley, more expansive than the organism concept in that it emphasized the totality of biotic and abiotic relationships: "In an ecosystem the organisms and the inorganic factors alike are *components* which are in relatively stable dynamic equilibrium. Succession and development are instances of the universal processes tending towards creation of such equilibrated systems."[13]

The more fundamental challenge to Clementsian theory came from American botanist Henry Gleason. Gleason had started with a Clementsian view of plant communities, but his field research and rudimentary statistical analysis did not support Clements's holistic view. He simply could not find any empirical evidence or statistical analysis to support the idea that plant communities developed in a predetermined and self-organizing fashion. Variation in plant associations, Gleason argued, was determined by environmental factors at a specific site and the plant species available after disturbance. Plant communities were essentially artifacts of history, not of biological predestination.[14] They were, he argued, governed by the laws of probability, not by some metaphorical genes encoded in a particular association of plants. When he published his critiques in 1917, 1926, and 1939, plant taxonomists who worked with an individualistic lens were sympathetic, but ecologists rejected it outright.[15]

1950s and 1960s

In the 1950s, ecologists took a second look at both Tansley's ecosystem concept and Gleason's individualistic perspective, and Clements's organismic theory, though not his ecological holism, faded with a few prominent exceptions.[16] Ecosystem ecology integrated Tansley's work with other areas of systems thinking, affirming Tansley's insight that ecosystems were fundamental units of study. Whereas Clements had studied plant communities as organisms, ecologists such as George Hutchinson, Eugene Odum, and Thomas Odum studied ecosystems as machines with complex circuitry that provided feedback mechanisms and self-correction over time. Whereas Clements had looked to individual organisms for his models, the new systems ecologists looked to guided missiles and early computers for inspiration.[17] Population ecologists, by contrast, focused on dynamic population levels and were rooted more directly in evolutionary theory. They confirmed Gleason's conclusion that populations were governed as much by the contingencies and probabilities of history as by genetics and other predictable factors.

While these two streams of ecology—ecosystem and population—differed significantly in their models, they shared a number of common traits, most notably a growing emphasis on mathematical rigor, which they viewed as the key to carrying ecology beyond its descriptive roots in natural history to a more powerful, predictive science. "Mathematization," one ecologist wrote in 1940, "is to be regarded as a normal phenomenon in the course of the development of any science."[18] More specifically, mathematics provided a bridge between the individualistic perspective of Gleason and the holistic perspective of Clements and Tansley. By representing individuals or individual relationships mathematically, ecologists could start with an individualistic emphasis yet still explain how whole populations or ecosystems were greater than the sum of their parts.

Ecosystem ecology emerged in the 1950s out of systems thinking, including cybernetics. George Hutchinson at Yale University led this movement in the late 1940s, but the rise of ecosystem ecology is most often associated with Eugene and Thomas Odum. Indeed, the textbook that Eugene Odum published in 1953 and that the two brothers coauthored through numerous editions, *Fundamentals in Ecology*, trained a generation of ecosystem ecologists.

As Eugene Odum described it, ecosystem ecology of the 1950s and

1960s was revolutionary. Influenced by geochemistry as well as biology, it took a functional view of ecosystems, emphasizing the movement of matter and energy among their constituent parts. Hutchinson and the Odum brothers argued that nutrient flows were circular within an ecosystem and that this provided critical feedback loops that stabilized the system: "The new ecology," Odum wrote in 1964, "is thus a systems ecology . . . the new ecology deals with the structure and function of levels of organization beyond that of the individual and species."[19] An ecosystem was a system in which myriad components functioned individually in great chains, particularly chains of energy transfer. The goal of the new ecology was to quantify these flows, at least at an abstract level. At the same time, the Odums assumed that many ecosystems exhibited stability, homeostasis, and self-regulation, emergent properties that were not explained entirely by the behavior of individual components.[20] By applying Tansley's ecosystem concept, modeled after a complex machine and understood through detailed mathematical calculations, they carried Clements's holism through a reductive and quantitative reform.[21]

The systems ecology that the Odums advanced trained not only ecosystem ecologists in the 1950s and 1960s but the public as well. In the atomic age, when real and potential human impacts on the environment appeared catastrophic, an ecology of equilibrium found a receptive audience. The Odums argued that nature was balanced and self-regulating and that ecology provided tools with which scientists could measure how much human population growth and technology were disrupting nature's balance. In particular, it could highlight and measure the degree to which humans were disrupting the fundamental processes that produced ecological stability, and it could guide the restoration of those processes. Popular understandings of ecosystem ecology then drove Congress in the 1960s and 1970s to pass sweeping environmental legislation designed to protect, as NEPA puts it, "productive and enjoyable harmony between man and his environment."[22]

While ecosystem ecology redefined holism, population ecologists working with Gleason's individualist focus offered a more foundational critique of Clements's work as metaphor-rich natural history rather than rigorous science. As they saw it, empirical tests and mathematical population models falsified Clements's concept of a monoclimax plant association completely. Robert Whittaker, for example, wrote in 1951 that earlier methods of testing the association hypothesis assumed its validity from the outset, which is why Clements did not find counterevidence.

In the Smoky Mountains, Whittaker ran transects along various environmental gradients—moisture and elevation—specifically testing the association hypothesis and finding it wanting. What he found was "a complex continuum of plant populations," marked by complexity and variability rather than a metaphorical organism marching toward a fixed climax.[23] Whittaker's was an ecology built on evolutionary biology and Mendelian genetics rather than on teleological assumptions.[24]

1970s, 1980s, and 1990s

The transition in ecology, both ecosystem ecology and population ecology, accelerated rapidly in the 1970s based primarily on growing computation power. Field ecologists could gather and analyze vastly greater data sets, and their research reinforced Gleason's conviction that ecological patterns were not entirely predictable. Ecologist Robert Colwell wrote in 1984, "The experimental evidence does not often reveal the well-ordered, equilibrium world of MacArthur and Levins [or Odum], but rather, a sometimes turbulent and bewildering place where disturbance, natural enemies, biochemistry, life histories, and behavior play leading roles, along with the original cast of competitors."[25] Ecological patterns were, at best, probabilistic.

Ecologists once again claimed that a "new ecology" had emerged, but unlike the new ecology that Eugene Odum had declared, this had little room for ideas of balance and homeostasis at the ecosystem scale. "Economy-of-nature, and balance-of-nature," Robert McIntosh wrote in 1980, "are excellent catch phrases, and strong on metaphorical (and I would add, normative) content, but have been difficult to transform into theory in ecology."[26] It was this "new ecology" that historian Donald Worster has referred to as the discovery of disorder, and it was this ecology that Robert Colwell described as ecology finally discovering evolutionary biology.[27]

Ecosystem ecologists remained committed to general systems theories, but their work increasingly emphasized dynamism, complexity, disequilibrium, and chaotic behavior at the ecosystem scale rather than the equilibrium or homeostasis that the Odums had initially found.[28] On the one hand, they continued to see order and pattern in ecosystems for which they sought causal explanations.[29] On the other hand, Frank Egler concluded, an ecosystem "is not only more complex than we think. It is more complex than we can think."[30]

To reconcile observations of order and complexity, ecologists focused on hierarchical organization with complex systems, drawing from general systems theorists such as political scientist Herbert A. Simon. He argued that complex systems, whether "artificial" systems such as computer programs and bureaucracies or ecosystems, had large numbers of individual parts interacting in nonsimple ways with feedback systems that enabled them to adapt to changing conditions.[31] Though all individual parts of a system might relate to one another, Simon and other theorists explained, the strength of those relationships varies. Strong sets of relationships cluster together hierarchically within a complex system, such that the system as a whole is comprised of various subsystems.[32] One of the metaphors that Simon used was that of Chinese boxes or Russian dolls, except that in this case, "opening any given box in a hierarchy discloses not just one new box within, but a whole small set of boxes."[33]

Hierarchy theory had important implications for the study of ecology.[34] If one accepted that this was an accurate description of ecological systems, then the key to understanding them was not to understand and predict every relationship but to identify the most important clustering of relationships analogous to the boxes in Simon's metaphor. Or as ecologist Richard Levins explained it, "the general strategy for analyzing complex systems is through some decomposition into subsystems and their interactions."[35]

Throughout the 1970s, 1980s, and 1990s, a number of paths emerged in ecology, and in each one ecologists struggled with the central problems of pattern and scale and with problems of relating ecological parts and wholes. What held them together was a common embrace of complex systems theory that recognized both irreducible complexity and sufficient organization and feedback mechanisms to allow for ecological adaptation and persistence. On the one hand, Simon Levin writes, interactions among organisms and their environment "can result in temporally constant spatially nonuniform patterns, or spatially constant temporally nonuniform ones, or spatiotemporal mosaics."[36] On the other hand, ecosystems demonstrate a "hidden order" that can be expressed in probabilistic terms.[37]

Understanding ecosystems probabilistically might have been a step forward for ecology, but it was difficult to apply in the legal and political context of federal land and resource management. Indeed, land and resource law, including new environmental legislation in the 1970s, reflected popular beliefs in nature's stability and balance more than the cutting edge of complex systems theory; and federal land and resource agencies promised sustained resource outputs more than they promised to continually adjust land uses to complex changes in ecological systems.

One of the leading figures working in the 1970s, 1980s, and 1990s to bridge the gap between science on the one hand and law and politics on the other was C. S. Holling. He and his colleagues argued that federal land and resource management was rooted in an outdated scientific framework. "Our traditional view of natural systems," he wrote, "might well be less a meaningful reality than a perceptual convenience. . . . An equilibrium centered view is essentially static and provides little insight into the transient behavior of systems that are not near the equilibrium. Natural, undisturbed systems are likely to be continually in a transient state."[38] Using this outdated framework, federal agencies continued to treat their lands and resources as static and predictable, setting them up for management failure.

Holling and others argued that federal agencies needed to reframe their planning and management around the vastly more complex realities of ecological systems. They needed to stop looking for stability and balance in the ecosystems they managed and look instead for the remarkable fact that key ecological characteristics persist *despite* routine disturbance and change. In other words, they persisted in a state of disequilibrium. Holling called this persistence "resilience." A resilient system was one that could absorb disturbance and change without losing its fundamental characteristics: patterns, processes, functions, etc.[39]

Holling and others in the resilience school offered a general theory of the "adaptive cycle." Ecosystems, they argued, do go through predictable stages: (1) growth or exploitation, (2) conservation, (3) collapse or release, and (4) reorganization. As systems go through this adaptive cycle they invariably change, but they can change in very different ways. A resilient system will reorganize with persistent relationships and functions, while a less resilient system will lose basic relationships and functions.[40] So unlike Clements, who predicted particular outcomes of ecological development, Holling and his colleagues were predicting a process of ecological development, the outcomes of which could vary widely based on a number of factors.

The challenge for Holling was to explain the thresholds of resilience. Here Holling followed the insights of hierarchy theory. Some groups of relationships within an ecosystem, he explained, were of a higher order than others in determining long-term characteristics and functions: "the complexity of living systems . . . emerges not from random association of a large number of interacting factors rather from a smaller number of controlling processes."[41] Ecological change, he wrote,

> Functions at each of a number of scales—small and fast ones involving plant and soil processes of physiological ecology, through slower patch dynamics controlled by plant competition, through still coarser and slower ecosystem dynamics controlled by mesoscale processes like fire and human harvesting, to topographic properties shaped by geomorphological processes. Rare events, such as hurricanes, or the arrival of invading species, can unpredictably shape structure at critical times or at locations of increased vulnerability.[42]

Disturbances occurring at smaller spatial and temporal scales might appear chaotic, yet processes at broader spatial and temporal scales may not be affected and may therefore provide inertia that maintains the overall characteristics of the system. The opposite is also true. If change occurs at a sufficiently broad scale, the ecosystem will reorganize with fundamentally new characteristics. Thus a wildland fire, on its own, often will not fundamentally alter an ecosystem; indeed, it may play a vital role in maintaining ecological diversity. However, climate change will alter the fundamental character of an ecosystem.

Holling applied resilience theory directly to land and resource management, showing just how much it challenged earlier ideas of scientific management. The problem with traditional management, he and others argued, was that it "implicitly assumes that the problem is well-bounded, clearly defined, relatively simple, and generally linear with respect to cause and effect. But when these same methods of control are applied to a complex, nonlinear, and poorly understood natural world, and when the same predictable outcomes are expected but rarely obtained, severe ecological, social, and economic repercussions result."[43] In treating ecological systems as if they were simple and clearly understood, traditional management tended to reduce natural variation and fluctuation in an ecosystem in order to render it predictable, efficient, and stable. This "pathology of natural resource management," they argued, is inevitably destructive because it tends to reduce the natural variation in an ecosystem and thereby undermine its resilience.[44]

The resilience view, Holling wrote in 1973, led to radically different management strategies. It began with a "recognition of our ignorance," an emphasis on heterogeneity and fluctuation, and an expectation of the unexpected. A resilience-based management approach would emphasize "the need to keep options open, the need to view events in a regional rather than a local context, and the need to emphasize heterogeneity."[45] The golden rule of management, Holling and Gary Meffe later explained, should be to "retain critical types and ranges of natural variation in ecosystems. That is, management should *facilitate* existing processes and variabilities rather than changing or controlling them. By so doing, ecosystem resilience and the organizing processes and structures of ecosystems will be maintained."[46]

Holling, along with Carl Walters, Lance Gunderson, Gary Meffe, and a range of other ecologists began calling for "adaptive management" to respond to the reality of complex ecological systems. "Uncertainty," Walters and fellow ecologist Ray Hilborn wrote in 1978, "is a pervasive feature of ecological management problems. Rarely is it possible to predict even the short-term effects of major interventions."[47] Yet managers often employed deterministic models that presumed full knowledge of the populations or systems they managed. In adaptive management, by contrast, managers would acknowledge the various uncertainties they faced and treat management projects as experiments. In other words, managers would treat their understanding of ecological systems as working hypotheses to test through management activities, and data from those activities could be used to support or falsify the hypotheses. "Adaptive approaches as defined here," they wrote, "imply a process of learning about system responses through experience."[48]

As should be clear, principles of adaptive management directly challenged the confident sustained yield goals of silviculture and range management, which rested on the assumption that managers could reduce and ultimately control natural variability. If agencies could not "claim to predict with certainty either the ecological effects of the activities, or the efficacy of most measures aimed at regulating or enhancing them," then they needed to stop promising predictable sustained yields of various resources.[49] Managers, they argued, needed to admit the limits of their predictive power publically and redefine management not as a system of *control* but as a system of adaptation to the dynamic changes in ecological systems.

Reframing land and resource management around the insights of contemporary ecology was no small challenge, and existing legal and political pressures did little to support such reform. Resilience was, after all, hard to prioritize over concrete management goals such as board feet of timber or animal unit months of forage. But in the 1980s, a new concept that captured public attention provided a more concrete focus for management reform: biodiversity. For ecologists, biodiversity was a complex idea rooted in hierarchical systems ecology that highlighted the need for adaptive management. The public and judges interpreting the ESA might not have had the same nuanced, scientific understanding of biodiversity, but they certainly recognized that biodiversity was an important value threatened concretely by species extinction and habitat destruction.

In 1979, biologist Elliot Norse was asked to analyze "the status of life on Earth" in light of reports that tropical rainforests were being systematically eliminated by human development. "Knowing no existing term that encompassed all that was being lost," Norse later wrote, "we called it 'biological diversity.'"[50] For Norse this term encompassed genetic, species, and ecosystem diversity simultaneously. It referred, in other words, "to *all* forms of life" and "extends over 'all levels of organization' that humans perceive in the living world."[51]

The public could understand the concrete problems of species and habitat loss and increasingly understood them as indicators of an underlying biodiversity crisis: the fundamental unraveling of Earth's ecosystems and therefore the dynamic equilibrium that supported all life.[52] Paul and Anne Ehrlich, for example, wrote in 1981 that species were like the rivets holding an airplane together. The wings would remain securely fastened after losing one or two rivets, but if more and more rivets were removed the wings would at some point fail catastrophically. No one, they admitted, could predict the threshold of this catastrophe, but the current rate of species loss was clearly sending "spaceship earth" hurtling in that direction.[53] Thus, the concept of biodiversity both translated the insights of complex hierarchal ecosystem theory into concrete terms—species extinction and habitat loss—and connected them to the fate of Earth.[54]

This sense of crisis not only galvanized a segment of the public; it led to the development of a new subfield called conservation biology in which researchers themselves openly blended the descriptive insights of ecology with the normative commitment to protect biodiversity. A wide range of scientists played prominent roles in establishing and popularizing the field—Paul Ehrlich, E. O. Wilson, Thomas Lovejoy,

Jerry Franklin, etc.—but the scientist most commonly credited for its development was Michael Soulé, who co-organized the first major conference on conservation biology in 1978 and helped found the Society of Conservation Biology and its journal, *Conservation Biology*, in 1985 and 1987, respectively. Soulé explained that conservation biology was not a traditional research subfield; it was a crisis discipline in which "it is sometimes imperative to make an important tactical decision before one is confident in the sufficiency of the data. In other words, the risks of nonaction may be greater than the risks of inappropriate action."[55] Researchers in most areas of science were concerned primarily with type one errors—the error of a false positive—so they preferred to remain silent until they were confident that they had all the data they needed to have full confidence in their conclusion. Researchers in conservation biology, by contrast, argued that the crisis nature of biodiversity loss made such a posture irresponsible. The greater danger in their minds was of making type two errors—the error of the false negative. In this way, their posture was much closer to that of land managers for whom "the concern is failure to act when action is needed."[56]

The rise of ecological theory such as Holling's and the emergence of biodiversity as a normative conservation language set an interesting stage for ecosystem management in the late 1980s and 1990s. As explained in the next chapter, the ESA (1973) imposed strict preservation requirements for species that the interior secretary determined were threatened with or in danger of extinction, and it required the relevant federal agencies to base their assessment and recovery efforts on the best available science. Ecologists and conservation biologists therefore needed to find ways of fulfilling the ESA's requirements for predictable preservation outcomes while using sciences that emphasized uncertainty and complexity.[57]

One of the most politically significant tools they developed was the "population viability assessment" (PVA) in which scientists determined "minimum viable populations" (MVP) for a species. One of the pioneers of this tool was Mark Shaffer, whose doctoral research in the late 1970s focused on establishing a viable minimum population of grizzly bears in the Yellowstone region. In a 1981 *BioScience* article, he defined a minimum viable population (MVP) of a species as "the smallest population having a 99% chance of remaining extant for 1000 years despite the foreseeable effects of demographic, environmental, and genetic stochasticity, and natural catastrophes."[58] Though conservation biologists often wrestled with the exact timeframe and probabilities of a viable population, they generally agreed that viability must be understood as a very high probability of survival over a relatively long period of time. Built into this idea was the assumption that a population needed to have a margin of safety that went beyond the minimum number required to survive under "average conditions." The population needed to have a sufficient (1) size, (2) distribution, (3) structure, and (4) genetic variability to withstand a range of perturbations and surprises.[59]

Not surprisingly, then, "estimating MVPs is complex. Efforts to consider all four effects jointly and, beyond these, to consider the consequences of habitat fragmentation, have been slow to develop."[60] The four effects, Michael Gilpin and Michael Soulé explained in 1986, were interrelated in ways that made their full extent difficult to predict, particularly as general rules. Some species were more sensitive to stochasticity in one or another of the key areas, so MVP was necessarily species specific. What is more, because of the irreducible complexity and uncertainties in determining a MVP and applying it to management decisions, the final determinations essentially came down to expert or professional judgments. All models used to study viability involved simplifying assumptions that represented the best professional judgments of scientists in the field, and the statistical probabilities of viability offered by the models were always working hypotheses.[61]

Remarkably, despite its lack of precision, the MVP concepts gained legal force through the Forest Service's 1982 regulations implementing the National Forest Management Act (NFMA). The regulations interpreted NFMA to mean that "fish and wildlife habitat shall be managed to maintain viable populations of existing native and desired non-native vertebrate species in the planning area. For planning purposes, a viable population shall be regarded as one which has the estimated numbers and distribution of reproductive individuals to insure its continued existence is well distributed in the planning area."[62] These regulations essentially required the Forest Service to include population viability assessment and risk management in national forest planning and pushed wildlife biologists in the Forest Service and other agencies to embrace many other tenets of conservation biology by the mid- to late 1980s.[63]

Based on an understanding of biodiversity risks, including those identified through population viability assessment, conservation biologists challenged traditional jurisdictional boundaries and agency planning practices. The problem with much conservation work, conservation biologist Reed Noss charged, was that it was often focused on spatial and temporal scales that did not match the key ecological determinates of biodiversity. Biodiversity, Noss explained, "can be conceived of as a nested hierarchy of elements at several levels of biological organization."[64] Managers were often good at focusing on smaller spatial and temporal scales that matched their jurisdictional boundaries and budgetary horizons, but they often ignored and needed desperately to address the landscape, or regional, scale. "A major lesson," Noss wrote, "is to consider each phenomenon in a broader, ultimately global, context; that is, to 'think big.'"[65] Or as Forest Service ecologist Jerry Franklin put it, "larger-scale approaches—at the levels of ecosystems and landscapes—are the only way to conserve the overwhelming mass—the millions of species—of existing biodiversity."[66]

As a corollary to large-scale planning, conservation biologists admonished managers to protect ecological patterns and processes across the landscape rather than focusing on individual resource outputs. After all, if one assumed that ecosystems were hierarchical, complex systems, one did not need to manage everything; one needed to manage in ways that protected the key, determinative patterns and processes at particular scales. Of particular importance, many of these processes were disturbances such as fire and flooding, that ecologists recognized as "organizing elements in the composition and structure of biotic communities."[67]

It was, however, very difficult to study and manage processes. Therefore, a landscape ecologist explained at one Forest Service workshop, "ecological processes generate patterns, and by studying these patterns we can make useful inferences about the underlying processes. There is an implicit concession here that it is actually the processes we are most concerned about, but these are often too difficult to study directly."[68] To translate concern about ecosystems into manageable research projects and management tasks, then, scientists and managers would use surrogate measures, such as viable populations of key species, that they believed served as indicators of overall ecological integrity.

Whether it was planning for grizzly bear protection in Yellowstone or, preferably, planning for dozens of species simultaneously in the Pacific Northwest, conservation biologists offered a number of key principles around which conservation planning should be organized. Some of the common principles included:

- Species well distributed across their native range are less susceptible to extinction than species confined to small portions of their range.
- Large blocks of habitat, containing large populations of a target species, are superior to small blocks of habitat containing small populations.
- Blocks of habitat close together are better than blocks far apart.
- Habitat in continuous blocks is better than fragmented habitat.
- Interconnected blocks of habitat are better than isolated blocks.
- Blocks of habitat that are roadless or otherwise inaccessible to humans are better than roaded and accessible habitat blocks.[69]

As these principles suggest, conservation biologists accepted that conservation efforts would be less than ideal, given the existing pattern of human land and resource use. What they demanded, though, was that federal agencies look at the landscapes they managed across jurisdictional lines through the needs of indicator species and manage accordingly.

The science of biodiversity protection was indeed a radical challenge to traditional federal management, but conservation biologists did not have a free hand to reform federal land policy and management. Existing resource statutes, congressional appropriations, and political divisions all provided enormous inertia for the traditional, fragmented approach to management. Yet buried within literally thousands of federal statutes governing the public lands, three carried the challenge forward with considerable force. NEPA (1969) ensured that the agencies considered new science, including conservation biology, in their land-use planning. They didn't need to adopt the conclusions that biodiversity protection was more important than resource outputs, but they were required to respond to those conclusions and the science behind it. More importantly, the ESA and NFMA both contained strict mandates that the courts interpreted through the new ecology and conservation biology. These two laws, with requirements for federal agencies to protect endangered species, habitat required to sustain endangered species, and all vertebrate species in the national forest system, were the regulatory hammer that required the agencies to look across their jurisdictional boundaries and protect not only individual plants and animals but larger ecological dynamics as well.

Understanding the demands of ecology and conservation biology in the late twentieth century helps significantly to explain the rise of federal ecosystem management in the 1990s. The concepts and rhetoric of ecosystem management furthered the scientific emphases on managing along ecological rather than political boundaries and managing for ecological processes and functions. But drawing on elements of both preservation and multiple-use conservation, ecosystem management simultaneously promised that these new emphases would allow agencies to accommodate substantial resource development activities. Ecosystem management, in other words, promised to incorporate new scientific emphases without turning federal land and resource management over to conservation biologists.

Political Science

Political science, particularly public administration, went through a shift in the second half of the twentieth century that paralleled ecology in important ways.[70] As scholars struggled to demonstrate that they were doing rigorous science, their fundamental unit of study shifted from the institution to the individual, they increasingly viewed administrative systems as open and dynamic rather than closed and stable, and they challenged the basic ideas of administrative management. Just as ecologists argued for adaptive management with feedback loops, public administration scholars argued for participatory management in which public participation held agencies accountable and helped them adapt to the changing landscape of public values.

As noted above, the shift in public administration did not directly reshape federal planning and management. Fewer federal managers had formal training in public administration than in the natural sciences, and no federal laws required the agencies to address public administration scholarship directly in their decision making. Nonetheless, the shift in public administration scholarship is significant for two reasons. First, policymakers and agency leaders drew on public administration scholarship for a variety of tasks, from drafting legislation to agency reorganization. Second, and perhaps more importantly, public administration scholarship reflected and encouraged the loss of public trust in administrative autonomy and discretion and a growing demand for democratic accountability in administrative decision making. Therefore, the evolution of political science scholarship in the twentieth century explains one of the central commitments identified in most definitions of ecosystem management: collaborative and participatory planning.

Early Public Administration Scholarship

Traditional public administration emerged in the late nineteenth century more as a reform movement than an empirical science. For most of the nineteenth century, agencies in the executive branch were staffed by political appointees at all levels, providing members of Congress and, to a lesser extent, the president, with a significant means of patronage. Following the Civil War, support for a more professional, merit-based civil service grew, and President Garfield's assassination by a disgruntled patronage-seeker in 1881 helped galvanize support for the Civil Service Act of 1883, known commonly as the Pendleton Act. As the Civil Service Commission reported, the act's primary intent was "to remove from American politics the degrading influences of the patronage system. Certainly, during the last sixty years, no other one cause has been so potent in tending to degrade American politics."[71] Traditional public administration emerged in the 1880s as part of an ongoing struggle to realize the act's goals in federal administration.

Public administration scholar Herbert Kaufman has argued that the field was driven by three primary values: representativeness, neutral competence, and executive leadership. Through this lens, the practice of public administration in most of the nineteenth century could be understood as the era of representativeness, where public administrators represented the members of Congress who appointed them. The field of public administration that emerged in the late nineteenth and early twentieth century could be understood as an effort to make federal agencies professional organizations defined by their politically neutral expertise and their clear executive leadership.[72]

The early leader in public administration scholarship, Woodrow Wilson, wrote in the midst of the civil service battles of the 1880s. For Wilson, the first normative principle of public administration was the necessity of separating administration from politics: "Most important to be observed is the truth . . . that administration lies outside the proper sphere of *politics*. Administrative questions are not political questions. Although politics sets the tasks for administration, it should not be suffered to manipulate its offices."[73] Agencies, he argued, were not themselves the policymakers; they existed to administer the policies established by Congress and the president.[74]

The second and third normative principles were efficiency and economy. With administration protected from politics and policymaking, the quality of public administration was to be judged by how efficiently and inexpensively the agencies fulfilled their mandates. For Wilson, ensuring efficiency and economy hinged on a clear hierarchical organization with a single source of authority—what he called the unity of command. In the case of the federal government, he argued that Congress was the singular source of authority, and agencies should be organized so that congressional authority flowed down through the ranks in a clear chain of command. Frank Goodnow, another early public administration scholar, provided a more thorough explication of this principle at the turn of the century, and President Teddy Roosevelt advanced it during his administration.[75] Indeed, the Forest Service that Gifford Pinchot organized during the Roosevelt administration stood for decades as a model of professional administration.

Organizational structures advanced during this period followed Max Weber's description of bureaucracy, which emphasized hierarchical lines of authority and control.[76] Ideas about management efficiency and economy, however, drew heavily from business management, particularly the work of Frederick Taylor. Taylor, like Wilson, Goodnow, Roosevelt, and others of the Progressive Era, shared an evangelical belief in what historian Samuel Hays has called "the gospel of efficiency."[77] Taylor wrote in his seminal work, *Principles of Scientific Administration*, that his goal was nothing less than answering Teddy Roosevelt's call for increased national efficiency: "We can see our forests vanishing, our water-powers going to waste, our soil being carried by floods into the sea; and the end of our coal and our iron is in sight. But our larger wastes of human effort, which go on every day through such of our acts as are blundering, ill-directed, or inefficient . . . are less visible, less tangible, and are but vaguely appreciated."[78]

Taylor argued that the most collaborative workers—factory workers who produced pieces of a larger product—deliberately worked below their capacity because they feared working at full capacity would give their managers unrealistic expectations for future productivity. Taylor believed that the first way to avoid underproduction was to break down each task to its simplest components and determine "scientifically" the most efficient way to perform that task. Based on these studies, rather than the testimony of workers, a company could then determine how long each task should take and establish clear incentives to make their workers achieve optimal productivity. For Taylor, businesses were analogous to the machines that many of them produced. He viewed organizations, then, through the lens of mechanical engineering and sought to define and craft—or in this case train—each part so as to optimize overall efficiency.[79]

Even with this brief description of Wilson and Taylor, it is possible to see some general connections to Clementsian ecology and its promise of control over nature. Although Wilson and Taylor had a mechanistic model rather than an organismic model of organizations, they shared with Clements the belief that it was possible to predict the efficiency, economy, and productivity of an organization by understanding its basic environmental determinants. And with Clements, they believed that an organization properly organized would lead invariably to optimal outcomes.

The significance of public administration, both as a field of scholarship and a practice, grew dramatically during the 1930s, as the size and scope of the federal government expanded rapidly. Luther Gulick, one of the leading figures in traditional public administration during this period, explained, "The New Deal is the decision of the Nation to have the government become the super-holding company of the economic life of America."[80] This posed new challenges for public administration, Gulick argued, because the growing scope and complexity of federal administration and the increasingly broad delegations of authority that Congress granted to administrative agencies made it impossible to isolate politics in Congress. Agencies required discretion to fulfill their broad mandates, and discretion made room for political values.

Yet Gulick modified rather than abandoned Wilson's politics/administration dichotomy. He continued to emphasize the importance of neutral competence, but he shifted the focus of control from Congress to the president. As with any complex company, he argued, the federal government had many different offices and departments, each with its own separate function. The key to organizing such a company was to ensure that the chief executive was the principal authority who ensured proper administration. What public administration scholarship needed to do, then, was "introduce into government a mechanism of central planning. This mechanism will be needed to draft the master plan, to elaborate its major details, to watch its operation, and to keep itself informed with regard to the changing needs of the Nation on every front."[81] As Vincent Ostrom explains, "The simple concept of hierarchy is replaced by a complex 'fabric of organizational interrelationships' resembling a 'jungle gym' but controlled by a single chief executive."[82]

Gulick's contributions were more than academic. He served on President Roosevelt's Committee on Administrative Management, and the committee's report in 1937 recommended significant changes in administrative organization designed to strengthen presidential control. Gulick offered numerous "principles of administration" that he argued would ensure proper administrative organization and, in turn, optimal efficiency and economy. These included principles of homogeneity (that administrative units needed to be organized around specialized tasks to be efficient), span of control (the idea that a supervisor could not effectively manage more than five or six subordinates), unity of command (a subordinate should only receive orders from one supervisor), etc.[83]

In 1951, Gulick published a study of federal forest management that outlined both the marked achievements of professional administration and new challenges that it posed. On the positive side, he argued that through the Forest Service and the rise of professional forest administration, the American people had "got what they wanted": steady wood supply, increased economic stability, watershed protection, etc. Directed by clearly identified goals, professional administration grounded in applied science had been effective and efficient. And, he declared, "the competence and loyalty of the foresters throughout the land, and the high esteem in which they are held, are a direct result of the educational and professional program set in motion by men like Gifford Pinchot and B. E. Fernow at the beginning of the forest program."[84]

At the same time, Gulick observed, the professional bureaucracy had increasingly displaced Congress and the president as policymakers, creating new challenges for political accountability. Indeed, Gulick argued, professional administration

> does not "settle the problem," even though it may perform a very great public service. Government action tends to open up a wider and wider compass of problems, interrelations, and conflicting interests. These new questions are beyond settlement by science or engineering, and therefore by the professional. It would seem that the political problems of government are inescapably extended; that new pressure groups are generated both among the

> professionals and the bureaucrats, and among the affected economic and administrative units; and that the political task thrown on the political machinery of government is thereby rendered both more difficult and more important.[85]

For Gulick, this was a cautionary lesson that reinforced the need for clear structures of command that could be traced back to the president.

Public Administration in the Postwar Period

As with Clementsian ecology, traditional public administration came under fundamental and sustained criticism in the mid-twentieth century, and the basic theories and models of the field changed dramatically. Scholars drew liberally on more established social sciences, ranging from psychology to economics, as well as on general systems theory and cybernetics. As with ecology, the new directions in public administration scholarship all suggested that administrative agencies were complex and dynamic and could not be predicted or controlled in any simple sense.

The new directions in public administration could be categorized in different ways. Writing in 1976, H. George Frederickson could speak of four new models of public administration that challenged the traditional bureaucratic model: "the neo-bureaucratic model, the institutional model, the human relations model, and the public choice model."[86] Rather than outlining all four of these in detail, it is sufficient to highlight some of their overlapping criticisms.

The pioneering challenge, which Frederickson called the neo-bureaucratic model, came from Herbert A. Simon and his colleagues in "the Carnegie School," who drew on cognitive psychology to explain administrative behavior. Writing in the mid-1940s, Simon voiced a number of concerns about traditional public administration. First, he argued that traditional public administration was more art than science, calling Gulick's "principles of management" little more than a collection of contradictory proverbs.[87] Each proverb might be true in a particular situation, Simon wrote, but taken together they were self-contradictory.[88]

Simon argued that the primary failure of traditional public administration that inhibited its development as a rigorous science stemmed from its mechanistic assumptions that treated organizations as simple systems and employees as interchangeable parts or cogs within the administrative machine and from its misunderstanding of administrative rationality: "First," Simon and his colleague James March wrote, "there is a tendency to view the employee as an inert instrument performing the tasks assigned to him. Second, there is a tendency to view personnel as a given rather than as a variable in the system." As a result, traditional public administration "largely ignored factors associated with individual behavior and particularly its motivational bases."[89] To truly understand an organization and its behavior, Simon insisted, one needed to focus on individuals within the organization and the real context of their decision making. "The construction of an efficient administration," Simon wrote in 1947, "is a problem in social psychology."[90] It is not difficult to hear in Simon and March's words echoes of the critique Whittaker leveled at Clementsian ecology.

Simon and March rejected many of Gulick's principles, arguing that it was not possible to determine a fixed "span of control" or to maintain "unity of command" in complex organizations. The real work of administration, they argued, was decision making, and the individuals making decisions were driven both by formal direction and incentives as well as by less formal psychological factors such as their sense of belonging to the organization and believing in its mission. In contrast to the machine model of Taylor and Gulick, then, Simon and his colleagues advanced a behavioral model: "a theory of motivation." An organization, they argued, was "a system of interrelated social behaviors of a number of . . . *participants*."[91] The purpose of organization, then, was to establish "bounded rationality" and certain "zones of acceptability" that limited the choices of each member of the organization. As Vincent Ostrom puts it, "An organization thus might reflect a composite of command networks rather than a single line of authority or chain of command."[92] As with his contributions to ecology, Simon was challenging the notion that bureaucracies could be understood or controlled as simple systems and must therefore be treated as complex hierarchical systems that can never be fully understood nor controlled.

Simon's critique went to the root faith of traditional public administration, namely that proper organization and control could ensure *optimal* decision making that met high standards of efficiency and economy. First, as just outlined, Simon argued that this faith rested on a naïve view of employee motivations. Second, and equally important, it rested on an unrealistic epistemology. The traditional model of rational decision making, he argued, assumed that the decision maker could gather all relevant information, compare all alternatives, and select the one that produced optimal results. Such a decision-making process was not only impossible but horribly inefficient. In reality decision makers worked with limited information and time in a dynamic context, and they most often made satisfactory rather than optimal decisions: "Most human decision making," Simon argued, "whether individual or organizational, is concerned with the discovery of satisfactory alternatives; only in exceptional cases is it concerned with the discovery and selection of optimal alternatives."[93] As Simon put it, decision making was typically a process of "satisficing."

The transition from Gulick's emphasis on administrative principles to Simon's emphasis on administrative motivation was reflected in a widely cited book on the Forest Service: Herbert Kaufman's 1960 study, *The Forest Ranger.* Kaufman studied under Gulick, who encouraged his interest in Simon's work despite its clear criticism of traditional public administration.[94] Because of Gulick's work on forest policy, Kaufman wrote his dissertation on the Forest Service, studying the mechanisms that allowed the agency to be both highly decentralized and highly consistent in its decision making. He found that the agency's hierarchical organization was significant, but that this did not adequately explain the consistency of the decisions that forest rangers made. In his initial, limited study, and in the extended research he later conducted for *The Forest Ranger,* Kaufman found that the most powerful influence on forest rangers was not their chain of command but their enculturation in the agency: "a common set of values, outlooks, and understandings."[95] As Terrence Tipple and Douglas Wellman reflected thirty years later, "*The Forest Ranger* painted a portrait of a field-level executive who . . . acted almost automatically in the organization's best interest, or in Simon's terms, within the service's 'zone of acceptance' [and who] efficiently and economically carr[ied] out public policy in an organizational context where the authority and power flows down in a clear, linear fashion."[96]

In his writing during the 1960s, Kaufman continued to emphasize this more complex picture of administrative behavior that he grounded in Simon's pioneering work. "Fundamentally," he wrote in 1964, "coordination is accomplished by two processes: central direction, which means that the activities of the elements of a system respond chiefly to cues and signals from some common source, and reciprocal relations, which means that the elements respond to cues and signals from each other. Every system employs some blend of the two processes."[97]

For Kaufman, this led to an organic picture of organizations that could not be reduced to a set of simple determinants. "Seen from this angle," he wrote, "what would the world of organizations look like? Probably crowded, continually changing, difficult to analyze, viscous."[98]

The other new models in public administration—the institutional model, the human relations model, and the public choice model—also rejected key principles of traditional public administration, attacking in various ways the belief in neutral competence and centralized executive leadership. What is more, they expanded the focus beyond the organization's internal relationships, describing organizations as complex open systems that were driven by relationships with their extra-organizational environment.

The institutional model was perhaps best articulated by sociologist Philip Selznick, beginning with an influential study of the Tennessee Valley Authority (TVA) in 1949: *TVA and the Grass Roots.* The TVA was a unique organization, created to be a public/private hybrid of sorts and given broad discretion to engage in development projects that would benefit people in the region. In other ways, though, the TVA was like a growing number of federal agencies to which Congress gave only broad, discretion-filled mandates. Selznik showed that the TVA was driven by what he called cooptation, where the agency satisfied new demands or interests as a way of maintaining its own stability. Some of this cooptation was formal and visible, but much of it was informal and therefore less visible.[99]

The political scientist Charles Lindblom, in a 1959 essay that enjoyed two decades of heated debate, also rejected the hope that proper organization and management could yield optimal administrative work, calling administrative decision making "the science of muddling through." Employing the institutional model, he rejected what he called the "rational-comprehensive" model of decision making. In this idealized picture of decision making, administrators first consider and rank all relevant values to determine the goals of decision making. Next they consider all possible alternatives and their consequences, and undertake a systematic comparison of the alternatives for the optimal decision. "For complex problems," Lindblom wrote, this model "is of course impossible. Although such an approach can be described, it cannot be practiced" for two reasons.[100] First, it was unrealistic to assume that the values, or ends, of decision making can be settled in advance. Second, "rational-comprehensive analysis leaves out nothing important. But it is impossible to take everything important into consideration unless 'important' is so narrowly defined that analysis is in fact quite limited."[101]

As a result, Lindblom argued, administrators made more modest, less comprehensive decisions aimed, as Simon suggested, at finding satisfactory choices for a given context. This was not, he argued, a lamentable failure of administration but a practical and defensible response to the complex realities both of the problems that administrators faced and the dynamics of administrative organizations. But perhaps the most important assertion of Lindblom's muddling thesis was that because administrators did not expect to achieve optimal outcomes, they tended toward incremental decisions that could be revisited "as conditions and aspirations changed and as accuracy of prediction improved."[102]

When applied to federal land and resource agencies, particularly the Bureau of Land Management (BLM), the new institutional model revealed a disturbing picture of federal administration. In the 1960s, scholars such as Philip Foss and Grant McConnell advanced the clientele capture theory of administrative behavior. They argued that in order to build and maintain supportive clientele that would ensure their own survival, agencies were coopted by and ultimately served their clientele rather than the public more broadly. One of the principal examples they offered was the BLM, which was locked in an "iron triangle" with ranchers and western members of Congress. This extremely stable set of relationships ensured, they argued, that the agency served western livestock interests rather than the public interest.[103]

The human relations model, led by scholars such as Rensis Likert and Chris Argyris, challenged the very notion of authoritarian control in organizations. They argued, based on empirical work in organizational psychology, that high levels of productivity were driven not by authoritarian oversight but by employee satisfaction. The organizational development they advocated, then, was aimed at nurturing employees and increasing communication between employees and their clientele. Scholars in this field were the most explicit in applying open systems concepts, describing organizations as open systems that were affected directly by their broader environment. This model did not necessarily have the most significant impact on public administration as a field, but in the 1960s it was certainly employed by consultants working with federal land and resource agencies.[104]

The last new model—public choice theory—was perhaps also less influential in public administration in the mid-twentieth century, but it became an increasingly dominant theory in later decades. Advanced most influentially by James Buchanan and Gordon Tullock in the 1950s and 1960s, public choice theory adopted an economic model. Individuals, they argued, were self-interested and developed optimizing strategies to meet their interests. This is true of a public employee as much as an individual citizen in the marketplace. As Vincent Ostrom explains, "Tullock's 'economic man' is an ambitious public employee who seeks to advance his or her career opportunities for promotions within a bureaucracy. . . . Tullock suggests that the limits on control in very large public bureaucracies will engender 'bureaucratic free enterprise' when individuals and groups within an organization proceed to formulate their own missions with opportunities for side payoffs, including graft and corruption."[105]

For public choice theorists, including Buchanan, Tullock, Vincent Ostrom, and Elinor Ostrom, an economic view of individuals did not necessarily condemn federal agencies to failure, but it rejected outright the Weberian model of bureaucracy and the hope that an agency could be effectively and efficiently controlled by Congress or the president. Public choice theorists went further than other new directions in rejecting the administration/politics dichotomy, arguing that administration was always political and needed to be treated as such. "The constitution of public enterprises must depend instead on the development of political mechanisms such as voting, representation, legislation, and adjudication for people to express their interests by signaling agreements and disagreements as the basis for ordering their relationships with one another. Such mechanisms provide essential means for informing public entrepreneurs about their strategic opportunities and limitations."[106] In other words, rather than trying to strip politics out of administration, federal agencies needed to have effective political mechanisms that tied them to the communities they served.

In rejecting Wilson's unity of command concept as ineffective and counterproductive, public choice theorists developed alternative models of political economy that were explicitly polycentric. They advocated for multi-organizational arrangements with many different nodes of power. Such an arrangement, they insisted, would provide more checks and balances and make organizations more flexible and responsive. Vincent Ostrom wrote that rather than seeing this polycentric, multi-organizational arrangement as chaotic, "political economists perceive a pattern of ordered relationships being sustained among diverse public enterprises. Each different public enterprise is accountable to its relevant community of interests and functions essentially as a public firm in a much larger industrial complex."[107] Indeed, he wrote, "industry characteristics of multiorganizational arrangements functioning in a public economy can be realized only when diverse public agencies are able to develop different economies of scale in response to varying communities of interest."[108] It shouldn't be surprising that one can find connections between public choice theory and ecological theory at this point, since both treat their subjects of study as complex hierarchical systems.

While the various movements in public administration during the 1940s through the 1960s differed in important ways, they collectively replaced the traditional public administration of Wilson and Gulick. Treating bureaucratic organizations as complex rather than simple systems, the new models undermined any faith that bureaucracies could be effectively controlled by a single authority such as Congress or the president. As with ecologists, public administration scholars worked to identify key determinative processes within administrative activity. And given their skepticism that any single authority could understand or control bureaucracies adequately, they developed new arrangements for administrative accountability.

The New Public Administration

Two important shifts occurred in the late 1960s and early 1970s: the emergence of "new public administration" and the emergence of "policy sciences." The former took an explicitly normative stance rather than the more positivist approach of scholars in the 1950s and 1960s, who were determined to demonstrate their scientific rigor. The latter retained the goal of scientific rigor but chose to study administration within the broader context of the policy process.

The "new public administration" was led by a loose coalition of scholars, particularly Dwight Waldo, Frank Marini, and H. George Frederickson. They rejected the politics/administration dichotomy emphatically, arguing that administrative decision making was inherently political and that administrators were not value neutral. They also rejected the possibility of Wilson's "unity of command," accepting that administrators were driven by a wide range of motivations and could not be controlled by a single authority. Thus, as Frederickson wrote, "we are witnessing a revolt against these values accompanied by a search for new modes of representativeness."[109] But perhaps most importantly, new public administration went beyond the earlier goals of efficiency and economy, arguing that social equity ought to be one of the central objectives of public administration. This wasn't simply meant to be a policy goal formulated by Congress but a commitment inculcated in public administration education and demonstrated by public administrators as they made and implemented policy.[110]

New public administration scholars echoed the concern voiced most clearly by public choice theory, namely that the virtue of a Weberian bureaucracy, its ultra-stability, was in fact a vice. Fixed bureaucratic patterns, they argued, reinforced the status quo, which was often rife with social inequality. What federal bureaucracies needed, then, was not a perfected form that resisted change but flexibility to adapt to change and to respond to new social needs and concerns. Much of the literature in new public administration focused on education, poverty, and other social programs, but the same critique could have been made of the Forest Service, which had once been seen as the epitome of professional administration. The agency's stability and consistency, explained by Kaufman in *The Forest Ranger*, had become a vice because the Forest Service was still emphasizing utilitarian timber production when a growing number of Americans were demanding protection of nonconsumptive values in the national forests. Somehow, agencies like the Forest Service needed to become more responsive to changing public values.

Thus, new public administration scholars called for dramatic shifts in representativeness: "Decentralization, devolution, termination, projects, contracts, evaluation, organization development, responsibility, expansion, confrontation, and client involvement are all essentially counterbureaucratic notions that characterize new public administration. These are the concepts that increase possibilities for social equity."[111] This was in some ways a redefinition of public administration. Rather than administration by professional expertise, public administration was to be a project that involved professionals, elected officials, interest groups, and individuals joined together to achieve broad common goals. Indeed, Frederickson predicted that "we will gradually come to define the organization to include citizens, clients, elected officials, and the like."[112]

New public administration scholars were decidedly pragmatic, emphasizing the limits of human knowledge. They followed Simon, Lindblom, and others in an effort to provide more realistic models of rational decision making. To Simon's "bounded rationality" and Lindblom's "incrementalism," Frederickson added "buffered rationality." By this he meant a more modest or limited understanding of administrative rationality that avoided the tendency of public administrators to "apply more than they really know."[113]

In the late 1960s and early 1970s, another development in the academy helped to reshape public administration. Generally falling under the title policy sciences, the new subfield treated administration as simply one aspect of the broader policy process. The new area of study was led by scholars such as Harold Lasswell, Theodore Lowi, and Lynton Caldwell. In the beginning of *Politics: Who Gets What, When, How,* Lasswell defined the study of politics as "the study of influence and the influential."[114] The focus, in other words, was not necessarily on formal organizational structures or legal statutes as much as on the many formal and informal sources of influence that shaped policy formation and implementation.

Unlike the new public administration, the goal of the policy sciences was more predictive than prescriptive. The emphasis was on the discovery of "interdeterminative relationships" that explained the politics and outcomes of public policy. Theodore Lowi, for example, argued that the politics associated with various policies was not explained by the subject area or by the particular bureaucratic structures involved but by the type of policy it was. Regulatory policy, whether it was in education or environmental protection, would produce a common set of political dynamics, while patronage policy would produce another. The politics of a given policy was determined by whether the government used positive or negative sanctions and whether it applied to individuals or broad categories of people.[115]

For Lowi, the preeminent issue in politics was the state's coercive power, and he was highly critical of Congress's tendency to delegate that power through broad and vague statutes. In his most widely cited book, *The End of Liberalism,* he observed with apprehension the wave of "*interest representation, cooperation, partnership, self-regulation, delegation of power, local option, creative federalism, community action, maximum feasible participation . . . and participatory democracy*" that had swept through Washington.[116] He agreed with new public administration that federal agencies needed more accountability in the exercise of coercive power, but he argued that the proper remedy was clear delegation of authority that made agencies more accountable to elected officials rather than greater public participation and devolution. This critique, however, was lost in the general tide of "democratic administration."

The Public Administration Context of Ecosystem Management

Scholars working in public administration and public policy continued to develop more complex and dynamic models throughout the remainder of the twentieth century, drawing not only on psychology and economics but on evolutionary biology as well. Public choice theory emerged as the dominant framework in public administration, but in the policy sciences it competed with other approaches, such as the "advocacy coalition framework" advanced by Paul Sabatier and Hank Jenkins-Smith and the "punctuated equilibrium" theory that Frank Baumgartner and Bryan Jones adapted directly from the work of Niles Eldredge and Stephen Jay Gould in evolutionary biology.[117] Indeed, toward the end of the twentieth century, the scholarship in ecology and in public administration/public policy merged at important points, such as ecologists drawing on Herbert Simon's work on complexity and hierarchy and those in political science and law using "complex adaptive systems" explanations from ecology.[118]

Whereas traditional public administration had assumed that order and control could be designed and perfected, the newer approaches had more modest goals of explaining how relative stability and order could emerge out of such complex and chaotic political systems: "Because organizational behavior is inherently complex," one scholar wrote, "it sometimes results in chaotic variations that cannot be controlled by hierarchical command structures, no matter how efficient and 'in control' supervisors appear to be, and even if employees are diligent."[119] And after a comparative study of community forestry in the United States and Canada, geographer James McCarthy concluded that the first lesson to draw from the study "is that superficially minor differences can lead to sharply divergent outcomes."[120]

The Clinton administration turned directly to a new approach in the 1990s in its effort to reinvent government, including ecosystem management. The "new governance" approach drew on the new public administration and new developments in business management theory, essentially affirming a strong federal agenda without challenging the anti-government and anti-bureaucratic ideology of the Reagan and

Bush administrations before it. As Frederickson explains, "What people want is less government and more governance."[121] Under this banner, the Clinton administration promised less rigid and static bureaucratic administration and more participatory, collaborative, and adaptive administration.

In many ways, new governance was a way of doing public administration in a context where scholars and the public alike had an aversion to concentrated authority. Harlan Cleveland had written in the 1970s that in the future of federal administration, the older, hierarchical models would give way to

> systems—interlaced webs of tension in which control is loose, power diffused, and centers of decision plural. "Decision-making" will become an increasingly intricate process of multilateral brokerage both inside and outside the organization which thinks it has the responsibility for making, or at least announcing, the decision. Because organizations will be horizontal, the way they are governed is likely to be more collegial, consensual, and consultive. The bigger the problems to be tackled, the more real power is diffused and the larger the number of persons who can exercise it—if they work at it.[122]

New governance scholars in the 1990s were essentially working to realize this vision of complex, adaptive administrative systems in the absence of most forms of centralized control. As public administration scholar Walter Kickert, who interestingly first studied "fuzzy set theory" in physics, explains, "Public 'governance' is the directed influencing of societal processes in a public policy network of many other co-directing actors." In this context, the federal government's role is "steering at a distance."[123]

New governance scholars and advocates called for the most dramatic democratization of public administration in the twentieth century, arguing that government needed to apply principles of business and the private sector to government, making it more "customer-oriented" and collaborative. "Instead of relying exclusively on government to solve public problems," new governance advocate Lester Salamon later wrote, "a host of other actors are being mobilized as well, sometimes on their own initiative, but often in complex partnerships with the state. In this new setting, traditional notions of public and private responsibilities are being turned on their heads and traditional conceptions of public administration rendered largely obsolete."[124] Indeed, another scholar wrote, "new governance seeks to combine new and old ideas into a comprehensive approach, centered on collaboration, flexibility, results, and engaging citizens . . . which in its totality would represent a new way of doing the public's business."[125]

The National Performance Review, initiated by the Clinton administration and headed by Vice President Al Gore, followed this basic philosophy, emphasizing things such as "Putting Customers First," "Serving the American People: Best Practices in Customer-Driven Strategic Planning," and "Business-Like Government." The very notion of bureaucratic efficiency was now based on the model of business or market efficiency, where employees are "empowered" to be responsive to their customers and satisfy their demands.[126] Certainly the federal government had certain legal obligations and authority to manage federal lands and resources, but rhetorically, the new governance model suggested that it would be steering from a distance.

Conclusion

Although the transitions in ecology and political science differ in important ways, their similarities outweigh their differences. Early ecology rested on teleological analogies to organisms, while early public administration rested on teleological analogies to machines. In both cases, scholars assumed that the object of study was designed for a predetermined purpose and would necessarily develop toward that purpose if properly organized and directed. In the second half of the twentieth century, however, ecologists and political scientists both increasingly emphasized complexity, chaos, and stochastic relationships, thereby undermining some of their predecessors' predictive confidence and normative weight.

This shift challenged the very foundations of federal land and resource management, the understanding of management as control over land, resources, and administrative agencies. In the newer ecological and political models, complete control was impossible. Ecologists focused instead on maintaining complex ecological processes and "historic ranges of variability," so that ecosystems could remain resilient in the face of dynamic changes. Public administration scholars focused on flexible, responsive, and collaborative administration held accountable by democratic participation. Thus, the overarching model of management at the end of the twentieth century was not control; it was adaptation or coping. One might call this "ecological satisficing," where the agency's expertise would not be demonstrated by producing final, optimal decisions that stabilized ecosystems but by learning and adapting to ecological change.[127]

While the conceptual models changed dramatically, legal and political demands that agencies faced, as well as their own administrative cultures, retained strong commitments to management as control. Whether legislation requiring species protection, political demands for timber output communicated through the appropriations process, or administrative programs focused on particular resource outputs, the pressures on federal agencies were still for predictable management outcomes.

It was in this context that ecosystem management looked so promising. It affirmed the revised concepts of scientific management and democratic governance and offered new ways of defining scientific and political authority in federal land and resource management; yet it still promised predictability in sustaining resource outputs and maintaining ecological protection. In the natural sciences, biologists and ecologists acknowledged the depth of ecological uncertainty that the agencies faced and promised that planning and managing at broader ecological scales on a more experimental and adaptive basis would nonetheless provide ecological protection that laws such as ESA and NFMA demanded. In public administration, new governance promised to mitigate deep public suspicion and distrust of the state, particularly a bureaucratic state defined by lines of concentrated political authority and power. Collaborative and participatory management, they argued, redistributed political power in a more egalitarian way, reinvigorating American civil society and avoiding the errors of earlier authoritarianism.[128] These two lines of thinking—ecology and public administration—thus merged in the rhetoric of ecosystem management, providing some of the central concepts and commitments of the new paradigm.

2 | The Policy Context of Ecosystem Management

Introduction

The intellectual shifts in the mid- to late twentieth century produced countervailing pressures on the four main federal land agencies. Ecology increasingly emphasized the complexity and dynamism of ecological systems as well as the importance of protecting biodiversity, suggesting the need for adaptive management guided by scientific expertise. At the same time, public administration provided an evolving critique of professional expertise, celebrating the importance of collaborative, participatory administration to increase agency accountability.

New concepts from ecology and public administration might not have had a direct impact on the federal agencies in the mid-twentieth century, but they were translated by popular discourse and written into landmark environmental legislation of the 1960s and 1970s. The environmental movement pressed Congress to apply an ecological lens, albeit one that still emphasized stability and balance, to federal lands and resources, and it pressed Congress for direct access to agency decision making. Congress responded with a flurry of legislation, adding the new public demands without repealing or even significantly amending the basic legal structure of federal land and resource management. Thus early federal law, ongoing appropriations, and agency culture continued to emphasize the protection or production of specific resources in discrete locations, even as Congress directed the agencies to adopt a more complex, ecological, and participatory lens to their work.

At the end of the twentieth century, the concepts and rhetoric of ecosystem management appeared promising to the Clinton administration and agency leaders precisely because they were new administrative tools to acknowledge and address the tensions that had grown in federal land and resource management: preservation and use; scientific authority and democratic authority. But building ecosystem management on the foundation of accumulated federal land and resource law meant that the new paradigm could not ultimately resolve these tensions in any clear sense; rather it led to the two very different ecosystem management models outlined above: one substantive model rooted in the Endangered Species Act (ESA) and one procedural model rooted in the National Environmental Policy Act (NEPA). Understanding the rise of ecosystem management therefore requires a firm grasp of this legal landscape.

Federal Policy in the Early Twentieth Century: Producing and Protecting Specific Resources

The federal government began building the nonmilitary federal lands system in 1872, when it reserved Yellowstone National Park as "a public park or pleasuring-ground for the benefit and enjoyment of the people."[1] Then, in 1891, Congress authorized the first forest reserves; in 1903, President Roosevelt created the first federal wildlife refuge; and in 1934, Congress finally accepted some management responsibility for the hundreds of millions of acres of unreserved public lands and passed the Taylor Grazing Act. Beginning in the early twentieth century, Congress or the president created the four main federal agencies that manage federal lands today—US Forest Service, 1905; National Park Service, 1916; US Fish and Wildlife Service, 1940; Bureau of Land Management, 1946—giving the system greater definition and building the constituencies that have supported and challenged its management.[2]

Two things are particularly important to note about these foundations. First, Congress and the president did not begin with a comprehensive plan for the federal land system. Rather, they reserved specific areas of land for specific purposes such as scenic protection and recreation, watershed protection and timber production, livestock grazing, wildlife protection, etc.; and the earliest laws that the four agencies implemented reflected these different purposes.

Second, the early policies that guided federal land and resource management were decidedly nonecological in the contemporary sense. The word *ecosystem* was not coined until a year after Congress passed the Taylor Grazing Act in 1934, and the ecology that did influence land and resource management was Clementsian, emphasizing the predictability of ecological processes and supporting a resource management orientation. As a result, federal land management in the first half of the twentieth century was guided by an agricultural or zoological model of resource production and protection. All federal agencies, including the Park Service, focused on producing particular resources or protecting specific objects that their constituencies valued.

The US Forest Service (1905)

In 1905, Congress transferred management of the forest reserves from the Interior Department's General Land Office to the Agriculture Department's Bureau of Forestry, which was renamed the Forest Service. Its charismatic first chief, Gifford Pinchot, pushed the transfer for political reasons, but the Forest Service was clearly at home in an agricultural context. Overton Price, Pinchot's assistant chief, argued in 1903 that "Forestry is a component part of agriculture,"[3] and years later Pinchot argued, "Wood is a crop. Forestry is tree farming."[4]

The forest reserves' enabling legislation, the Organic Act of 1897, established two main purposes for forest management: protecting "favorable water flows" from the forests and providing "a continuous supply of timber for the use and necessities of the citizens of the United States."[5] After Congress transferred the forest reserves to the new Forest Service, the agriculture secretary reiterated this mission, emphasizing the forest reserves' essential role in economic development:

> The continued prosperity of the agricultural, lumbering, mining, and livestock interests is directly dependent upon a permanent and accessible supply of water, wood, and forage, as well as upon the present and future use of their resources under businesslike regulations, enforced with promptness, effectiveness, and common sense. In the management of each reserve . . . the dominant industry will be considered first.[6]

Clearly, the Forest Service mission was sustainable resource production to support western settlement and economic development.

Pinchot was a masterful politician with close ties to President Theodore Roosevelt, and his wealthy family founded a school of forestry at Yale, through which Pinchot trained a cadre of professionals for the young Forest Service. He popularized "conservation" as the organizing concept for the Forest Service and for federal land management more broadly. Conservation, Pinchot argued, followed an agricultural model of maximum resource production within natural limits that would not jeopardize the forest's future productivity. Contrasting conservation to the preservation impulse behind the national park system, Pinchot explained, "The first great fact about conservation is that it stands for development . . . Conservation demands the welfare of this generation first, and afterward the welfare of future generations."[7] Just as importantly, Pinchot believed that maximum, sustainable production could be achieved because natural processes were predictable and open to scientific delineation.

In the Forest Service's first four decades, timber production on private lands in the Pacific Northwest and elsewhere was high, and demand for timber resources through the Depression was low. The agency focused less on maximum sustained yield of timber and more on custodial work, such as replanting cut over lands in the East and waging an all-out war on wildland fires.[8] But after World War II, a housing boom as well as general economic growth increased timber demand substantially, and the Forest Service shifted to a truly industrial agricultural model of forestry. Clear cutting became the dominant method of timber harvesting because it was efficient and because it allowed the agency to convert older, slower growing forest, into young, vigorous stands of a single desired species.[9]

At the same time, a growing number of Americans had the money and the time to enjoy the national forests' amenity values, and the number of recreational visitors in the national forests exploded. The Forest Service increasingly found itself caught between the demands of timber production and recreational use, and it promoted a policy of "multiple-use management" to assure Americans that it could do both. In 1960, Congress codified this framework in the Multiple-Use Sustained Yield Act. The national forests, Congress declared, "are established and shall be administered for outdoor recreation, range, timber, watershed, and wildlife and fish purposes."[10] The Forest Service was driven by a "conspiracy of optimism," believing that it could say yes to each new demand that Americans placed on the national forests, yet it continued to privilege timber production over other uses.[11]

The National Park Service (1916)

Congress established the National Park Service in 1916 to manage more than a dozen existing national parks and about twenty national monuments.[12] Whereas the Forest Service was oriented to resource development, Congress charged the Park Service with managing land for recreational enjoyment. In its organic act, Congress directed the agency, "to conserve the scenery and the natural and historic objects and the wild life therein and to provide for the enjoyment of the same in such manner and by such means as will leave them unimpaired for the enjoyment of future generations."[13] The early Park Service, answering to the general attitudes of the period, interpreted this mission as something akin to that of a zoo or museum; the agency's task was to preserve objects—scenic natural features, historical artifacts, and wildlife—so that present and future generations of visitors would have the opportunity to see them. The Park Service, no less than the Forest Service, was focused on individual resources that the public valued rather than on preserving "nature" more broadly construed.

The Park Service mission contained one clear conceptual tension—preservation and use—that did not initially trouble the agency.[14] Park visitation in 1916 was still relatively low, and the agency's first director saw his primary challenge as increasing park use and building a supportive constituency. Stephen Mather was a member of the Sierra Club and a staunch advocate of scenic preservation, but his legacy for the Park Service grew out of his background and expertise in advertising. As soon as Congress established the Park Service, he and his assistant Horace Albright mounted a remarkable campaign to promote the agency and its lands, particularly to the growing number of Americans who owned automobiles.[15] Railroad companies shared some of the marketing burden as they worked to fill their tourist lines to Yellowstone, Yosemite, and other national parks.[16] And Albright, who himself served as Park Service director from 1929 to 1933, built support in Washington, where the echo of Senator Joseph Cannon—"not one cent for scenery"—could still be heard.[17]

The Park Service's focus on auto tourism had profound effects on the national park system. Mather and Albright developed plans for a "park-to-park" highway that had complete circuits of national park units. They even proposed some national parks and monuments specifically to break up long drives between existing units, ensuring that tourists could drive conveniently through the park system. And once visitors got to the parks, Mather and Albright wanted to make sure that they had adequate roadways that would take them to some of the most scenic spots. By the late 1920s, roads were the agency's top budget priority. This emphasis also shaped the agency's staffing choices, leading to a Park Service staffed heavily by engineers and landscape architects but very few biologists.[18] Thus, the Park Service mission in the early twentieth century was not so much to understand and interpret nature but to make objects of natural beauty and intrigue accessible to visitors.

US Fish and Wildlife Service (1940)

Both the Forest Service and the Park Service were established with relatively clear, if somewhat complicated, mandates. The other two main federal land agencies—the Fish and Wildlife Service (FWS) and the Bureau of Land Management—came from more complicated lineages and were established without clear mandates, though they also followed a model of agricultural production and conservation.

President Franklin Roosevelt formed the FWS through administration reorganization in 1940, giving it responsibility for 193 management units that he officially renamed refuges. The units, beginning with Theodore Roosevelt's designation of Pelican Island in 1903, were a hodgepodge created by presidents and by Congress, each with its own mandate. What held them together was a general interpretation that these were places where it was "unlawful to hunt, trap, capture, willfully disturb, or kill any bird or wild animal . . . except as permitted by law or by rules and regulations of the Secretary of the Interior."[19]

The new agency was formed through a merger of the Commerce Department's Bureau of Fisheries and the Agriculture Department's Bureau of Biological Survey, which had both been moved to the Interior Department a year earlier, and it retained its commercial, agricultural orientation. The Bureau of Fisheries began as the US Fish Commission in 1871, organized to address declining Atlantic fish stocks. From its beginning, the agency's primary task was boosting fish production, and one of its first actions was to build a fish hatchery on the McCloud River in California. The agency adopted an agricultural, or in this case aquacultural, approach to fish management, which it implemented throughout the United States.[20]

The Bureau of Biological Survey, which itself had antecedents going back to the 1880s, was officially created in 1905. Like the US Geological Survey at the time, the Biological Survey's goal was to support agriculture and economic development. It researched ways to control agricultural pests and conducted extensive field surveys of various birds. But its significance for federal land management stemmed from two main activities. First, it managed most of the wildlife areas established by the president or Congress. Second, it became the nation's primary predator extermination service. The Biological Survey and the Forest Service conducted reviews of predator problems between 1905 and 1907, and beginning in 1906, the bureau served as an information center for state bounty systems, through which state and local governments paid hunters to exterminate predators.[21] But by 1915, the bureau had begun to wage an all-out war on predators, fought by its own army of professional hunters and trappers. Between 1915 and 1947, the bureau helped extirpate wolves from most of their range, and it killed almost two million coyotes.[22]

The bureau's most effective tool was poison. In 1920 it established the Eradication Methods Laboratory, which developed a wide range of poison compounds and techniques. The most common technique was to lace an animal carcass with strychnine and leave it out as bait for unwanted predators. And when predators learned to identify and avoid strychnine for its bitter taste, the lab fought back by putting strychnine in capsules. This work continued unabated after the bureau was absorbed by the FWS, which announced in 1945 an even more effective poison. Compound 1080 was odorless, tasteless, and water soluble, making it far more effective in the field.[23]

The treatment of predators is illustrative of broader attitudes toward wildlife in the early twentieth century. It followed an agricultural model, served the agriculture and livestock industries, and, as Donald Worster explains, reflected "a peculiarly intense moralism that . . . assigns every species to an absolute ethical category: good or bad. A few wild animals, songbirds chiefly, have been pronounced good; everything else is of use only for target practice."[24]

The FWS had distinct constituencies for each category of wildlife—bad and good. The livestock industry supported the agency's predator control efforts and ensured that Congress appropriated funding each year for the program. Sportsmen supported wildlife refuges politically and financially as production areas for favored game species. Through the federal duck stamp program initiated in 1934 and expanded in 1958, the agency could use money that hunters spent on migratory bird hunting licenses to purchase additional land for the refuge system.[25] Thus, while the taxpayers shared the cost of predator control equally, hunters paid disproportionately to protect game species.

In 1956, Congress finally gave the FWS statutory footing in the Fish and Wildlife Act. It gave the agency broader discretion to "take such steps as may be required for the development, management, advancement, conservation, and protection of wildlife resources through research, acquisition of refuge lands, development of existing facilities, and other means."[26] The act increased the agency's authority, but it did not change its basic orientation toward agricultural production.

The Bureau of Land Management (1946)

The Bureau of Land Management (BLM) was formed by administrative reorganization in 1946 through the merger of the Interior Department's General Land Office (GLO) and the US Grazing Service, and it initially pursued nineteenth-century land and resource disposal policy as much or more than twentieth-century conservation policy. Indeed, two of its primary jobs in 1946 were land disposal (moving land from federal ownership to state and private ownership) and mineral disposal by patent or lease. Its management of renewable resources—forage and timber—followed the agricultural model that the Forest Service had adopted, although it did not have the political strength to regulate grazing effectively by Forest Service standards.

The GLO was created in 1812 to liquidate the federal government's massive, and growing, estate. Over the next century, Congress created literally hundreds of land and resource disposal laws to divest over a billion acres of land and rich reserves of hard rock minerals and coal. The GLO's primary work was to survey the federal lands and handle the administrative work of transferring ownership through sale or grant; it had virtually no land *management* responsibilities.[27]

But the arid lands west of the 100th Meridian were poorly suited to agricultural settlement, and laws such as the Homestead Act of 1862 were simply not practical in this region. As a result, land disposal slowed to a crawl, and it became clear that the federal government would not be able to sell or give the public lands away under the terms of existing law. In 1929, even as Congress continued to build the national forest and national park systems, President Hoover attempted to resolve the arid lands issue once and for all. He wrote to the western governors that these lands "bring no revenue to the federal government. The federal government is incapable of adequate administration of matters which require so large a matter of local understanding. . . . Therefore, for the best interest of the people as a whole, and people of the western states and the small farmers and stockmen by whom they are primarily used, they should be managed and the policies for their use determined by the state governments."[28] He offered to transfer ownership of virtually all of the unreserved public lands managed by the GLO to state ownership, but the states rejected the offer because it did not include the subsurface mineral rights.

In 1934, Congress passed the Taylor Grazing Act, which gave the interior secretary two primary mandates: stabilize the western livestock industry and improve range conditions. The secretary first created a Division of Grazing in the GLO (1934–1939) and later the independent US Grazing Service. The agencies had considerable success regulating who could graze livestock on the public lands through grazing permit and grazing lease programs, but they exercised little control over livestock numbers. The Division of Grazing's first director was suspicious of the conservation movement and of expert, scientific management, and he advocated a system of "split control" in which the ranchers themselves administered the grazing program through advisory committees. While later directors challenged this model of "home rule on the range," Congress was careful to protect it.[29]

The BLM was formed in 1946 in the middle of one such challenge. The Grazing Service director from 1944 to 1946 attempted to professionalize range management and exert greater federal control over range management decisions. Western members of Congress punished the agency with budget cuts so steep that it lost 80 percent of its staff just as it was being merged into the new BLM.[30] The BLM spent the next fifteen years essentially rebuilding the range program's budget and staff and making very modest adjustments in range and forest management.

This brief sketch captures an essential aspect of federal lands history: each of the four main federal land agencies was established with a distinct mandate or a catalog of mandates, and each had distinct constituencies that supported or directed its work. Congress clearly did not intend to provide a comprehensive or common paradigm for land and resource management and certainly did not design the federal lands system to protect large-scale ecological systems. Federal lands management in the first half of the twentieth century focused explicitly on producing or protecting specific resources.

The Ecological Turn

In the 1960s and 1970s, however, environmentalists began to apply an ecological perspective to federal land and resource policy and found it wanting. They criticized the degree to which federal agencies managed for "separate things," sometimes at odds with one another, rather than working to sustain "webs of relationships," and they demanded that Congress rethink federal land and resource management.[31]

The rise of this ecological perspective in the 1960s was driven by two central factors. The first was that the median family income doubled in the two decades after World War II, and with it a growing number of Americans began to see the natural environment as central to a high standard of living.[32] For example, outdoor recreation on federal lands exploded, overwhelming federal agencies and their limited facilities.[33] Visits to the national park system increased from 33 million in 1950 to 79 million in 1960, and visits to the national forest system increased from 27 million to 93 million in that decade.[34] The Park Service responded with Mission 66, a $1 billion infrastructure improvement program from 1956 to 1966, so that the parks could accommodate millions more visitors each year. The Forest Service responded with its own infrastructure program, Operation Outdoors, from 1957 to 1962.

When Americans were not outside enjoying nature, they were inside learning about the natural world through books, television, and films. Disney released its popular film *The Living Desert* in 1953, following the next year with *The Vanishing Prairie.* One of the longest-running television nature programs, *Wild Kingdom,* began in 1963. This genre disseminated a popularized version of the new ecology described in the previous chapter, and a growing number of Americans became concerned about how human development was interfering with the balance of nature.

The transformation of American perceptions of nature in the mid-twentieth century was dramatic. Perhaps the most significant illustration is public perceptions of wolves and other predators. Though still despised by many hunters and ranchers, a growing number of Americans came to view the wolf not as a symbol of evil but as a vital symbol of wild nature. "Instead of the 'despicable,' cowardly wolf, ready to turn on its wounded packmates or devour its own dead," Thomas Dunlap writes, "we are told of wolves living in a natural world where all species have a function, where the death of prey animals is 'the predators' gift to the land.' . . . The coyote has been rehabilitated too. The sly, sneaking 'varmint' is now 'God's dog,' a valued part of the ecosystem."[35] For a growing number of Americans, then, it was not enough that Congress protect scenic views in the national parks; they demanded that Congress protect pristine nature, including predators.[36]

Congress responded to these demands in the 1960s, beginning most notably with the Wilderness Act of 1964, which it passed "to secure for the American people of present and future generations the benefits of an enduring resource of wilderness." Congress defined a wilderness as "an area where the earth and its community of life are untrammeled by man, where man himself is a visitor who does not remain."[37] Congress then required the Forest Service, Park Service, and FWS to survey their lands for areas with wilderness characteristics and began to build a wilderness preservation system that is now well over 100 million acres. (Congress did not extend the Wilderness Act to BLM lands until 1976.) Congress followed the Wilderness Act with other statutes aimed at protecting remnants of "wild nature" for Americans to enjoy, such as the National Trails System Act and the Wild and Scenic Rivers Act of 1968.

The second factor that drove the environmental movement was a growing fear that the same science and technology that created unprecedented prosperity in the United States were simultaneously threatening human health and undermining the very fabric of nature. In 1950, radioactive isotopes from nuclear testing in the southwest were found in New York rain; in the 1950s, researchers linked leukemia to strontium-90 in milk; in 1957, the FDA found traces of a potential carcinogenic pesticide on cranberries just before Thanksgiving; and in the late 1950s, a sedative taken by pregnant women, thalidomide, was found to cause horrific birth defects. Beyond human health, Americans learned that pesticides and herbicides marketed after World War II, which built the green revolution and expanded American agricultural production dramatically, simultaneously threatened wildlife.[38]

It is important to emphasize that this growing fear of toxic threats, perhaps more than the growing interest in wild nature, taught a generation of Americans to see the world not as a collection of discrete objects but as webs of relationships that spanned the entire nation and even the globe. After all, if nuclear testing in Nevada could increase cancer rates in New York, and if spraying DDT to eliminate mosquitoes in suburban neighborhoods at the headwaters could kill birds in the estuary, human actions clearly could have cascading effects over great expanses of time and space. Furthermore, this fear for human health, more than interest in wild nature, gave the environmental movement its breadth and moral force.[39]

Throughout the 1960s, scientists and public figures warned that these chemical threats were not isolated technical problems but symptoms of an overarching environmental crisis. In 1962, the popular science writer Rachel Carson published *Silent Spring*, a chilling account of the impacts that pesticides, or as she called them, "biocides," had on the environment. She explained to her readers how pesticides bioaccumulated in food chains and became concentrated enough at higher trophic levels to decimate bird populations.[40] Carson was a significant catalyst for the environmental movement, but she died in 1964 just as the movement was coalescing. The most influential scientist to carry Carson's message forward and develop it politically and socially was Stanford biologist Barry Commoner. In *Science and Survival*, published in 1967, Commoner warned that citizens could no longer trust science because its advances were producing so many environmental and human threats.

In 1970, the year of the first Earth Day, Barry Commoner appeared on the cover of *Time Magazine*. The feature article described him as "the Paul Revere of ecology" who had "a classroom of millions." Commoner lived up to his title as one of the "new Jeremiahs" of the environmental age the next year with his best-seller, *The Closing Circle*. Here, Commoner argued that Western society and the Western economy needed to be reorganized around four fundamental principles of ecology if humanity were to survive: "Everything is connected to everything else; everything must go somewhere, nature knows best, and there is no such thing as a free lunch." These were not, obviously, scientific laws; rather they were mantras and norms for society, providing segments of the American public with a language for environmental debate that drew on the perceived authority of science. As environmental historian Michael Egan argues, Commoner reflected a significant expansion of the environmental movement, which went beyond the growing middle-class and upper-class interest in natural amenities. Carson, Commoner, and others reframed environmental debate in ways that embraced concerns about human health and social justice.[41]

The environmental movement demanded political action to address both individual concerns and what a growing number of Americans saw as a fundamental environmental crisis. Congress responded with sweeping legislation that transformed the federal government into the nation's primary environmental regulator, including the Clean Air Act of 1970, Clean Water Act of 1972, Federal Insecticide, Fungicide, and Rodenticide Act of 1972, Safe Drinking Water Act of 1974, Resource Conservation and Recovery Act of 1976, Toxic Substances Control Act of 1976, the Comprehensive Environmental Response, Compensation, and Liability Act of 1980, and many others.

What is striking is that demands for federal action came at a time of growing public distrust in the government, and this only increased through the Vietnam War and President Nixon's resignation. The federal government, after all, had funded the science of atomic weapons and nuclear power; the Forest Service, an agency committed to conservation, was accelerating clear cutting, which looked environmentally destructive; and the BLM appeared to serve rather than regulate the livestock and mining industries. Environmentalists wanted action from a government that they didn't trust, which meant that they wanted Congress to pass laws on their behalf *and* they wanted mechanisms through which they could help direct agency discretion.

And environmentalists were joined in the 1960s by others who argued that the sheer size and scope of federal administration placed it beyond meaningful democratic control. Even presidents in the 1960s saw the executive branch as a behemoth that was difficult to control and struggled to reorganize and reform the executive branch so that it could more effectively direct federal agencies. As Arthur Schlesinger Jr. wrote in 1965, "No doubt the mid-century Presidents could blow up the world, but at the same time they were increasingly hemmed in by the growing power of the executive bureaucracy."[42] If elected officials could not control the federal bureaucracy, environmentalists and other critics argued, then Congress had an obligation to make the agencies themselves more directly accountable to the people.

While Congress did not make agency decisions fully democratically, environmental legislation in the 1970s was shot through with requirements for public dissemination of decision-making materials and public participation in the decision-making processes themselves. Looking back over federal land and resource legislation in 1979, resource scholars Gail Achterman and Sally Fairfax put it this way: Congress had created a "responsibility . . . to seek input from the general public, the mass public, the so-called man in the street. The mandate is no longer to provide an opportunity for those who will be affected by a decision to be heard. Under the new theory, the goal is to involve everybody."[43]

Federal land agencies that earlier might only have consulted with grazing permit holders and local officials in its range management decisions now had to open the process to a national audience, including environmentalists from distant cities.

Within this larger ecological turn, Congress took another look specifically at federal land and resource policy, beginning in earnest with the eighty-eighth Congress from 1963 to 1964. The "Conservation Congress," as it is often called, engaged in a fundamental debate over the future of federal lands. This was not, as it became in the 1980s and 1990s, primarily a partisan debate, since Democrats controlled both the legislative and executive branches of government. Rather, it was a debate between those interested in protecting traditional natural resource development and use, and those attempting to redefine conservation and preservation through ecological and aesthetic lenses.

The Public Land Law Review Commission and Multiple-Use Management

To address the tensions between traditional resource development and newer recreational and environmental interests, Congress established the Public Land Law Review Commission (PLLRC), led by Wayne Aspinall (D-CO). Between 1964 and 1970, the commission prepared a remarkably detailed review of the federal estate, including the thousands of disparate and at times conflicting statutes that guided federal land agencies. Its report in 1970, "One Third of the Nation's Land," fell far short of a new comprehensive vision; the commission made 405 recommendations regarding federal lands and resources that to some degree reaffirmed the complex and disjointed legal structure of their administration.

What the commission found in its work was that the number of demands on the federal estate had grown dramatically since the last comprehensive review from 1929 to 1931 and that virtually all of these demands were legitimate.[44] The committee's challenge, then, was to affirm these diverse uses while recommending tools to limit their conflict. The committee also recognized an "ever-growing concern by the American people about the deterioration of the environment," which existing public land law simply didn't address.

In making recommendations for navigating these challenges, the commission essentially adopted a version of the multiple-use management paradigm:

> Statutory goals and objectives should be established as guidelines for land-use planning under the general principle that within a specific unit, consideration should be given to all possible uses and the maximum number of compatible uses permitted. This should be subject to the qualification that where a unit, within an area managed for many uses, can contribute maximum benefit through one particular use, that use should be recognized as the dominant use, and the land should be managed to avoid interference with fulfillment of such dominant use.[45]

The commission's framework was guided by two key assumptions. First, the commission assumed that public lands should provide the "maximum benefit for the general public." Second, it assumed that Congress could not provide a singular definition of either "maximum benefit" or "general public" and needed therefore to endorse something like interest group adjudication in federal land and resource decisions. There were, it noted, many different "publics" that Congress needed to serve: the national public, the regional public, the federal government as sovereign, the federal government as proprietor, state and local government, and the users of public lands and resources.[46] By determining the interest of each group in a given policy area, such as livestock grazing, Congress could look for the decision that would provide the greatest satisfaction to the most groups.

The commission further endorsed delegating this interest group adjudication to federal agencies within broad guidelines. In one sense, it reaffirmed the Forest Service's multiple-use mission as codified in the Multiple-Use Sustained Yield Act of 1964. At the same time, it rejected the Progressive Era view of professional administration as something that needed to be shielded from direct political engagement. The commission, then, recommended that Congress mandate a participatory planning process that would actually insert democratic deliberation into agency decision making:

> Public land management agencies should be required by statute to promulgate comprehensive rules and regulations after full consideration of all points of view, including protests, with provisions for a simplified administrative appeals procedure in a manner that will restore public confidence

in the impartiality and fairness of administrative decisions. Judicial review should generally be available.[47]

To be clear, public participation was not new, since the agencies already consulted with specific land users and local governments. What was new was the scope: "all points of view, including protests."

The report contributed to significant legislative action dealing specifically with the two multiple-use agencies. In 1976, Congress updated the Forest Service's organic legislation in the National Forest Management Act and provided the BLM with its first organic legislation in the Federal Land Policy and Management Act. Both laws made multiple-use management official policy for the agencies. NFMA directed the Forest Service to "provide for multiple use and sustained yield" and "include coordination of outdoor recreation, range, timber, watershed, wildlife and fish, and wilderness."[48] Likewise, Congress instructed the BLM to manage "in a manner which recognizes the Nation's need for domestic sources of minerals, food, timber, and fiber from the public lands."[49] Congress did not provide a ranking of these resources, which it listed alphabetically. Instead, Congress mandated extensive planning processes for both agencies, the hallmarks of which were to be interdisciplinary science and public participation.

The Environmental Statutes that Drove Ecosystem Management

Though the PLLRC's report shaped legislation for specific agencies, it clearly did not address the environmental movement's evolving demand for a comprehensive response to the environmental crisis. The commission was formed in 1964, after all, just as the environmental movement's ecological lens was taking shape, and it was chaired by a representative who, as one scholar puts it, wanted to make sure that federal land and resource policy was not "unduly influenced by the emotional environmental binge."[50] The PLLRC's report focused on balancing *resource uses* rather than on defining and mandating some broader ecological vision of environmental protection.

So even as the PLLRC worked from 1964 to 1970, Congress faced growing pressure to establish a national environmental policy. The biggest problem with environmental law, political scientist Lynton Caldwell lamented in 1963, was that it lacked any clear, overarching purposes:

"Our national tendency," Caldwell argued in 1963, "is to deal with environmental problems segmentally. . . . The public decision-maker . . . must deal with environmental questions without the help of a general body of environmental policy to which he may turn for authoritative guidance." And the consequences, he insisted, were clear: "No massive research is required to document the inadequacy of our environmental decision-making. . . . Segmental thinking, segmental decision making—the 'practical' approach to practical problems—has again and again produced some very impractical results."[51]

Writing specifically about federal land and resource policy in 1968, Caldwell explained that it "is based upon a set of historically derived assumptions—legal, economic, and political—that provide no means for taking the fundamental ecological context of land use into account." By contrast, he opined, "An ecosystems approach to public land policy assumes a scope of policy that embraces all land regardless of its ownership or custody under law. The metes and bounds of ecosystems are determined by physical and biological forces. Men may impose their own arrangements on natural systems, but engineers, surveyors, and lawyers neither amend nor repeal the so-called laws of nature." An ecosystems approach would be holistic and scientific.[52] And rather than advancing an ecosystems approach, he later complained, the PLLRC's report had simply accepted the segmented reality of federal land law: "a multiplicity of redundant, incongruent and conflicting laws and regulations; and underlying all of these, widely differing assumptions in philosophies with respect to Uncle Sam as a landowner."[53]

What Caldwell and other environmentalists demanded was more than the commission's multiple-use recommendations. They wanted comprehensive, substantive statements of environmental protection. And it was in this context that Congress passed the two most powerful statutes that, as overlays on top of NFMA, FLPMA, and other agency-specific legislation, drove ecosystem management: the National Environmental Policy Act of 1969 (NEPA) and the Endangered Species Act of 1973 (ESA).[54] What is more, Congress passed both statutes almost unanimously, demonstrating both that the environmental movement had achieved considerable political strength and the fact that most members of Congress did not anticipate the power these laws would have.

NEPA

Precursors to NEPA date back to the Resource Conservation Act, which Senator James Murry (D-MT) introduced in 1959. In it, he proposed an overarching policy of environmental protection and the establishment of an environmental advisory council, but Congress rejected this kind of comprehensive approach at the time. As public demands shifted during the 1960s, Congress reconsidered, and in 1968 Senator Henry M. Jackson (D-WA) and Representative George Miller (D-CA) organized a joint House-Senate colloquium on the environment. One participant, Governor Laurence Rockefeller, echoed Caldwell's lament: "We do not have a clearly stated national attitude toward the environment. . . . agencies need better guidelines."[55] Caldwell, who was heavily involved in these debates, argued that "some new facility will be needed to provide a point at which environmental policy issues cutting across jurisdictional lines of existing agencies can be identified and analyzed, and at which the complex problems involved in man's relationship with his environment can be reduced to questions and issues capable of being studied, debated, and acted upon by the President, the Congress, and the American people."[56]

For Jackson, the key to environmental protection was broad-scale land-use planning. A broad enough scale could allow planners to find better opportunities to balance economic development, environmental protection, and social equity. "Intelligent land-use planning and management," Jackson later told the Senate, "provides the single most important institutional device for preserving and enhancing the environment, for ecologically sound development, and for maintaining conditions capable of supporting a quality life and providing the material means necessary to improve the national standard of living."[57] Broad-scale planning was not itself novel; indeed, the New Deal had emphasized broad conservation planning. The novelty was Jackson and Caldwell's ecological orientation: "'Conservation' as a concept," Caldwell allowed, "has been helpful principally as an intermediary proposition, midway between unrestricted competition among resource users and an ecologically based view of public responsibility for the self-renewing capabilities of the ecosystem."[58] The new system of land-use planning that Jackson and Caldwell envisioned would be reframed through an ecological lens and would apply to both public and private lands.

Given the legal and political differences between public and private lands, Jackson first introduced a bill in February 1969 to deal with federal policy and planning: the National Environmental Policy Act. As amended and passed on December 23, 1969, its intent was substantive. Its general purposes were to "encourage productive and enjoyable harmony between man and his environment; to promote efforts which will prevent or eliminate damage to the environment and biosphere and to stimulate the health and welfare of man; to enrich the understanding of the ecological systems and natural resources important to the Nation; and to establish a Council on Environmental Quality." It included additional substantive goals of attaining "the widest range of beneficial uses of the environment without degradation, risk to health or safety, or other undesirable and unintended consequences" and assuring "all Americans safe, healthful, productive, and aesthetically and culturally pleasing surroundings."[59]

These aims presented a legislative challenge. On the one hand, it was impossible for Congress to predetermine and therefore mandate all of the substantive decisions necessary to achieve such lofty goals. Agencies needed enormous discretion if they were to even attempt to reach them. But without some strict mandates that were legally enforceable—action-forcing mechanisms—agencies could use that discretion to thwart the law's intent. Late in the process, then, Jackson added a procedural requirement that agencies engage in an interdisciplinary and public review process to determine and potentially avoid the negative environmental impacts of their actions.

Jackson's amendment, which became Section 102 of the act, mandates that federal agencies "utilize a systematic, interdisciplinary approach which will insure the integrated use of the natural and social sciences and the environmental design arts in planning and in decision making which may have an impact on the environment." Specifically, for every "major Federal action significantly affecting the quality of the human environment," the responsible federal agency was required to prepare what became known as an environmental impact statement (EIS). As the courts later interpreted it, NEPA required an EIS to be exhaustive, providing, among other things, a detailed account of "(i) the environmental impacts of the proposed action, (ii) any adverse environmental effects which cannot be avoided should the proposal be implemented, (iii) alternatives to the proposed action [and] (iv) the relationship between local short-term uses of man's environment and the maintenance and enhancement of long-term productivity."[60] What is more, as interpreted by the Council for Environmental Quality and the courts, NEPA required extensive public disclosure and public participation in federal decision-making processes that involved an EIS.

Showing how far the political climate had changed in ten years, NEPA faced little resistance. Given its sweeping policy declarations and binding procedural requirements, it clearly threatened the status quo, yet Congress passed NEPA over only fifteen opposing votes in both houses. Indeed, political support for the act and for a national environmental policy was so strong that President Nixon chose to sign it on January 1, 1970, as a symbol of Republican commitment to environmental protection: "The nineteen-seventies," the president declared, "absolutely must be the years when America pays its debt to the past by reclaiming the purity of its air, its waters and our living environment . . . the decade of the seventies will be known as the time when this country regained a productive harmony between man and nature."[61]

It is important to see in the legislative history that the authors of NEPA intended it to have a substantive impact on federal decision making by linking the procedural planning requirements of Section 102 and the broad, substantive statements of purpose.[62] It is just as important to see that the courts have interpreted NEPA differently. In the landmark decision *Calvert Cliffs' Coordinating Committee, Inc. v. US Atomic Energy Commission* (1971), the D.C. Circuit Court ruled that the substantive goals were flexible, which "leaves room for a responsible exercise of discretion and may not require particular substantive results."[63] Judge J. Shelly Wright ruled that an agency's decision might be struck down on substantive grounds, but only if it reflected such an imbalance of costs and benefits that it was arbitrary and capricious. The strictly enforceable mandate was to prepare environmental impact statements outlined in Section 102. This, the judge determined, bound all federal agencies and permitted no discretion or deviation. This procedural interpretation has frustrated the law's original authors and environmentalists. Caldwell later wrote, "To regard the action-forcing provision of Section 102 (the so-called NEPA Process) as the essence of the Act is to misinterpret its purpose."[64]

Nonetheless, NEPA's procedural requirement transformed federal decision making in a number of critical ways that did push federal agencies to apply an ecological lens in their planning and helped lead to ecosystem management. First, NEPA set new requirements for the use of both interdisciplinary science and public participation in agencies' decision making. Whereas range, forest, and park managers might have made significant decisions in the past based on their professional expertise in a particular field and after consulting with parties they believed were affected, under NEPA, those managers were required to gather interdisciplinary scientific data from a full range of disciplines and to make the EIS process open to full public participation. This forced the agencies to consider new ecological science and to encourage broad public participation, which were both later central to ecosystem management.

Second, and just as importantly, NEPA and later statutes such as NFMA and FLPMA made fundamental changes to the composition of agency staff. Prior to the 1970s, it was generally true that foresters ran Forest Service decision making, range conservationists ran BLM decision making, etc. But in order to meet the requirements set forth by NEPA regulations, the agencies had to hire biologists, archeologists, sociologists, and a host of other "ologists." This opened the door for a whole new professional cadre within the agencies, many of whom applied an ecological lens to federal lands and resources. Without these new professionals, for example in the Forest Service research station, the agencies would not have had the capacity to develop and implement ecosystem management.

ESA

Just a few years after NEPA, Congress passed what is probably the single most powerful federal conservation statute in American history: the Endangered Species Act of 1973. And whereas the courts have read NEPA as imposing procedural requirements, they have ruled that the ESA places strict substantive requirements on both federal agencies and private landowners. NEPA and ESA, then, are fundamentally different laws with different outcomes and different politics, and they have proved to be very different mechanisms for driving ecosystem management.

Early state and federal wildlife policy had focused primarily on protecting a narrow range of "good" species. In the 1960s, though, as a growing public applied an ecological lens to the biotic landscape, they came to see a vastly broader range of plant and animal species as "good" because of the presumed role they played in their ecosystems. What is more, they found that a wide range of species had either gone extinct or were in serious decline. The issue was not, they argued, simply a set of isolated problems: extinction of the passenger pigeon in 1914, the existing population of whooping cranes hovering near extinction with only forty-four individuals remaining, etc. Rather, species extinction was a burgeoning crisis that would have cascading effects ecologically.[65]

In 1966, Congress responded with the Endangered Species Preservation Act. It required the Interior Department to compile and maintain a list of endangered species, and it mandated that Interior, Agriculture, and Defense agencies work to protect those species "insofar as is practicable and consistent" with their missions. The act also gave statutory footing and definition to the National Wildlife Refuge System that had been growing steadily since 1903, and it prohibited the "taking" of endangered species within the refuge system. In total, the act was an important statement of interest in endangered species protection; it was also, two scholars write, "a broad but toothless policy."[66]

In 1969, Congress passed the Endangered Species Conservation Act, which expanded specific provisions of the 1966 act without altering its basic approach, and it passed a number of other statutes that provided more enforceable protection for specific species. In 1971, for example, Congress passed the Wild Free-Roaming Horses and Burros Act unanimously, which places strict limits on agency management of feral horses and burros on federal lands; and in 1972, Congress passed the Marine Mammal Protection Act, which established a detailed regulatory program for charismatic marine mammals such as whales and sea lions.

Support for broader and stronger legislation protecting endangered species continued to grow. President Nixon, struggling to keep Democrats from claiming political ownership of environmental protection, complained in 1972 that "the most recent act to protect endangered species . . . simply does not provide the kind of management tools needed to act early enough to save a vanishing species."[67] In 1973, Congress responded with the Endangered Species Act, which differed dramatically from the earlier acts of 1966 and 1969 in both its scope and its regulatory power. The act had two critical purposes: "to provide a means whereby the ecosystems upon which endangered species and threatened species depend may be conserved [and] to provide a program for the conservation of such endangered species and threatened species."[68] Unlike NEPA, however, the ESA provided both procedural *and* substantive requirements to ensure that these purposes were met: Sections 4, 7, and 9. Section 4 requires the interior secretary to list *any* species that it determines is threatened with or in danger of extinction based solely on the best available scientific and commercial data, regardless of its economic implications.[69] Section 4 also requires the secretary to designate critical habitat necessary to protect the species and develop a recovery plan that will protect the species and lead to its removal from the endangered and threatened species list. And whereas the earlier acts had applied only to wildlife, the ESA extended protection to all animals and plants in danger of extinction.

Section 7, innocuously entitled "Interagency Cooperation," places a strict requirement on federal agencies to "insure that any action authorized, funded, or carried out by such agency . . . is not likely to jeopardize the continued existence of any endangered species or threatened species."[70] What is more, the agencies are not permitted to make this determination on their own but must "consult" with the FWS or the National Marine Fisheries Service. Gone from the ESA was the earlier qualifier from 1966 and 1969 that the agencies needed to protect listed species to the extent *practicable.* This new requirement was unqualified.

Section 9 prohibits *anyone* from "taking" a listed species unless specifically authorized to do so, and the act defines "take" as "harass, harm, pursue, hunt, shoot, wound, kill, trap, capture, or collect, or attempt to engage in any such conduct."[71] The earlier acts had established the modest prohibition against "taking" listed species within national wildlife refuges; the ESA, by contrast, recast this as a blanket prohibition applicable to all public and private land in the United States.

Looking back after forty years, what is perhaps most remarkable about the ESA's passage is the fact that like NEPA it faced little opposition in Congress. Indeed, the final bill that came out of conference passed in the Senate unanimously and in the House by a vote of 345 to 4, and President Nixon was effusive in his praise of that act during the signing ceremony: "Nothing is more precious and more worthy of preservation than the rich array of animal life with which our country has been blessed."[72] In fact, the only real point of debate in Congress was whether or not the ESA infringed on states' rights to determine their own wildlife policy.

The legislative history makes it clear that neither Congress nor the president anticipated the scope or power of Sections 4, 7, and 9. As Nixon suggested in his reference to the "rich array of animal life" and "wildlife," he and many members of Congress were not thinking of protecting *all* threatened and endangered species, even though this is precisely what Section 4 requires.[73] Both Congress and the president seemed to have a modest interpretation of Section 7 as well. One senator explained that an agency was required to consult over their actions but would be free to make a final determination and take actions that jeopardized listed species if they believed it was necessary.[74] And both Congress and the president clearly underestimated the impact of Section 9, which they seem to have interpreted to mean only that a landowner could not directly kill or trap a listed animal. But the ESA included the word "harm" in its definition of take, which the courts have interpreted to mean not only direct harm to a listed species but also indirect harm caused by habitat modification.[75]

In general, then, Congress grossly underestimated the impact that the ESA would have on public and private land use, and certainly members of Congress such as Bob Dole (R-KS), Jesse Helms (R-NC), Bob Packwood (R-OR), and Mark Hatfield (R-OR) later regretted their support for the act. This misunderstanding of the act reflected Congress's scientific understanding of the time. They were working with an older assumption that the primary threat to species was overexploitation, and they were also working with a more static view of the natural world that could be managed through relatively static law.[76]

The ESA's full implications and force became apparent, however, soon after its passage. The Supreme Court ruled in 1978 that the ESA required the Tennessee Valley Authority to stop construction of the nearly completed Tellico Dam in order to protect the three-inch snail darter; and it found in 1995 that the prohibition against "harming" a listed species can be read legitimately as prohibiting the modification of a species' habitat on public or private land.[77] As the ESA's full implications and force became apparent, Congress amended the act several times. It established a committee often called "the God Squad" with authority to exempt particular actions from the ESA if they are essential to the public good, and it authorized "incidental take permits" and "habitat conservation plans" (explained in subsequent chapters) to make the act more flexible. Opponents have also tried, unsuccessfully, to repeal the act.[78]

The ESA's strict regulatory provisions and political durability were the very foundation of ecosystem management in the late 1980s and 1990s. During the 1980s, ecologists and conservation biologists emphasized the importance of "biodiversity" not simply as diversity of species but also as diversity ranging from genetic variability to the diversity of broad ecological systems. And as federal agencies matched this science with the rigid regulatory provisions of the ESA, the act forced much broader preservation activities that threatened well-established resource uses. Ecosystem management, then, emerged out of the struggle to comply with the ESA and maintain as much traditional land use as legally possible.

The Ecological Turn for Federal Agencies

The popular ecological turn of the 1960s and sweeping environmental legislation of the 1960s and 1970s had profound impacts on all four federal land agencies, including the introduction of significant new tension in federal land and resource management. As a general rule, the agencies continued their traditional programs while struggling to add science, public participation, and species protection to their decision making.

US Forest Service

Of the four federal land agencies, the Forest Service faced the greatest challenges in adapting to the new political and legal context. The Multiple-Use Sustained Yield Act of 1960 had affirmed a broad range of forest values and the agency's broad discretionary authority. Between 1969 and 1973, the Forest Service faced unprecedented criticism for clear cutting in the Monongahela National Forest, massive terracing projects in the Bitterroot National Forest, and a disproportionate emphasis on timber production. Rarely had the Forest Service, thought to be the paragon of public administration just a decade earlier, faced such widespread criticism.[79]

Many of these reports accused the Forest Service of betraying its own multiple-use, sustained-yield mission. One prominent report in 1970 found that "Multiple use management, in fact, does not exist as the governing principle on the Bitterroot National Forest," and it accused the agency of engaging in "timber mining" rather than sustained-yield management. The report concluded, among other things, that "the heavy timber orientation is built in by legislative action and control, by executive direction and by budgetary restriction. It is further reinforced by the agency's own hiring and promotion policies and it is rationalized in the doctrines of its professional expertise."[80] Ironically, perhaps, the report came out just as the Nixon administration was ramping up timber harvest levels even further.

The agency also faced litigation over clearcutting in the Monongahela National Forest. In that case, the Izaak Walton League argued that clearcutting violated the agency's original organic act from 1897, which authorized the Forest Service to harvest "dead, matured or large growth of trees," which "shall be marked and designated, and shall be cut and removed." The courts ruled that the act's plain meaning barred clearcutting as a harvest method: "We are not insensitive to the fact that our reading of the Act will have serious and far reaching consequences, and it may well be that this legislation enacted over seventy-five years ago is an anachronism which no longer serves the public interest. However, the appropriate forum to resolve this complex and controversial issue is not the courts, but the Congress."[81] The case threw the Forest Service back to Congress and opened a vigorous public debate over the agency's multiple-use management paradigm.

In 1976, Congress passed the National Forest Management Act (NFMA), which represented a compromise between timber and environmental interests. NFMA reaffirmed the Forest Service's multiple-use mission, and it explicitly authorized clearcutting under certain guidelines. In that way, it ensured that the agency could maintain its vigorous timber program. But NFMA also mandated two things that over time reshaped the agency's mission.

First, NFMA mandated a revised Forest Service planning process that required the agency to prepare a comprehensive forest plan for each national forest every fifteen years. The impacts of this process are inseparable from those of NEPA, since the agency quickly determined that the forest plans were "major Federal action[s] significantly affecting the quality of the human environment" under NEPA and merged its forest planning and EIS preparation accordingly. The new planning process required extensive use of scientific data as well as extensive public participation, giving environmentalists greater access to Forest Service decision making and, more importantly, forcing the agency to hire more ecologically minded specialists. This elevated environmental considerations in Forest Service planning.[82]

Second, if NFMA contained planning requirements that overlapped with elements of NEPA, it also contained a mandate that echoed the ESA. The act included a provision requiring the Forest Service to "provide for diversity of plant and animal communities based on the suitability and capability of the specific land area in order to meet overall multiple-use objectives."[83] Even more importantly, the agency's first set of regulations implementing NFMA, which it issued in 1979, interpreted that to mean

> Fish and wildlife habitat shall be managed to maintain viable populations of existing native and desired non-native vertebrate species in the planning area. For planning purposes, a viable population shall be regarded as one which has the estimated numbers and distribution of reproductive individuals to insure its continued existence is well distributed in the planning area. In order to insure that viable populations will be maintained, habitat must be provided to support, at least, a minimum number of reproductive individuals and that habitat must be well distributed so that those individuals can interact with others in the planning area.[84]

This commitment increasingly forced the agency to look at national forests through an ecological lens, because it had to look at forests through the perceived needs of various species.

As explained in chapters 4 and 5, the commitment to maintain viable populations challenged the Forest Service's commitment to high levels of timber harvesting in the 1980s and 1990s. Newer work in ecology and conservation biology showed that fulfilling the viability mandate required far more than the agency had anticipated. Viable population analysis took into account a range of risk factors for each species, including habitat fragmentation and modification, which increasingly suggested the need for dramatic reductions in timber harvesting. The Forest Service, then, was stuck between its ongoing commitment to timber production, which Congress endorsed each year in the appropriations process, and the legal requirements of the new environmental statutes. For the Forest Service, ecosystem management promised a new rhetoric and a new set of tools to balance these two competing demands.

National Park Service

The National Park Service was not impacted as dramatically as the Forest Service by new environmental legislation, but it was impacted by the popular ecological turn of the 1960s. As the public understanding of "nature" became more ecological, the Park Service's mission to preserve and interpret nature shifted as well. A growing number of Americans, the agency discovered, wanted to experience "wild nature" rather than a collection of zoo specimens. As a result, the agency found that it could no longer simply preserve natural objects for public viewing but needed to preserve these objects *and* the broader ecological processes that sustained them.

Significant change can be traced to the Interior Department's Special Advisory Board on Wildlife Management, which Secretary Stewart Udall established in 1962. Chaired by Aldo Starker Leopold and including a number of prominent scientists, the board released its report, "Wildlife Management in the National Parks," the following year. The Leopold Report, as it is generally called, was significant because it merged an ecological perspective with the Park Service's traditional preservation mission: "As a primary goal," the report began, "we would recommend that the biotic associations within each park be maintained, or where necessary recreated, as nearly as possible in the condition that prevailed when the area was first visited by white man. A national park should represent a vignette of primitive America."[85] Recommending that the Park Service preserve a vignette of primitive America reaffirmed the agency's existing preservation mandate, but the board reinterpreted that vignette in broader ecological terms of maintaining ecosystems.

One notable aspect of the report was its emphasis on nature's dynamism and complexity, which precluded a zoo or museum approach to preservation: "The major policy change which we would recommend to the National Park Service is that it recognize the enormous complexity of ecological communities and the diversity of management procedures required to preserve them." Indeed, authors of the Leopold Report acknowledge that they "are calling for a set of ecologic skills unknown in this country today."[86]

Generally overshadowed by the Leopold Report, the National Academy of Sciences (NAS) released a separate, and damning, study of the Park Service's research program in the same year. The committee found that whereas comparable agencies spent roughly 10 percent of their budgets on research, the Park Service had spent less than 1 percent from 1960 to 1962. It was, for scientists on the committee, "inconceivable that property so unique and valuable as the national parks . . . should not be provided adequately with competent research scientists in natural history as elementary insurance for the preservation and best use of the parks."[87] The NAS report found the lack of research particularly appalling because of how they interpreted the main purpose of national parks. The purpose was not, they wrote, to attract more visitors and provide more amenities; "the purpose of the national parks should be the preservation of nature, the maintenance of natural conditions, the avoidance of artificiality, with such provisions for accommodation of visitors as will neither destroy nor deteriorate the natural features which should be preserved for the enjoyment of future visitors who may come to the parks."[88] And the NAS report went even further than the Leopold Report in emphasizing the complexity and dynamism of nature. Parks, they argued, "are dynamic biological complexes with self-generating changes. To attempt to maintain them in any fixed condition, past, present, or future, would not only be futile but contrary to nature."[89]

The challenge for the Park Service was that the more it viewed the parks and monuments through an ecological lens, accounting for the complexity and interconnected character of ecological systems, the clearer it became that their boundaries rarely corresponded to ecological boundaries. Indeed, maintaining "biotic assemblages" and "biological complexes" was difficult or impossible without support from surrounding landowners. Threats to park resources, they realized, were as often external and therefore beyond the agency's jurisdiction as they were internal and therefore redressable through park management.

By the 1970s and 1980s, the issue of external threats to the parks had gained political traction. In a 1980 report, the Park Service catalogued over 2,000 external threats to the park system, and in 1981, the agency's "State of the Parks" report contained this sobering assessment:

> Because neighbors of federal lands often have different land-use objectives and expectations, conflicts can be formidable. Service actions to protect and preserve the irreplaceable park resources of national significance must be forged in the crucible of the political process as attempts are made to determine the appropriate relationships between park lands and lands of neighbors.[90]

As political scientist John Freemuth explains, people increasingly viewed the parks "as islands surrounded by the rising tide of civilization."[91] And by the end of the 1980s, scholars were addressing this issue through the language of ecosystem management, showing how protection of parks and wilderness areas required protection of the broader ecosystem processes that sustained them.[92] For the Park Service, then, ecosystem management was less about balancing internal tensions in its mandate and more about addressing external threats to parks and monuments.

US Fish and Wildlife Service

Like the Park Service, the FWS's mission was reinterpreted in the 1960s. From its roots in the US Fish Commission and Bureau of Biological Survey, the FWS had already evolved significantly by mid-century. At the same time, the agency still lacked a clear identity and purpose and was often caught among at least three significant constituencies. Preservationists pressured the agency to privilege the protection of rare and endangered wildlife species within the national wildlife system, emphasizing the Migratory Bird Conservation Act of 1929's mandate to establish certain refuges as "inviolate sanctuaries" for wildlife; sportsmen pressured the agency to provide recreational opportunities for hunting and fishing, seeing the refuges essentially as fish and game parks; and the livestock industry pressured the agency to maintain its active predatory control program, including the extensive use of poisons.[93] To the extent that it worked to satisfy all three sets of demands, the agency was pursuing three very different management paradigms: preservation, conservation, and agriculture, respectively.

During the 1960s, Congress passed two significant statutes to address these pressures. The Refuge Recreation Act of 1962 authorized a wide range of recreational activities, but "only to the extent that is practicable and not inconsistent with . . . the primary objectives for which each particular area is established."[94] The Refuge Administration Act of 1966 moved closer to a comprehensive organic act for the agency. It consolidated management units into the National Wildlife Refuge System and charged the FWS with the task of "conserving, protecting, restoring, and propagating selected species of native fish and wildlife that are threatened with extinction."[95] These acts essentially codified aspects of the agency's existing, fragmented mission without providing a unifying mandate.

But by the late 1970s, the agency had been transformed both by shifts in public attitudes and by environmental legislation. In 1960, the agency was still involved in predator control programs supported by the livestock industry, but its Division of Predator and Rodent Control (PARC) faced mounting criticism, both external and internal. In 1964, the Special Advisory Board on Wildlife Management, which had reviewed national park management the year before, issued a report on the agency's predator control program. The program was, the board pronounced, neither necessary nor based on sound science. Rather, it rested "on the subjective judgment of the PARC field men or supervisors in conference with livestock operators and agricultural officials." The program, they continued, was "an end in itself and no longer is a balanced component of an overall scheme of wildlife husbandry and management."[96] The board found in particular that PARC's poisoning techniques were ineffective in protecting livestock and killed wildlife indiscriminately.

Environmental organizations found predator control itself objectionable, but they were particularly outraged by PARC's use of poisons, or toxicants, such as Compound 1080. At a time of growing public concern over toxic chemicals in the environment, PARC's continuing use of indiscriminate poisoning techniques seemed particularly ill conceived.[97] In 1970, Defenders of Wildlife sued the agency, arguing that NEPA required it to prepare an EIS for its predator control program. This put enormous pressure on the administration to address the issue more directly. It responded by commissioning another report, usually called the Cain Report. The 1972 report called for an end to the use of toxicants in predator and rodent control. The report did not condemn predator control as such, recommending only that the agency target individual animals that preyed on livestock. In 1972, President Nixon issued an executive order that banned the use of Compound 1080, strychnine, cyanide, and thallium on federal lands, bringing forty years of systematic federal predator control to an end.[98]

If the agency changed slowly in response to public attitudes towards wildlife, it changed swiftly and dramatically in response to NEPA and the ESA. From a small agency that ran fish hatcheries, managed poorly funded and staffed wildlife refuges, and killed predators and rodents, the FWS quickly became the nation's preeminent plant and wildlife regulator.

As a procedural law, NEPA did not alter the agency's substantive mandates, but it transformed the agency's workload almost overnight. In 1971, two scholars observe, the agency anticipated the following responsibilities:

> Reconnaissance and feasibility studies for 50 Bureau of Reclamation projects, 140 Corps of Engineers projects, and 410 other federal water projects; the review of 6,000 Department of the Army permit applications for work on navigable waterways, leading to the preparation of 400 reports;

> 130 nonfederal applications for the Federal Power Commission, 50 for the Atomic Energy Commission, and 16 for the Forest Service and the Bureau of Land Management. In addition, the Service expected to review at least 1,200 environmental impact statements during the first half of 1971.[99]

Thus, an agency that had already struggled to fulfill its obligations in refuge management, research, and outreach was suddenly asked to serve as an authority in government-wide NEPA implementation.[100]

And even as the agency was drowning in a sea of NEPA paperwork, it became the nation's top wildlife regulatory agency, with responsibility for all terrestrial and freshwater threatened and endangered species. The magnitude of this responsibility is staggering, both in its scope and volume and in its regulatory power. Consider just the following responsibilities. First, the agency is required to review petitions to list species as threatened or endangered, making its final determination after a review of all relevant scientific and commercial data. Second, the agency is required to designate critical habitat and prepare a recovery plan for each listed species. Third, it is required to "consult" with other federal agencies under Section 7 of the ESA to determine whether their actions are likely to jeopardize listed species. The time and money it takes to fulfill all of these responsibilities grossly exceeds the agency's capacity.

As a result of the FWS's workload and resource base, it has been perpetually behind in ESA obligations. Environmental organizations have sued the agency repeatedly and successfully for failing to meet deadlines for listing decisions, critical habitat designation, and recovery planning. Legally, the agency has ninety days to review listing petitions, and if it finds that listing is warranted, it has another twelve months to complete the listing process. But the agency is so limited in funding and staff that it often determines that listing a species is "warranted but precluded," which means that scientific evidence warrants listing but that the agency does not have the resources to complete the listing process. This stops the twelve-month listing clock.[101] And once the agency has listed a species, it often cannot meet its subsequent obligations. In January 1995, for example, the threatened and endangered species list included 944 species, but the FWS had prepared only 415 recovery plans.[102]

NEPA and ESA both pushed the FWS to adopt ecosystem management in the 1990s, but its reasons were different from the other land management agencies. Certainly, like the Park Service, the FWS was concerned with external threats to refuge wildlife, particularly migratory birds. But the real push for an ecosystems approach was the agency's inability to meet the ESA's strict requirements. As the primary regulatory agency implementing the ESA, along with the National Marine Fisheries Service, the FWS supported ecosystem management in the hopes that it would lead to multi-species and landscape-level approaches to protecting vulnerable species, thereby providing more efficient means of protection than the strict species focus of the ESA.

The Bureau of Land Management

Going into the 1960s, the BLM had very different challenges than the other three federal agencies, so the environmental movement and environmental legislation impacted it differently as well. Unlike the Forest Service, which received the Multiple-Use Sustained Yield Act in 1960, the BLM had no statutory footing for its multiple-use management. Indeed, in 1960, the BLM had no organic act at all. It was still managing western rangelands under the Taylor Grazing Act's qualification, "pending its final disposal," and it still had not established the kind of professional authority that the Forest Service had enjoyed for half a century. It was, in many ways, an agency in search of a mission.

Congress gave the agency temporary authority under the Classification and Multiple-Use Act and the Public Land Sales Act of 1964, which were part of the legislative compromise package that created the Public Land Law Review Commission. These acts granted the BLM broader authority and discretion in land classification, disposal, and management, hinting that public lands the agency managed were, in fact, the nation's fourth major federal land system rather than simply leftover lands on the auction block.

In the 1960s, three factors began to reshape the agency. First, with administrative support and temporary legislation, the BLM began to professionalize and articulate a multiple-use mission even in the absence of any overarching legislation. Second, the agency faced growing criticism for its failure to stand up to the livestock and mining industries it served, becoming a favorite example for scholars of an agency "captured" by its clientele. Third, outdoor recreation introduced the agency to a new, and often hostile, constituency that had largely ignored public lands in the past. Congress had never authorized recreational use or funding on the public lands, so the BLM was entirely unprepared for the recreation boom. These three factors pushed or supported the BLM in an incremental move toward the Forest Service model of professional multiple-use management.[103]

In the 1970s, two laws reshaped public lands' politics and management. As with the other three federal agencies, NEPA changed the agency's decision-making process. The BLM had been working toward comprehensive planning in the 1960s, but it had not employed the kind of interdisciplinary, environmental focus that NEPA required. Furthermore, while BLM managers consulted with local land users and governments in their planning, they had not cultivated the kind of broad public participation outlined in NEPA. The BLM therefore faced significant NEPA litigation in the 1970s, leading to a number of landmark decisions. In one of these cases, *NRDC v. Morton* (1974), the court ordered the BLM to prepare 212 site-specific range management EISs, creating 212 requirements for scientific review and public participation in a decision-making process that had historically involved only BLM managers, grazing permit holders, and local governments.[104]

The other statute that significantly shaped the BLM was the Federal Land Policy and Management Act of 1976 (FLPMA), which gave the BLM its first comprehensive, multiple-use mandate, formally ended the policy of wholesale public land disposal, and formalized a comprehensive planning system that the agency had been struggling to develop for over a decade. As with the Forest Service, the combination of NEPA and FLPMA forced the agency to hire a much wider range of "ologists," adding new voices and perspectives to its planning and management.

One of the unique challenges that the BLM faced and faces is that the public lands it manages lack the kind of spatial definition of national parks, national forests, and national wildlife refuges. Their only organization is the administrative districts the BLM has created for planning and management purposes. In some places, the BLM manages every other square mile of land; in other places, the BLM manages isolated 40- or 80-acre tracts surrounded by private land. This has made cohesive management, ecological or otherwise, extremely difficult. While ecosystem management was perhaps embraced weakly by the BLM, the new framework spoke very clearly to the agency's collaborative management orientation.

Conclusion

When President Reagan took office in 1981, his first Secretary of the Interior, James Watt, promised dramatic changes in federal land and resource policy to counteract the rise of modern environmentalism and modern environmental law. He promised, in particular, to reduce environmental regulation and increase natural resource production. Furthermore, President Reagan took office during the Sagebrush Rebellion, when many western states claimed ownership of federal lands within their boundaries. Secretary Watt mollified these demands considerably, not by transferring ownership but by giving western resource interests greater political control over the federal estate. If NEPA, ESA, and other environmental laws were centripetal forces helping to unify federal land policy to a limited extent, Watt and the Reagan administration more broadly strengthened the centrifugal forces that disaggregated federal lands and resources for various uses to avoid what it saw as efforts by environmental organizations to turn federal lands into one large national park.[105]

Despite the administration's success in reducing funding for environmental regulation and boosting production of certain resources, it could not modify or eliminate the new environmental mandates Congress passed in the 1970s. Throughout the 1980s, then, federal land agencies continued to struggle in balancing use and preservation, resource development and resource protection, and they faced growing public criticism and adverse legal rulings that called their expertise into question. All of this set the stage for federal agencies to embrace ecosystem management in the late 1980s and early 1990s, because it offered them a new language and a new framework for fulfilling the conflicting promises that Congress and the agencies themselves had made.

3 | Yellowstone: The Ecosystem Management Laboratory

Introduction

Ecosystem management, the previous two chapters explained, grew out of important conceptual, political, and legal shifts in the 1950s through the 1980s. The paradigms that had guided federal land and resource management—preservation and multiple-use conservation—simply couldn't keep pace. Indeed, one can understand the formal adoption of ecosystem management in the 1990s not as the embrace of a new, fully developed management system but as an admission that preservation and conservation were inadequate. And one can understand intense conflict over ecosystem management in the 1990s as a fundamental disagreement over this very point. Defenders of particular resource outputs or inviolate preserves understood that ecosystem management's complexity threatened their interests.

To illustrate this, one has only to look at the history of the nation's first national park: Yellowstone. Naturalist Paul Schullery explains that Yellowstone has been "the site of something . . . dynamic in human culture, a kind of perceptual experiment that will never end. We are embarked on a great journey of discovery in Yellowstone, one of the most exciting and instructive adventures we have undertaken in our quest to understand this exasperatingly elusive thing we call nature."[1] In addition, Yellowstone has been the site of an experiment in what became known as ecosystem management. Here, federal agencies have looked for ways to provide more integrated management of what is often called the Greater Yellowstone Ecosystem (GYE) (see Fig. 3.1). This experiment succeeded in certain respects but also faced strong resistance from entrenched commitments to traditional preservation and conservation management.[2]

The history of this experiment is instructive because in many ways it prefigured the rise and fall of ecosystem management as an overarching paradigm for federal lands and resources. It began because existing

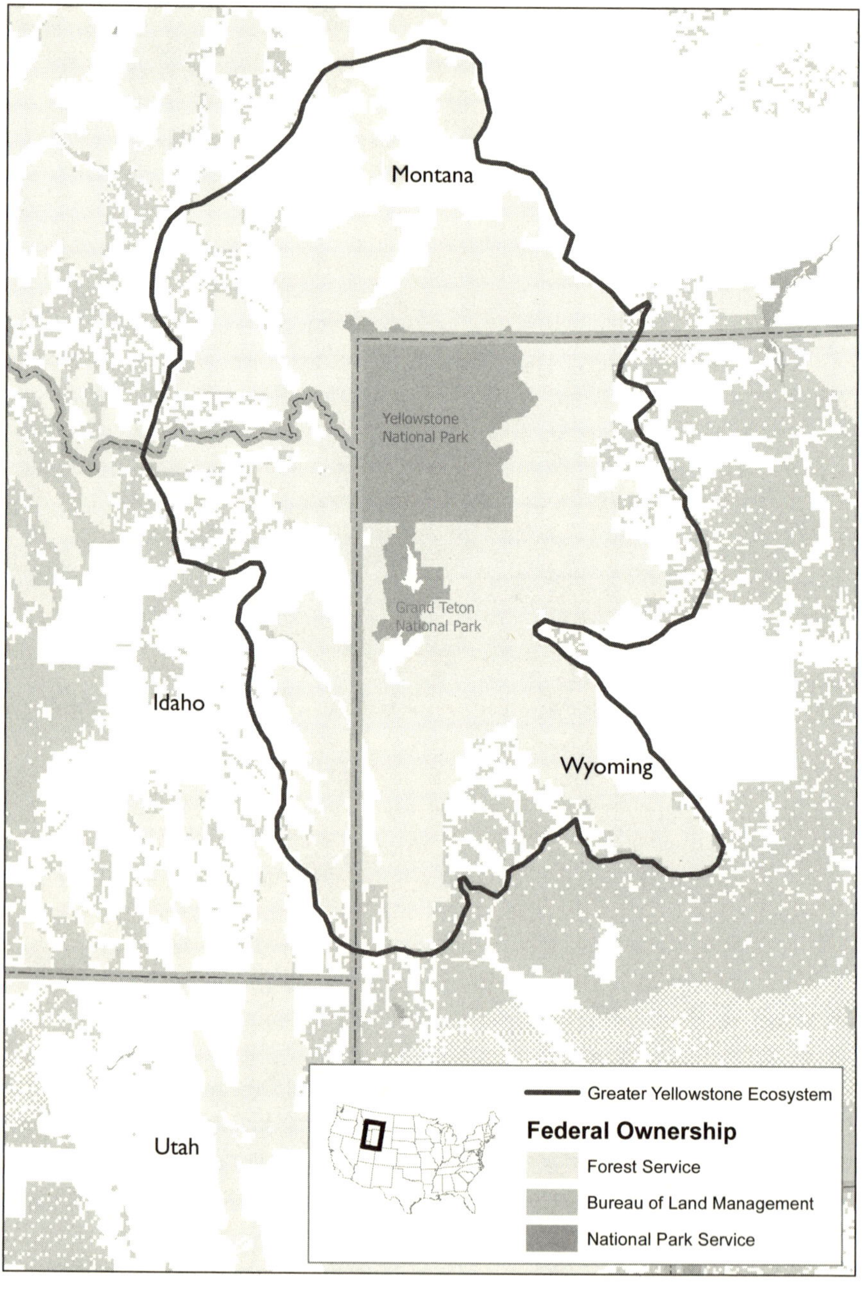

Figure 3.1: The Greater Yellowstone Ecosystem. Map prepared by Cam Patterson.

political boundaries, relatively static goals for resource outputs or scenic preservation, and autonomous professional management within these constraints could not meet important political and legal demands on federal agencies. It declined because these demands were not unanimous and powerful opposition remained to cross-boundary, adaptive, and collaborative management. What happened in Yellowstone at the end of the twentieth century is that certain goals and practices advanced under the banner of ecosystem found space within the complex fabric of federal land and resource management, while the banner itself fell from favor. Here, more specifically, goals and practices required by the Endangered Species Act (ESA) and other federal law found purchase, while more comprehensive visions of ecosystem management faded from view. Or understood differently, the ecosystem management paradigm that agencies embraced was a transparent layer placed on top of existing management practices, and traditional preservation and multiple-use paradigms still shine through brightly.

Yellowstone National Park: 1872–1960

Congress established Yellowstone National Park in 1872 "as a public park or pleasuring-ground for the benefit and enjoyment of the people." Congress clearly had in mind enjoyment of Yellowstone's unusual geological features and remarkable wildlife, and wildlife protection was a major concern at the time.[3] Park managers invested considerable resources feeding elk and bison to increase their populations, and they drove out hunters and killed predators to protect elk and bison herds. Indeed, Schullery explains, the early park was essentially "an unfenced zoological garden for the enjoyment and enrichment of visitors who rarely saw such animals elsewhere."[4]

Elk

Yellowstone National Park was established also at a time when elk populations were declining in many parts of the country, including extinction of the Eastern Elk in 1880. Early park advocates therefore emphasized elk protection and argued as early as the 1880s that the park's legal boundaries should be redrawn to include the full extent of elk habitat in the region. In 1898 and again in 1919, they mounted serious, though unsuccessful, political campaigns to expand park boundaries around elk winter range south of the park and therefore protect a larger amount of what they called the "Greater Yellowstone Area."[5] Thus, even before the National Park Service was formed in 1916 and the term *ecosystem* was coined in 1935, park advocates were trying to draw boundaries for the park as defined by the needs of a particular animal species, intimating an ecological approach to management that would later take hold.

Without changes to the park boundaries, the Park Service was content to manage elk as an exhibit for park visitors. One park historian writes that "the Park Service sought to present to the public an idealized setting of tranquil pastoral scenes with wild animals grazing in beautiful forests and meadows bounded by towering mountain peaks and deep canyons. . . . preserving the scenic façade of nature, the principal basis for public enjoyment."[6] And certainly park visitors in the late nineteenth and early twentieth century came with the expectation of seeing herds of elk and bison.

With this preservation lens the Park Service created an ironic dilemma. By exterminating predators, banning hunting within the park, and feeding elk to reduce winter mortality, the Yellowstone elk herds grew significantly. So even as park managers continued to feed elk, they began to express growing fear in the 1920s that the elk herds exceeded Yellowstone's carrying capacity and were destroying the range. George Wright and other prominent biologists concurred. In a 1934 report, "Fauna of the National Parks," Wright and Ben Thompson argued that elk overpopulation had degraded the northern range rapidly in the previous five years, which threatened both the elk herd and other range-dependent wildlife. The elk population, they said, had to be reduced.[7] Park managers in Yellowstone began culling the elk herd in 1935, planning to reduce the herd to what they believed was the park's carrying capacity of 5,000 animals, and continued reductions until 1967. In 1935, over 3,000 elk were killed, and that number grew to a high of over 4,500 in 1962.[8]

The approach to elk management that park managers adopted was steeped in Clementsian thinking about ecological succession and carrying capacity coupled with extremely limited scientific data. They assumed that the park had a fairly stable carrying capacity, and that by feeding elk in the winter to avoid "unnecessary" mortality and culling elk in the summer to keep the population stable, they could maintain ecological equilibrium within the park. When episodic events, such as the droughts in the 1930s, reduced Yellowstone's range carrying capacity, the Park Service and state game agencies saw this as an aberration from the park's "natural" carrying capacity and simply increased their feeding programs to maintain elk populations at the "natural" level. As Schullery puts it, "the elk were not supposed to die of anything except bullets. This was asking a lot of the elk."[9]

During the thirty years of elk reduction, the Park Service faced growing, though hardly overwhelming, opposition from biologists who complained that the agency lacked a scientific basis for its elk population target. The Park Service did not, however, seem particularly concerned. In 1943, for example, when Aldo Leopold proposed studying the northern elk herd to reassess the scientific basis for elk management, the park superintendent declined, explaining that this would divert funds from the park's essential programs and that it wasn't necessary in order for park staff to make appropriate management decisions.[10]

Bison

The story of Yellowstone bison follows a similar trajectory. Congress established the park at a time of rapid extirpation of bison from a pre-European contact population of perhaps 30 million to less than 1,000 animals by the end of the nineteenth century, 200 of them in Yellowstone.[11] And even in the park, which prohibited hunting, poachers continued to reduce the bison population from 200 in 1889 to 25 in 1894. Responding to growing public alarm, Congress increased the penalty for killing a Yellowstone National Park bison in 1894 to a $1,000 fine, two years in prison, or both, and in 1902, President Roosevelt appointed a game warden in Yellowstone, who began an aggressive recovery program. The game warden, Buffalo Jones, essentially turned the park into a bison ranch, and the bison herd became a significant tourist attraction for the park.[12]

In the late nineteenth and early twentieth centuries, park managers fed bison in the winter to reduce regular winter mortality and bred them to increase the herd.[13] But as the bison population grew rapidly, park managers became increasingly concerned that they had more bison than the park could support. They initiated a bison reduction program, and between 1908 and 1944, park managers shipped 691 live bison from the park and slaughtered 1,863, selling or donating the meat to Indian reservations.[14]

By the late 1930s, however, scientists began to challenge the Park Service's bison management. In the same 1934 report where Wright and Thompson supported elk reduction, the two scientists called for more "natural" approaches to bison management. They explained that intensive management had been appropriate when the bison was threatened with extinction but that by the 1930s the herd was large enough to permit a less intensive program. The Park Service, they argued, should be "devoted to the task of returning the Yellowstone herd to the wild state insofar as the inherent limitations of the park will permit."[15] During the 1940s, park managers signaled a shift in bison management by closing the buffalo ranch.

The shift toward re-wilding Yellowstone bison sparked heated debate, not simply over bison management but over the Park Service's basic mission in places like Yellowstone. Former park service director Horace Albright argued passionately that the bison herd was a central park exhibit, and that the Park Service had an obligation to give visitors the experience of seeing a large herd of these majestic animals. The current director, Newton Drury, by contrast, argued that the mission of the national parks was to sustain natural areas, so "an artificial buffalo show is contrary to long established national park policy."[16] The debate, it should be emphasized, did not focus on whether or not the Park Service should protect nature; rather it focused on the definition of what was or wasn't "natural."

Predators

As noted above, wolf management was no longer necessary within the national park after about 1880, but control of other predators such as cougars and coyotes continued through the early 1930s. This, like the US Forest Service's war on fire, reflected the agricultural conviction that managers needed to eliminate all "wasteful" mortality so as to maximize the outputs of wildlife or timber.

But starting in the mid-1920s, biologists outside the Park Service began to challenge predator control. Organizations such as the American Society of Mammologists and prominent biologists such as Joseph Grinnell and Charles Adams condemned predator control, arguing that predators played important roles in their ecosystems and that the Park Service had absolutely no scientific argument to support continued predator control. The first scientific survey of wildlife in the national parks, "A Preliminary Survey of Faunal Relations in National Parks," appeared in 1933. In it, biologists George Wright, Joseph Dixon, and Ben Thompson went so far as to argue "that the rare predators shall be considered the special charges of the national parks in proportion that they are persecuted everywhere else."[17] And in a 1936 publication, Wright went even further, recommending that park managers allow natural predation to control game species such as elk.

Over the course of the 1930s, the mounting pressure from biologists reshaped the agency's official policy toward predators. In 1931, Park Service director Albright wrote that "predatory animals are to be considered an integral part of the wild life protected in national parks, and no widespread campaigns of destruction are to be countenanced," although he went on to say that predators would still be killed if they posed a demonstrable risk to key wildlife species.[18] And in 1935, the Park Service officially ended predator control outside of Alaska. The Park Service's new policy put it at odds not only with ranchers in the Yellowstone region but with the Forest Service and the Agriculture Department's Biological Survey as well.[19]

Bears

Bears, though shot along with other predators outside the park, were celebrated within park boundaries. Bears were so popular that in the 1920s and 1930s, the Park Service built bleachers next to some of the open dumps where concessioners disposed of food waste so that visitors could watch grizzly bears feeding. Park naturalists even gave evening lectures at the dumps, using them as tools for park education, and despite official policy to the contrary, both visitors and park personnel routinely fed black bears along the main roads in the park. Yellowstone National Park, it would seem, was a large zoo for bears. As Director Albright once wrote, "One of the duties of the National Park Service is to present wild life 'as a spectacle.' This can only be accomplished where game is abundant and where it is tame."[20] Yet as bear feeding continued, both at open dumps and by individuals, injury to visitors and their property increased.

As park managers looked for ways to reduce property damage and protect visitors, they reexamined bear management and began to emphasize the importance of keeping bears wild. In 1942, the biologist Olaus Murie argued that the main issue was that both bears and humans had lost their fear of one another and had too much direct contact. Responding to Murie and others, the Park Service closed the feeding shows and actively discouraged bear feeding by park visitors.[21]

Murie continued to study Yellowstone bears, adding immensely to the Park Service's paltry base of scientific knowledge about both grizzly bears and black bears. Based on his study of bear behavior, Murie became a leading authority and advocate for the protection of bears in their natural and wild condition. In 1945, he wrote to Director Drury, explaining that "natural food habits, reactions to environment, distribution as affected by the associated fauna and flora, and seasonal weather conditions, are all parts of the ecological story offered by an animal like the grizzly." It would violate the agency's mission, he went on, to "destroy these values in an animal that has become so scarce in the United States, simply for showmanship and the satisfaction of an artificially created human appetite."[22]

When taken as a whole, the evolution of park wildlife policy from 1872 to 1960 reveals hints of an ecological perspective on management; park managers moved haltingly away from the initial zoological model of wildlife protection and embraced to varying degrees a commitment to a more "natural" management approach. The ecological perspective was generally provided by research biologists outside the Park Service and by the few biologists on staff with the agency, who emphasized the importance of ecological relationships between individual species and their environment. The scientists called for the protection or restoration of natural conditions in Yellowstone as a key measure to protect wildlife, including the reintroduction of extirpated predator species. They also insisted that the Park Service needed to expand Yellowstone's boundaries so that they corresponded with habitat boundaries of the park's wildlife. It was, they argued, impossible to sustain wildlife while protecting only that part of its habitat located within the park. They argued that the agency needed to base its decisions on scientific research and that it should allow each species in the park to "carry on its struggle for existence unaided."[23] Yet park managers did not turn primarily to scientists for guidance, so they continued to emphasize visitor services and the visitor experience rather than radically redefining their mission in Yellowstone.

The Ecological Turn

As outlined in the previous chapter, the four main federal land agencies went through significant transitions in the 1960s and 1970s that were driven by a popular understanding of ecology and by statutes like the National Environmental Policy Act of 1969 (NEPA) and the ESA. For the Park Service as a whole, the transition began in earnest as a response to the 1963 Leopold Report's recommendation "that the biotic associations within each park be maintained, or where necessary recreated, as nearly as possible in the condition that prevailed when the area was first visited by white man. A national park should represent a vignette of primitive America."[24]

The report expanded some of the arguments that Wright and his colleagues had made in their 1934 report. First, it called on the Park Service to conserve *all* the ecological parts and processes and even to restore those that had been removed by past human action. Second, it also settled on the time of European contact as a baseline for proper ecological management. Finally, it recommended active management of wildlife—including culling—to restore their "natural" population sizes. As quoted in the previous chapter, authors of the Leopold Report understood that they were calling for a new set of skills and a new management framework.

The new ecological knowledge called for in the report grew through wildlife research in the 1960s. In 1959, biologists John and Frank Craighead had initiated a long-term grizzly bear study and had quickly become the world's leading experts on the species. Their landmark report in 1967, *Management of Bears in Yellowstone National Park*, was significant in part because it reflected new methods in the study of bear population dynamics. Placing radio tracking collars on dozens of bears to follow their movements, the Craighead brothers provided a new picture of Yellowstone's grizzly bears: "Yellowstone bears belonged not to the park alone, but to an entire ecosystem."[25]

Responding to the Leopold Report, the Craigheads' early research, and a number of other factors, the Park Service, Forest Service, and US Fish and Wildlife Service (FWS) formed the Greater Yellowstone Coordinating Committee (GYCC) in 1964, which provided a forum for the three agencies to discuss pressing issues that crossed jurisdictional boundaries. The committee addressed fairly circumscribed issues in its first twenty years, but one set of issues increasingly pushed the committee toward broader ecosystem planning: endangered species.[26]

Bears

In 1973 the interior secretary established an Interagency Grizzly Bear Study Team (IGBST), which initiated a long-term monitoring program. The team produced an impressive body of research that only reinforced the idea that grizzly bears did not belong to the park alone but to the entire ecological region. The team found, for example, that the bear's habitat range was larger than the Craigheads had estimated and that the home range of a male grizzly bear was on average 1,450 square miles. They also documented the bears' dependence on cutthroat trout, white bark pine nuts, winter-killed elk, and other resources spread throughout the region. Once again, the research showed that the "bears were not Yellowstone bruins, but creatures of a huge ecosystem."[27]

Because of disagreements over grizzly bear population trends, the interior secretary also requested review by the National Academy of Sciences. The NAS issued the "Report of the Committee on the Yellowstone Grizzlies" in 1974, which served as another indictment of Park Service research. Its authors argued that the agency simply did not have enough information to make reliable management decisions.

That same year, Fund for Animals petitioned to have the grizzly bear listed under the ESA. FWS biologists concluded that the grizzly bear was a threatened species, primarily because of timber harvesting and road construction that destroyed and fragmented its habitat and because those who viewed the bear as a "dangerous vermin" continued to shoot it outside the park. The agency also cited problems with existing bear management efforts, particularly the lack of detailed scientific information.[28]

Under its ESA obligations, the FWS then had to identify the grizzly bear's critical habitat. The agency released critical habitat designations in November 1976, which included millions of acres of national park and national forest land, and it finished a formal recovery plan in 1982. Implementing the recovery required significant coordination among the FWS, the Park Service, and the Forest Service, and the GYCC produced "Guidelines for Management Involving Grizzly Bears in the Greater Yellowstone Area" in 1979. The guidelines divided the region into five land-use categories, or management situations, including core grizzly bear habitat in which all land uses would be subordinate to grizzly bear protection.[29]

In response to the FWS's recovery plan for grizzly bears, in 1983, the interior secretary established an Interagency Grizzly Bear Committee with representatives from the relevant federal agencies as well as the relevant state wildlife agencies. The new committee differed from the study team, which continued its work, in that it was a management rather than a research committee, and it added a new layer of coordination in addition to that provided by the GYCC.

From 1975 to 1985, then, the grizzly bear served as a focal point for debate about management of the Yellowstone region, and it pushed the Park Service and Forest Service into more concrete management coordination. It also played a role in solidifying the idea and language of the Greater Yellowstone Ecosystem (GYE). Indeed, by the end of the 1970s, this phrase was well established among scientists, land managers, policymakers, and environmentalists. There were certainly other species that shaped this discourse, especially elk and bison, but the regulatory force of the ESA, with its attending obligations for grizzly bear recovery, was by far the most significant driver.[30]

Beyond the Bear

The Interagency Grizzly Bear Committee had advanced ecosystem planning, but its mandate was to consider one species. During the 1980s scientists and environmental organizations began pushing to expand this ecosystem approach beyond the bear. In 1983, a number of environmental organizations joined together to form the Greater Yellowstone Coalition (GYC), which by 1987 represented thirty-four regional organizations and ten national organizations. The coalition served as one of the most significant political forces pushing for the agencies to address protection of the ecosystem as a whole.[31] The Craigheads' boundary lines had corresponded to their grizzly bear habitat estimated at 5 million acres, roughly half of which were located inside national park boundaries. As the GYC looked beyond this single species and accounted for other significant bio-physical linkages, it argued that the boundaries encompassed roughly 19 million acres.[32] In 1984, it released a report entitled "Threats to Greater Yellowstone," which identified eighty-eight significant threats to the area. Many of them were threats specifically to the resources of Yellowstone and Grand Teton

National Parks.[33] The organization was particularly concerned with management of national forest lands surrounding the parks, where the Forest Service permitted logging, mining, and drilling activities, and it called for a buffer zone around the parks in which those activities would be limited.

In 1985, this report featured prominently in a congressional hearing hosted by house subcommittees on public lands and national parks and recreation. The hearing highlighted partisan battle lines around ecosystem management that would continue into the twenty-first century. Democrats would consistently use a popular ecological lens, emphasizing scientific authority and calling for more cohesive federal land management; Republicans would rebuff these calls as thinly veiled efforts to turn all federal lands into parks and wilderness areas. Subcommittee chairs John Seiberling and Bruce Vento, both Democrats, expressed their concern over inadequate protection for and management coordination in the GYE, while Republican representatives Dick Cheney, Larry Craig, and Ron Marlene challenged the very basis for this concern. As Cheney explained at the outset of the hearing, "I think the burden is . . . to show that the current system needs repair. . . . that somehow the current system with all the various requirements that have already been imposed is not functioning properly."[34]

The hearing is instructive, because it addressed most of the major concerns about an ecosystem approach to management. First, subcommittee members and witnesses discussed the difficulty of defining an ecosystem as a management area. The Park Service director, for example, stated that "the art of ecosystem management is in its infancy mainly because one must first ask, 'How is that ecosystem defined? By whom? From what perspective?'. . . We know what the ecosystem is of the grizzly, for example. We can define that. But when we use it as a general term . . . we are using it incorrectly scientifically."[35] Republican representatives emphasized this ambiguity throughout the hearing, pushing the debate toward management of specific resources rather than of a region.

Second, the hearing highlighted issues of remaining uncertainty about the connections within the GYE. This is where the GYC's report featured prominently. Republican representatives challenged the concern that drilling outside of Yellowstone National Park would impact the park's geysers or hydrology generally, and one even challenged the assumption that grizzly bears needed to forage outside of the national park boundaries. Cheney insisted that "Yellowstone is in fact in excellent condition . . . maybe what the committee ought to be looking at is not threats to Yellowstone, but threats to other national parks if we are going to pursue ecosystem legislation or some kind of management concept."[36]

Third, the hearing highlighted Republican concern about how the science and rhetoric of ecosystem management would impact agency discretion, particularly that of the Forest Service and the BLM. Republicans supported agency discretion, but they worked through the appropriations process to control it. They were suspicious that ecosystem management would reframe agency discretion, pushing the multiple-use agencies toward park-style preservation. Ron Marlenee (R-MT) asked at one point, "this whole hearing is about total control of issues . . . on land outside of the park by the Park Service?"[37] The multiple-use agency directors were also concerned about how their discretion might be altered in the GYE, concerned in particular that Congress might expand the park boundaries or in some other formal way provide the region with a unified mandate. Forest Service Chief Max Peterson insisted that "the Yellowstone area has been a model for interagency cooperation and management. We expect . . . that we be fully cognizant of the natural values of the area and exercise care and sensitivity in the decisionmaking"; and the BLM Director Bob Burford assured the subcommittees "that existing statutory authority provides an adequate mechanism to protect the GYE and the multiple values of the non-park federal lands in the area."[38]

After the hearing, the subcommittee chairs asked the Congressional Research Service (CRS) to gather more information and make recommendations on how the agencies could improve their management administratively. The CRS researchers first prepared a map of the GYE that intentionally followed federal property boundaries when possible, ultimately identifying the ecosystem as roughly 14 million acres and 90 percent federally owned.[39]

CRS researchers gathered a vast amount of data from federal and state agencies to determine how well they were managing the ecosystem. The report highlighted inadequate protection of grizzly bears, which it also identified as a prime indicator of ecosystem health in the region. It pointed to unmitigated disruption caused by the growing miles of roads built for resource extraction and recreation. The report also identified enormous weaknesses and gaps in the scientific information that the agencies used for coordinated management. The problem was both the lack of available data and the grossly inadequate data coordination and sharing. After receiving the report, the subcommittee chairs observed that there was a striking unevenness in the agencies' data for the region: "One realizes that each agency has its own special set of needs as well as constraints when it comes to collecting information. But there is also a need to look at the 'big picture,' and that has proved exceedingly difficult to do. It is precisely the inability to do so that has permitted some of the major ecosystem problems to build until today they seem nearly intractable."[40]

Even before the CRS report came out, the agencies were chastened. One month after the hearings, the Park Service and Forest Service created a dedicated planning team to jump-start a new coordination effort, and the next year the agencies signed a memorandum of understanding that redefined the GYCC's work. The new planning team, headed by Sandra Key from the Park Service and Jack Troyer from the Forest Service, served as the GYCC's primary staff on its next two major documents. The first document, "Greater Yellowstone Area: Aggregation of National Park and National Forest Management Plans" (1987), was a compilation and analysis of all existing general management plans for the national parks and national forests within the GYE. The planning team was generally positive in its evaluation of the existing plans, but it also pointed to gaps and inconsistencies in agency data and tensions among the various park and forest plans.[41] The "Aggregation" document, which confirmed many of the criticisms laid out in the 1986 CRS report, was largely noncontroversial. It was, after all, primarily a compilation of data rather than any statement of planning or management direction.

Next, however, the planning team in Billings began to work on a document intended to advance coordination and collaboration across the wide-ranging planning areas so that, among other things, Congress would not be tempted to "tinker with boundaries, and curtail multiple use."[42] The GYCC completed a draft document in 1990, "Vision for the Future: A Framework for Coordination in the Greater Yellowstone Area," which established overarching goals for national parks and national forests in the Greater Yellowstone Area. The authors were careful to emphasize that the vision document "is not a regional plan. . . . It does not make specific land allocation decisions, and does not seek to change the separate missions of the national forests or the national parks."[43] Therefore, the vision document was not subject to NEPA requirements.

Nonetheless, they also made clear that it was intended to provide meaningful goals that would direct future planning by the Park Service and the National Forest Service. It was intended, in other words, to direct agency discretion toward common ecosystem management goals.

The document itself deserves attention, because it shows how the agencies tried to balance the competing pressures to maintain the status quo in terms of basic agency missions, mandates, and management goals while simultaneously reframing them so that they would collectively improve protection of the Greater Yellowstone Area, a title that the agencies preferred to the Greater Yellowstone Ecosystem. On the one hand, the vision document focused only on the roughly 12 million acres of national park and national forest land in the region, assuring its readers that it would not impact privately owned land. Furthermore, its authors made clear that the vision document would not interfere with any existing rights to use federal lands and resources and that "opportunities for recreation and commodity development, including timber harvesting, grazing, and mineral development will be provided for on appropriate federal lands."[44]

On the other hand, the document also identified a new, central goal for the GYA that, if achieved, would be a major shift in the overall management of federal lands in the region: "The overarching goal is to conserve the sense of naturalness and maintain ecosystem integrity in the GYA through respect for ecological and geological processes and features that cross administrative boundaries. This will result in high-quality air and water, complete protection for fragile geothermal features, and maintenance or restoration of biological diversity and resulting fish and wildlife populations." And at the end of the document, the authors implied that they were pioneering a new land management approach: ecosystem management.[45]

In essence, the vision document promised that the agencies would fulfill all of their diverse resource mandates while simultaneously achieving ecological sustainability, and they would do it administratively through better coordination rather than through any change in law. The larger scope of integrated management would allow the agencies to better arrange land uses so as to avoid conflicts between environmental protection and resource use and development.[46]

The GYCC released the draft vision document in July 1990 to "an unprecedented, extensive, and explosively heated public comment period."[47] On the one hand, many comments reflected the fear that congressional Republicans had voiced in the 1985 hearing, namely that the vision document was essentially an attempt to manage the entire Greater Yellowstone Area like a national park.[48] One critic wrote to Interior Secretary Manuel Lujan lamenting, "my worst fears are being realized, and it appears that this truly is developing into a creation by an association, if not a conspiracy, of preservationist-minded individuals and groups."[49] More importantly, governors of the affected states along with a majority of the congressional representatives opposed the vision document vehemently. They turned immediately to Interior Secretary Lujan, demanding substantial revisions of the document: "The crux of our concerns rests with the worry that this document will ultimately be used as a 'master' management plan and will result in the wholesale amending of the individual forest management plans . . . that could undermine the concept of multiple-use on public lands."[50] Finally, popular opposition was evident through public protests in communities that depended on federal resource development.[51]

On the other hand, environmental organizations applauded the vision document for addressing the GYE and offered less vitriolic criticism that the document did not provide action-forcing mechanisms to ensure that the agencies prioritize ecosystem protection: the Greater Yellowstone Coalition complained that the "current draft clearly does not go far enough to rein in environmentally damaging resource management activities in Greater Yellowstone"; and the National Parks and Conservation Association argued that the final draft "should clearly state . . . that activities that impair Ecosystem resources or values will be precluded."[52] Essentially, the environmental organizations wanted to reduce agency discretion dramatically to ensure that ecosystem protection would trump resource development whenever they believed that the two conflicted.

The planning team leaders and other agency employees expressed surprise at the document's reception, largely because they had seen the document as a broad policy statement that would influence but not necessarily reduce agency discretion. But as one of the region's forest supervisors later reflected, the draft vision document was "much more than broad coordinating guidelines. It provided for very specific guidelines to be used in obtain [*sic*] the vision of the GYA. Many of our publics recognized this specificity, as we should have, and concluded that it was truly a planning document."[53]

The draft vision document alarmed the Bush administration, since it was essentially in the middle of an ecosystem management crisis with the northern spotted owl in the Pacific Northwest. There, the National Forest Management Act (NFMA) and the ESA were putting enormous pressure on the administration to privilege species protection over resource development, and administration officials were intent on avoiding another ecosystem management crisis in the Yellowstone region. Therefore, they took an intense interest in the document and its revision.

With direct oversight from regional offices and Washington, the GYCC revised the draft document and released it as "A Framework for Coordination of National Parks and National Forests in the Greater Yellowstone Area" in September 1991. The final document was much leaner—ten pages compared to seventy-four in the draft—and while it retained many of the basic goals found in the vision document, the framework document was fundamentally different. Goals in the document had been stripped of controversial environmental vocabulary. For example, the first goal, "Conserve the Sense of Naturalness and Maintain Ecosystem Integrity," was rephrased in more sterile language as, "Maintain Functional Ecosystems."[54] The short document also included affirmations of each agency's distinctive mission and elaborated very little on how the different agencies would achieve broad ecosystem goals.[55] As Jack Troyer, co-director of the Billings planning team, wrote during the revision process, "Expectations based on current direction and comment led to a final document . . . that absolutely states the separateness of agency missions, even more clearly disclaims the relationship of private lands, and replaces the words sense of naturalness."[56] The final draft was, in effect, written to assuage fears that the national parks and national forests were being subsumed in a single preservation plan.

Oversight from agency leaders and Interior Department officials led to charges of undue political manipulation and prompted a lengthy investigation by the House Subcommittee on the Civil Service. Soon after the framework document emerged, three senior Park Service and Forest Service staff who bore responsibility for the vision document received involuntary reassignments, and the planning team in Billings was disbanded. The House subcommittee and its bipartisan investigation team ultimately concluded "that the Department of Interior engaged in a politically motivated, underhanded operation to destroy the draft vision document because it was unacceptable to powerful and monied commodity and special interest groups."[57]

The GYCC continued to meet, but it focused once again on specific, largely noncontroversial issues, such as unified outfitter guidelines for the Greater Yellowstone Area and data management.[58] The Bush administration had nipped the first large-scale federal ecosystem management project in the bud in order to ensure that existing land and resource programs could be managed as separate things rather than integrated into a single program focused on ecosystem protection.[59]

Beyond the Vision and Framework

In the 1990s, the politics of the Greater Yellowstone Area made a comprehensive ecosystem management plan unlikely if not impossible. Even the Clinton administration, which championed such plans elsewhere, did not reinitiate a vision-type project. This does not mean that the agencies ended all efforts at integrated management of the ecosystem. They simply moved forward without a strong, integrating policy of ecosystem management. Some management issues were framed by ecologically oriented rhetoric and law, and some management issues were framed by economically oriented rhetoric and law, even within the broader arena of wildlife management. This can be seen most clearly in the management of wolves and bison.

Wolves: From Ecosystem Management to Ecosystem Restoration

As explained earlier, predators were significant indicators of changing public values. Once widely despised as varmints, wolves had become for environmentalists the very symbol of wild nature. One sociologist explains, "environmental activists argue that the wolf is the last remaining link in a chain of ecological restoration to a prehistoric time . . . wise use activists argue that wolf reintroduction is primarily a strategy to gain increased social control over private property."[60] Since wolves were listed as an endangered species, though, environmentalists had the upper hand. So even though the agencies did not have an overarching ecosystem management plan, it might be said that the agencies practiced ecosystem management, in this case with an emphasis on ecological restoration, of the wolf.[61]

The first, and perhaps most striking, aspect of wolf recovery is that it happened at all.[62] As noted above, wolves were systematically exterminated in the region, and officials trapped the last wolf in Yellowstone National Park in 1926. What is more, fear of wolves and support for their extermination continued to exist in the Greater Yellowstone Area throughout the twentieth century. Livestock growers in particular were grateful that wolves had been removed, and they staunchly opposed the wolf's return. Wolf recovery happened nonetheless because a substantial majority saw the wolf in ecological terms as a keystone species in the Yellowstone ecosystem that added significant value to Yellowstone National Park and its surrounding lands.[63]

Calls for wolf recovery in Yellowstone National Park began in the late 1960s, growing out of the Leopold Report and the Park Service's subsequent emphasis on preserving natural conditions as vignettes of primitive America. As early as 1966, biologists and conservationists recommended wolf reintroduction in Yellowstone as a way to restore natural conditions in the park and to check the growth of elk herds in the park, and a year later the FWS listed an eastern subspecies of wolf under the Endangered Species Preservation Act of 1966.[64]

Responding to the changing politics of wolf protection, the Interior Department held an interagency meeting in Yellowstone National Park in 1971 to discuss wolf management, both the possibility that wolves were naturally recolonizing the park and the possibility of reintroducing wolves, and it initiated a study to determine whether or not wolves were naturally repopulating the region. Before the wolf study ended, though, the fundamental legal framework for wolf protection shifted dramatically. Congress passed the Endangered Species Act of 1973, and the FWS listed the gray wolf, *canus lupus*, as endangered throughout the lower forty-eight states, with the exception of a Great Lakes population that were listed as threatened. This provided the first binding regulatory mechanisms for protecting the gray wolf and a much stronger justification for restoring them to parts of their historical range that would still support them.

Listing the gray wolf as an endangered species had a number of important consequences for wolf management in the GYE. First, after the interior secretary listed the wolf as endangered or threatened throughout its range in the lower forty-eight states the GYE became one of three core protection areas for the gray wolf in its northern Rocky Mountain range. Second, whereas earlier discussion of wolf policy had been at the administrative level, the ESA brought Congress and the courts directly into the debate and shifted the burdens of responsibilities among federal agencies. So even though the GYCC was well established and working on cross-boundary management challenges by the mid-1970s, it was not asked to manage wolf recovery. This responsibility fell to the FWS and to a dedicated Northern Rocky Mountain Wolf Recovery Team that focused exclusively on the endangered wolf.[65] What is more, the recovery team differed substantially in its membership representation from that of the GYCC. Rather than primarily a committee of Park Service and Forest Service staff, the committee was chaired by a representative of the Montana Department of Fish and Game and included members from the University of Montana, the National Audubon Society and other environmental organizations, and the livestock industry.[66]

Convened in 1975, the Northern Rocky Mountain Wolf Recovery Team issued a preliminary recovery plan in 1980. It identified three areas suitable for wolf recovery—central Idaho, northwestern Montana, and the Yellowstone area—but it stopped short of recommending wolf reintroduction. In 1982, Congress provided a new legal option for wolf reintroduction, the "experimental nonessential" population provision of the ESA. Under the new provisions, the FWS could introduce experimental populations of a species and afford them a lower level of protection than mandated for listed endangered species. In particular, it gave the service flexibility in permitting private property owners to "take" (kill, harass, etc.) wolves under defined circumstances.

In 1987, the recovery team issued a revised Northern Rocky Mountain Wolf Recovery Plan. The new plan offered a more specific timeline for wolf recovery, and by a margin of one vote the team recommended wolf reintroduction in the Yellowstone area. The recovery plan triggered the NEPA requirement for an Environmental Impact Statement that would analyze the impacts and alternatives to the plan. Here, the debate moved out of administrative hands and squarely into the congressional domain when Congress barred funding for the EIS.

In 1988, Congress once again blocked funding for an EIS, but approved $200,000 to fund a study called "Wolves for Yellowstone?" In 1990, the FWS and the Park Service released the first two volumes of "Wolves for Yellowstone?," reporting that wolf reintroduction was not likely to have an adverse effect on the livestock industry or on other wildlife in the park and that it might actually increase tourist revenue. The report's findings acknowledged significant uncertainties with respect to the impacts of wolf reintroduction, but it still provided strong support for wolf reintroduction by contradicting some of the more extreme predictions about wolves undermining the livestock industry or wiping out elk populations.[67] Despite the report, Congress remained divided and continued to bar funding for a full EIS on wolf reintroduction.[68] Certainly the political firestorm over the vision document did not put Republicans in a particularly good mood to support ecosystem restoration projects.

What broke the gridlock in Congress was not a shift in views on wolf reintroduction, but evidence that wolves were likely to repopulate areas of the northern Rocky Mountains naturally and thereby automatically receive full protection as a listed, endangered species. A growing number of those who opposed wolf protection overall therefore began to work on a proposal for reintroduction that would give states and private landowners more options for wolf control than they would have after natural repopulation. Senator James McClure (R-ID) became one of the most ardent supporters of this type of compromise approach and introduced a bill, the Northern Rocky Mountain Gray Wolf Recovery Act of 1990, which would authorize reintroduction with two key caveats. First, wolves would be reintroduced in designated "Core Zones" and natural repopulation would be permitted in a designated "Natural Recovery Area." The wolves would receive full protection under the law within those areas. In exchange, wolves outside of the designated "Core Zones" and "Natural Recovery Area" would be delisted, removing all ESA protection. This would appease those who wanted to see wolves back in Yellowstone while giving states and private landowners assurance that they could control wolves as they saw fit in other areas. Second, the bill would prohibit the interior secretary from ever relisting the wolf under the ESA, reserving that authority for Congress alone. The bill amounted to a species-specific amendment to the ESA.[69]

Testimony at the hearing for Senator McClure's bill followed a well-established script. Environmental organizations argued that wolf recovery was essential for the GYE. One witness explained it this way: "Do you all remember a few years ago when that lunatic smashed the face of Michelangelo's Pieta at St. Peter's in Rome with a hammer? That is what wiping out the wolf in Yellowstone seems like to me. We now have the opportunity, actually, to undo an act of savage vandalism against one of the world's greatest masterpieces and restore it to wholeness."[70] Witnesses from the agricultural and livestock organizations in Idaho, Montana, and Wyoming remained just as vehemently opposed to wolf reintroduction. One witness from the Idaho Farm Bureau Federation explained, "It is inconceivable for me to imagine, after the successful eradication of this plague and menace to our society in the mountain west, we are even thinking about the reintroduction of the wolf. . . . The reintroduction of wolves threatens livestock and the viability of the livestock operation."[71] And as with the hearings in 1985, Republican members of Congress and representatives of commodity groups expressed a deeper concern that environmental organizations were simply "using a species' status to gain control over policies and ministrations of the region."[72]

In the end McClure's provision for delisting the wolf outside of the designated areas undermined support for the bill. With legislative action unlikely, and at the suggestion of the FWS Director, Congress authorized a Wolf Management Committee and asked it to find a compromise solution that had eluded Congress. The new committee included just three representatives from federal agencies, three representatives from the states, and four representatives of the main interest groups involved. The law that authorized the committee stipulated that it should develop a consensus plan for wolf reintroduction supported by at least six of the ten committee members by May 1991.[73] After several months of tough negotiation and periods of gridlock, the FWS offered a proposal that received the requisite votes. The proposal recommended reintroduction with a nonessential experimental population designation. Furthermore, it recommended giving the states primary responsibility for wolf management and allowing landowners to shoot wolves in this population on sight.[74] As with Senator McClure's bill, the committee recommended a de facto, species-specific amendment to the ESA.

The report generated its own controversy, particularly over its amendment of normal ESA requirements, and it remained clear that the proposal would not be the template for wolf reintroduction. Finally, in November 1991, Congress authorized funding for a full EIS, and the Interior Department formed the Gray Wolf Interagency EIS Team to carry it out. Planning was once again returned to the agency level. The draft EIS appeared in 1993. It included alternatives ranging from implementing the Wolf Management Committee's recommendations to not reintroducing wolves at all, but its preferred alternative was to reintroduce wolves in Yellowstone and central Idaho as nonessential experimental populations. This meant that, unlike naturally repopulating wolves in the Glacier National Park area that received full protection as an endangered species, the preferred alternative could include compromises important to the three states involved and to the livestock industry. These compromises included nonlethal harassment of wolves to scare wolves away from private property and killing wolves caught killing livestock. The preferred alternative set the bar for delisting the wolf at ten breeding pairs in each of the three recovery areas. In the summer of 1994, the secretaries of interior and agriculture signed the record of decision on the final EIS, paving the way for wolf reintroduction.[75]

In January 1995 and again in 1996, the FWS imported wolves from Canada, and in 1996 it relocated ten wolves from northern Montana to Yellowstone. In all, the agency released seventy-six wolves in Yellowstone and central Idaho, and over the next five years, the wolf populations in all three recovery areas increased even faster than expected.[76] Livestock depredation did occur in the affected states, but not at the level predicted by the more ardent opponents of wolf reintroduction. Furthermore, in order to build political support for wolf recovery, Defenders of Wildlife established a trust in 1987 to compensate ranchers who lost livestock to wolves, and the Wolf Compensation Trust continues to pay ranchers for these losses. According to the organization, between 1987 and 2009, wolves killed a total of 3,832 animals (about 175 per year), and the trust provided a total of $1,368,043 to compensate livestock operators for these losses.[77]

Since wolf population numbers rose faster than expected, the biological criteria for recovery were met in 2002, and the FWS began the process of delisting the wolf in 2003. Legal conflicts lengthened the process considerably, but in May 2011 the FWS delisted the gray wolf in Idaho and Montana, and in August 2012, the FWS delisted the wolf in Wyoming. In announcing this final delisting, the FWS director pointed to wolf recovery as a major success of the ESA.[78]

The work of wolf recovery is an interesting case of what might be called, perhaps ironically, species-specific ecosystem management. The preferred EIS alternative included an ecological perspective in that it emphasized the natural boundaries established by wolves rather than pre-established political boundaries. It required collaborative efforts, although the primary collaborative work and political compromise took place at the congressional level rather than the administrative level. And it balanced the socioeconomic interests of ranchers with ecological values by utilizing the nonessential experimental designation and allowing ranchers to "take" wolves under certain conditions, while Defenders of Wildlife provided funds to compensate ranchers for livestock losses to wolf predation.

Bison

If wolf recovery was framed in ecological terms, albeit species-specific terms, bison management was framed primarily in economic and political terms. Here the question was not how federal and state agencies could work together to restore an ecological component of the GYE; rather the question was how federal and state agencies could collaborate to stop an ecological process, bison migration, from threatening domestic livestock. In the former case, the ESA mandated ecological restoration; in the latter, the ESA had no influence. Bison management, with its own constellation of legal authority, administrative participants, constituencies, and geographic areas, therefore lies even further along the spectrum away from the kind of holistic, integrated ecosystem management attempted in the vision document.[79]

As explained above, the initial impetus for controlling the number of Yellowstone bison was a concern that the herds would overgraze their range and damage park resources. Around the same time, though, an additional layer of concern emerged, this time a concern about the impact that Yellowstone bison could have when they migrated outside the park boundaries. In 1917, bison tested positive for exposure to the bacteria *Brucella abortus*, which can result in brucellosis, a contagious disease that causes spontaneous abortions in domestic cattle and undulant fever in humans.[80] Since 1934, the USDA Animal and Plant Health Inspection Service (APHIS) and state authorities have worked to eradicate brucellosis in domestic livestock and have spent over $3 billion in the effort.[81] Together, federal and state agencies have created a comprehensive program for domestic livestock. While the program has a number of important layers, the most significant for bison management is a classification program that ranks states according to the rates of brucellosis infection in their cattle. For states that are declared free of brucellosis, ranchers may ship their cattle across state lines with few restrictions. But as the rates of brucellosis infection increase, APHIS imposes costly requirements for testing and vaccination. Thus, eradicating brucellosis is a priority not only out of concern for individual herds of cattle but for a state's overall livestock industry.[82]

As initial efforts succeeded in reducing brucellosis in domestic cattle, federal and state officials expressed growing concern that Yellowstone bison threatened to undermine the eradication program by transmitting the disease to cattle grazing on federal and private land around the park. Park officials refused to test and slaughter bison carrying the disease, but in 1968, they agreed to contain the otherwise free-roaming herds of bison within the park to block their interaction with livestock grazing just beyond the park boundaries. They hazed bison trying to leave the park, set up fences to contain them, and shot bison that made it through these initial lines of defense. It is interesting that this took place at the same time that the Park Service made natural regulation official policy for elk, the Craigheads began their intensive study of grizzly bears, and some biologists began to call for the reintroduction of wolves. The growing ecological perspective applied to these other animals was not duplicated in bison management.

The stakes increased dramatically in 1985, when APHIS declared Montana and Wyoming brucellosis free, thus exempting ranchers in those states from testing requirements. The states therefore demanded that the Park Service stop bison from migrating into areas used by domestic cattle, and when the Park Service failed to act, officials from Montana's Department of Fish, Wildlife, and Parks (DFWP) began shooting bison at the park boundaries. In 1985, Montana created a hunting season for bison, allowing hunters to kill bison as they left the park. Over the next few years, state officials, hunters, and Park Service officials shot a growing number of bison: 84 bison were killed in 1984; 500 bison were killed the year after the Yellowstone fires of 1988.[83]

As the number of bison shot at the park boundaries grew, so did tension among federal and state agencies that had dramatically different mandates and priorities for bison management. The Park Service continued to maintain its policy of free-roaming bison herds within the park; state officials and APHIS continued to demand containment of bison within the park. Short of building a concrete and steel fence around the park—bison will walk through lesser fences without difficulty—some compromise was needed.

In 1989, the Montana DFWP and Department of Livestock, the Park Service, the Forest Service, and APHIS began negotiating a joint plan for managing bison and brucellosis. The efforts led to the formation of the Greater Yellowstone Interagency Brucellosis Committee in 1990 and an interim plan and environmental assessment in 1991. The interim plan called for public and state hunting to keep bison from leaving the park and interacting with domestic livestock. Still, tensions continued to escalate after APHIS, under pressure from state veterinarians, threatened to downgrade Montana's brucellosis-free status if bison that might carry brucellosis made it beyond the park boundaries into Montana. In 1995, the Montana governor sued APHIS and the Park Service on a number of grounds related to bison management, resulting in a settlement agreement that prevented APHIS from downgrading the state's brucellosis-free status as long as it complied with the interim bison management plan.[84]

The agencies developed new interim plans in 1995 and again in 1996, both of which upheld the basic strategy of hazing and, if that did not work, killing bison that left the park. After a particularly harsh winter in 1996–1997 that sent bison out of the park in search of food, the interim plan led state and park officials to kill over 1,000 bison. The scale of the slaughter caused a political firestorm that elevated questions of bison management to the cabinet level in Washington, where the interior secretary, the agriculture secretary, and the Council on Environmental Quality organized meetings with officials from the relevant federal and state agencies along with the states' congressional delegations. Following the meetings, APHIS withdrew its threats to downgrade the states' brucellosis-free status unless a brucellosis outbreak actually occurred in domestic livestock.[85]

In 1998, the agencies released their first draft EIS and Interagency Bison Management Plan (IBMP). The new draft plan made some minor changes, but it still emphasized zero-tolerance for brucellosis-positive bison leaving the park. The draft drew 67,000 public comments, over 70 percent of which opposed the agencies' plan. The draft EIS was modified slightly, and a final EIS and Joint Management Plan was issued by the Department of the Interior and the Department of Agriculture in 2000, over ten years after the first efforts to develop an interagency bison management plan. The final plan maintains its original emphasis on controlling bison movements and keeping bison that test positive for brucellosis and all pregnant females from leaving the park.[86]

The IBMP was a three-step plan, designed to be phased in over a five- to six-year period. In step one, bison would continue to be essentially blocked from leaving the park by hazing, capture, and killing. In step two, which was anticipated by 2002 or 2003, the agencies would allow 100 bison that tested negative for exposure to *Brucella abortus* to migrate into established zones just beyond the park boundaries. In step three, which was anticipated by 2005 or 2006, 100 bison could migrate into designated zones just outside the park without testing. Even when fully implemented, then, the plan would impose rigid boundaries for migrating bison, limiting them to zones just outside the park on national forest land and some private land. To this end, the agencies purchased land and conservation easements in these zones to limit conflicts with bison.[87]

The IBMP, which is a binding plan for the Park Service and Forest Service, proved very difficult to implement, and the agencies fell well behind schedule. In a 2008 report, the Government Accountability Office (GAO) explained the challenge of plan implementation this way: "The plan has two broadly stated goals: to 'maintain a wild, free-ranging population of bison and address the risk of brucellosis transmission.' The plan, however, contains no clearly defined, measurable objectives as to how these goals will be achieved, and the partner agencies have no common view of the objectives."[88] The report goes on to recommend that federal and state agencies work together to improve the process by "clearly defining measurable objectives, and to enhance the agencies' collaboration and accountability to the public on this issue."[89]

Looking back on the bison management controversy and plan, one can see how far it was from the kind of broad, ecosystem approach advocated in the vision document exercise and how different it was from wolf management under the ESA. Whereas the agencies worked under direction from the FWS to protect and manage wolves wherever they roamed, they followed an older approach of protecting bison only within a rectilinear preserve. In many ways, then, bison management returned Yellowstone National Park to its earlier role as a zoo or museum protecting a popular exhibit for tourists. There simply was no legal or political force capable of forcing the federal agencies to protect the bison's place in the larger Yellowstone ecosystem.

Conclusion

The GYE, with Yellowstone National Park at its center, was the first major laboratory for ecosystem management by federal agencies, and it remains one of the ecological icons of the United States. As such, it demonstrates some of the key motivations for ecosystem management, some of the management benefits of an ecological approach, and some of the most implacable challenges of trying to achieve various ecosystem management goals. Indeed, it is instructive precisely because it illustrates the practical complexities of applying ecosystem management theory as well as the intense political opposition to any new paradigm that threatens the status quo.

Ecosystem management of the Greater Yellowstone Area's federal, state, and private land began with the Park Service and its efforts to protect the most iconic species in Yellowstone and Grand Teton National Parks. In many ways, elk, bear, and wolves taught the Park Service what it meant to think about managing beyond park boundaries, because these animals demonstrated quite clearly that the park boundaries did not align with the boundaries of specific species' core habitats. Other concerns, such as the longevity of Yellowstone's famous geothermal systems, also forced the Park Service to look beyond park boundaries at external threats.[90]

This initial motivation for ecosystem management, as a way of dealing with external threats to resources in Yellowstone National Park, led to suspicion that the Park Service was simply trying to expand its preservation mission beyond park boundaries. This was expressed perhaps most clearly in the congressional hearings of 1985, when Republican members of Congress painted ecosystem management as essentially a land grab by the Park Service. They sought to reinforce clear distinctions between national parks, with their preservation mission, and national forests, with a multiple-use mandate that has traditionally privileged economic development of natural resources. What is more, they successfully resisted any meaningful, comprehensive management plan for the GYE.

Looking back, ecosystem management left an imprint on federal agencies in the GYE, though it certainly did not replace preservation and conservation entirely. The agencies learned to share information more effectively and learned to appreciate certain ecological processes that cross jurisdictional boundaries. That is, they learned to think of the Yellowstone region as one large ecosystem. And in certain cases, particularly where backed by the ESA, the agencies applied and continue to apply elements of ecosystem management to their work. At the same time, federal agencies can never work from a blank slate, and political as well as legal opposition prevented the agencies from turning the common principles of ecosystem management into a controlling mandate.

II | Adopting Ecosystem Management

The first three chapters explained three primary contexts or roots of ecosystem management: intellectual, policy, and management. New work in ecology and public administration challenged the Progressive Era framework of scientific management, calling for a more complex, adaptive, and deliberative approach to deal with the dynamic realities of both ecological and political systems. In both fields, scholarship suggested that the failure of earlier management had been to isolate resource outputs from their larger ecological context and isolate resource managers from their larger political context. Scientists and environmentalists also reframed land and resource issues with the concept of biodiversity, arguing that historical variability at the genetic, species, and ecosystem levels were all essential to the health and productivity of the biosphere. New environmental laws, passed in response to the growing political power of the environmental movement, almost unwittingly created the legal drivers of ecosystem management. Certainly, members of Congress knew that the National Environmental Policy Act (NEPA) and the Endangered Species Act (ESA) would advance environmental protection, but many of them did not anticipate the full scope and power that these laws would have. With their requirements to gather and/or apply the best available science, they forced or enabled federal agencies to grapple with the scientific concepts supporting ecosystem management. And new management challenges, largely created by new environmental law, led federal agencies to collaborate on issues that crossed jurisdictional lines.

In the late 1980s and the early 1990s, these three overlapping contexts or roots forced federal agencies, the courts, and Congress to reconsider federal land and resource management in fundamental ways. The balance that Congress had struck in the 1970s between resource development and environmental protection was not holding, and the multiple-use agencies in particular were caught to some extent between Congress, which tended to disaggregate land and resource values to please different constituencies, and the courts, which were demanding that the agencies take a more integrated approach to biodiversity protection.

Federal agencies adopted ecosystem management as official policy at the end of the George H. W. Bush administration and the beginning of the Clinton administration as a new rhetorical and practical framework for managing these challenges. The Clinton administration advanced ecosystem management as a way to break the gridlock, the "train wrecks," and the injunctions of the 1980s by reframing land and resource management in broader, integrated terms. It suggested that agencies deserved and needed discretion and flexibility to once again manage the federal estate. It was, in other words, a tool of agency discretion.

Understood this way, the debate over ecosystem management in the late 1980s and 1990s becomes clearer. It was, to a considerable extent, a debate over how the discretion implied by ecosystem management would be used. The debate was most intense in cases where ecosystem management appeared to be a substantive framework with predetermined goals, such as ecosystem protection. It was less intense when ecosystem management appeared to be a procedural framework that Congress and interest groups could control with established tools. This, though, returns to a central question of the book: was ecosystem management during the Clinton administration treated as substantive or procedural? The answer, as revealed in the following chapters, is that it was both, and it depended primarily on the laws through which agencies implemented ecosystem management projects. Two models, it is argued here, emerged in the 1990s, a substantive model driven by the ESA and a procedural model driven by NEPA.

4 | Resistance and Acceptance: Ecosystem Management in the Bush Administration

Introduction

The new science, new laws, and new management challenges of the 1960s and 1970s exposed federal agencies to a fierce criticism and endless rounds of litigation. The National Park Service faced ongoing criticism for its weak science program, and the US Fish and Wildlife Service (FWS) faced criticism and litigation for failing to meet its ESA obligations, particularly deadlines for species listing and critical habitat designation. But these challenges were not so much about the agencies' underlying mission; they were about mission failure. The two multiple-use agencies, and disproportionately the US Forest Service, faced more foundational criticism of multiple-use management, which critics said allowed the agencies to privilege timber, grazing, and mining over other land uses and values. They also faced perpetual litigation that slowed some of their management activities to a crawl. They were trapped between judicial interpretations of NEPA (1969), the National Forest Management Act of 1976 (NFMA), the Federal Land Policy and Management Act of 1976 (FLPMA), and the ESA (1973) on the one hand and congressional/administrative demands to increase resource outputs on the other.

The politics of this debate were shaped by the Reagan administration's backlash against 1970s environmental regulation, which the rapidly growing and professionalized environmental organizations defended. When elected in 1988, President Bush distanced himself from the anti-environmental rhetoric of the Reagan administration, promising to be the "environmental president." At the same time, the Bush administration retained a strong resistance to regulatory policy.

This tension proved important throughout the Bush administration, as key administration leaders sought to advance environmental protection while limiting regulatory policy and encouraging economic growth. It would be too simplistic to say that the Bush administration was disingenuous about its commitment to environmental protection, particularly in light of the president's leadership in amendments to the Clean Air Act that significantly reduced sulfur dioxide emissions in the 1990s. A better answer is that the paradigm for environmental protection of the "environmental president" framed land and resource questions in political and economic terms rather than ecological terms. This was clear in the administration's handling of wetlands regulation and protection of the northern spotted owl. In both cases, it supported protection, but it opposed the more expansive and ecologically framed protection that environmentalists demanded.

Yet in an interesting twist, the Bush administration directed the Forest Service to adopt ecosystem management as official policy as a strategy to deflect international criticism at the Rio Earth Summit in 1992. The United States had, for the last decade, criticized other countries for reducing biodiversity, and the administration knew that the same criticism would be turned back on the United States in Rio. Announcing that the Forest Service was adopting ecosystem management, the administration hoped, would silence those critics.

It is difficult to guess exactly what the new approach might have meant if President Bush had won a second term, since ecosystem management lacked clear definition. The Bush administration's record suggests that it might have celebrated discretion and flexibility over and against the strict regulatory requirements of the ESA and NFMA, but it would likely have worked to ensure that agency discretion was not guided by the ecologists and conservation biologists who most supported ecosystem management. This was demonstrated by the Bush administration in conflicts over wetlands delineation and the northern spotted owl.

George H. W. Bush: The Environmental President

The Reagan administration had waged an all-out war on environmental regulation, and regulation more broadly, from 1983 to 1989. President Reagan's top land and resource priority, administered initially by Interior Secretary James Watt, was to "get America moving" by cutting red tape and increasing natural resource production.[1] The administration's priorities were perhaps best reflected in its attempt to dismantle the Council on Environmental Quality (CEQ) and replace it functionally with Vice President Bush's Task Force on Regulatory Relief. The priorities were also evident in annual budget proposals, which included steep cuts in environmental programs, particularly those programs for environmental monitoring and regulatory compliance.[2]

The 1988 presidential election was therefore an important election for environmental policy. Campaigning amidst fires that burned vast swathes of Yellowstone National Park, growing conflict in the old-growth forests of the Pacific Northwest, and increasing problems with acid rain, Vice President Bush pledged himself to environmental protection. On the campaign trail, he positioned himself as a conservationist in the tradition of Gifford Pinchot and Teddy Roosevelt, and he insisted, "I am an environmentalist: always have been."[3] If elected, he promised, he would be "the environmental president."[4] Specifically, he said, he would reduce acid rain by cutting sulfur dioxide emissions, strengthen enforcement of toxic waste regulations, address climate issues, and establish a new goal of "no net loss of wetlands."[5] Bush insisted that he did not need to reject his economic and anti-regulatory positions in order to protect the environment: "We can have both economic growth and a cleaner, safer environment."[6]

After his inauguration, President Bush sent what environmental organizations saw as very mixed signals that affirmed both environmental and economic commitments. On the one hand, he appointed respected environmentalists William Reilly and Michael Deland to the top positions in the Environmental Protection Agency (EPA) and CEQ,[7] and he followed up on several of his campaign promises. In June 1989, the new president reiterated his pledge to protect wetlands: "You may remember my pledge, that our national goal would be no net loss of wetlands. . . . It's time to stand the history of wetlands destruction on its head. From this year forward, anyone who tries to drain the swamp is going to be up to his ears in alligators."[8] He also pushed significant amendments to the Clean Air Act, signed in 1990, which limited sulfur dioxide emissions through a cap and trade system and eventually reduced acid rain dramatically.

On the other hand, his appointments for land and resource issues raised concern for environmentalists. His chief of staff, John Sununu, was hostile to the agenda of most environmental organizations, and the National Wildlife Federation described the appointment of James E.

Cason as assistant secretary of agriculture as "just a shocking appointment . . . he [has] an absolutely abysmal track record on the environment."[9] Manuel Lujan, Jr., whom President Bush appointed as interior secretary, had just retired from Congress (R-NM). On the one hand, he had forged a very close working relationship with Morris Udall (D-AZ) and had provided strong support for clean air and water legislation. On the other hand, his record on land and resource policy reflected traditional Western interests in resource development.

Secretary Lujan, like the president, called for a "balanced" approach to economic development and environmental protection on the federal estate. This can be understood primarily as a reaffirmation of the Forest Service's multiple-use management paradigm, emphasizing the opportunity to manage various *resource uses* across large areas of federal land, separating them to avoid conflict. In other words, he and others in the Bush administration saw resources as discrete or separate things, while advocates of the new ecosystem management approach saw them embedded in "webs of relationships."[10]

The Bush administration alarmed environmentalists almost immediately by attacking the very law that had driven ecosystem management to the foreground and forced agencies to apply an ecological lens in Yellowstone: the ESA. "It's just too tough an act," Secretary Lujan argued, "We've got to change it. Do we have to save (an endangered species) in every locality where it exists? Do we have to save every species?"[11] Or as President Bush later put it, the ESA "was intended as a shield for species against the effects of major construction projects like highways and dams" not as a foundation for comprehensive ecosystem protection.[12] The ESA, both the secretary and the president argued, was intended to protect specific species, from specific threats, in specific places, for specific periods of time; it was not meant to serve as a paradigm for federal land and resource management. As Secretary Lujan explained in a draft letter to Representative Duncan Hunter (R-CA), "we do not wish to see the Endangered Species Act used as a means of solving broader questions of maintenance of biodiversity."[13]

Yet the act clearly framed the work of the ESA in ecological terms, making scientists the primary decision makers in species protection, and the new ecology of the late twentieth century supported a very broad interpretation of the ESA's habitat protection requirement. From a scientific perspective, informed by the clear goal of species and habitat protection, federal agencies could not first promise resource outputs and then work biodiversity protection around those goals. If they were to achieve the ESA's mandate, they would have to reverse that order. As a General Accounting Office (GAO) report on ecosystem management later explained, federal agencies "must no longer be focused primarily on individual uses on individual land units or on protecting individual natural resources." Instead they must coordinate "human activities across larger areas and over longer time frames so as to maintain or restore an ecosystem's integrity and function."[14] The Bush administration's opposition to both the ESA and ecosystem management stemmed from the conviction that this type of management approach was too open ended, relied too heavily on the judgment of scientists, seemed politically to favor environmentalists, and clearly threatened the tenure of public land and resource developers. It was, in other words, "just too tough."

Wetlands Redefinition

Two conflicts illustrated the Bush administration's idea of balanced conservation, framed in political rather than ecological terms: wetlands protection and the northern spotted owl. As noted above, President Bush's most specific environmental pledge during the 1988 campaign was a promise to stop the net loss of wetlands in the United States. During the campaign, he explained, "My position on wetlands is straightforward: All existing wetlands, no matter how small, should be preserved."[15] He reiterated this pledge later in the campaign and during the first year of his presidency: "You may remember my pledge, that our national goal would be no net loss of wetlands. And together, we are going to deliver on the promise of renewal, and I plan to keep that pledge. . . . Wherever wetlands must give way to farming or development, they will be replaced or expanded."[16] As one observer suggests, "The goal of no net loss probably was deemed a politically viable promise because it sounds flexible and absolutist at the same time . . . environmentalists tended to emphasize the 'no loss' part of the promise, while developers emphasized 'no *net* loss.'"[17]

Section 404 of the Clean Water Act, the primary regulatory provision governing wetlands, prohibits placement of dredged or fill material in a wetland without a permit from the US Army Corps of Engineers. Permit issuance is based on guidelines promulgated by the EPA in consultation with the Corps. Wetlands regulation was particularly difficult politically for the Bush administration, because nearly three-quarters of all remaining wetlands were in private ownership.[18] As a result, wetlands regulation could very easily be framed as environmental regulation versus property rights, exactly the framing that the administration was trying to avoid.[19]

One of the basic challenges of wetlands regulation, which the Bush administration addressed, was how to define and delineate wetlands. The challenge stems from the fact that there is no clear, universal definition of a wetland, and federal definitions of wetlands varied depending in part on their legal purpose.[20] The divergent definitions themselves would probably not have caused controversy, since they overlapped considerably and were too broad for direct application.[21] Controversy grew as these general definitions were operationalized and the agencies developed the criteria and indicators necessary to move from the general definition to precise and consistent delineation that could be justified and enforced by the courts.[22] This legal necessity conflicted with the general scientific understanding that wetlands boundaries represented a continuum that was difficult to pinpoint in many cases: "Unfortunately," wrote authors of the FWS's 1976 "Interim Classification of Wetlands and Aquatic Habitats of the United States," "legal definitions are usually based as much on facility and pragmatism as they are upon accuracy of meaning. Hence, legal definitions of wetland may bear little resemblance to the ecological concepts embodied in the term. There is no single, correct, indisputable, ecologically sound definition for wetland because the gradation between totally dry and totally wet environments is continuous."[23]

The Corps and the EPA shared the same basic definition of wetlands under the Clean Water Act: "Those areas that are inundated or saturated by surface or groundwater at a frequency and duration sufficient to support, and that under normal circumstances do support, a prevalence of vegetation typically adapted for life in saturated soil conditions."[24] Both agencies agreed, in other words, that wetlands were indicated by hydrology, vegetation, and soils. Yet the agencies initially produced separate technical manuals to operationalize the definition. In 1989, the agencies collaborated on a new, unified manual for wetlands delineation.[25] It maintained the basic definition of wetlands, but it offered a new delineation protocol that would have expanded the land area designated as wetlands and therefore regulated under the Clean Water Act. The 1989 manual therefore drew extensive criticism from those representing private property interests, who argued that it would expand the regulatory reach of the Clean Water Act to areas that were not technically wetlands.[26] In one 1992 congressional hearing, a farm representative argued, "farmers are not, in general, opposed to protecting wetlands; that is land that is *truly wet*."[27]

The 1989 manual put the Bush administration in a difficult position between the president's promise to stop the net loss of wetlands and the president's anti-regulatory credentials, so the administration created a separate task force to rewrite the manual.[28] In the summer of 1991, the task force proposed revisions to the 1989 "Federal Manual for Identifying and Delineating Jurisdictional Wetlands." In one sense, the revisions were simply clarifications that made the 1989 manual more precise and consistent by reducing inferences on the part of wetlands delineators. Yet the 1991 revisions clearly established a more restrictive protocol designed to reduce the total area of wetlands identified for federal regulation. For example, whereas the 1989 manual required inundation or saturation of the soil for seven days out of the growing season, the 1991 revisions required fifteen days of inundation or twenty days of saturation.[29]

The administration faced a firestorm of criticism from environmentalists and hunters, while developers and private property advocates praised the revisions as common sense changes.[30] For his part, the president argued that his administration was simply avoiding the demands of extremist environmentalists, shielding private landowners from out-of-touch bureaucrats, and simplifying the definition of wetlands. "I've got a radical view of wetlands," the president said, "I think wetlands ought to be wet."[31]

The administration's problem was that its "common sense" approach cut both ways politically. On the one hand, it would have freed private property from regulation by cutting the acreage of "jurisdictional wetlands" under the Clean Water Act by 30 to 50 percent. On the other hand, it also eliminated protection for publically recognized and celebrated wetland areas. The State Water Control Board of Virginia, for example, estimated that the new protocol would remove wetlands designation from 86 percent of the Great Dismal Swamp, and areas of the iconic Florida Everglades would lose wetlands status as well.[32] When these implications became clear, the administration scrapped the 1991 manual and began work on a new delineation protocol.

While the issue of wetlands delineation did not shape ecosystem management directly, it is instructive because it illustrates the potential tension that exists between science and law in ecosystem management. For environmentalists, the Bush administration's revision of wetlands delineation guidelines was a clear-cut case of crass politics trumping objective science.[33] Ecologists do not have a single definition of wetlands, nor do they have a single set of criteria and indicators that one would use to locate the boundaries of a wetland area. In ecology, multiple definitions of wetlands can exist without inhibiting the basic work of scientific research. One of the reasons for this ambiguity is that ecologists focus not so much on endpoints or fixed characteristics of wetlands but on their ecological relationships, functions, and processes. This is why for many wetlands ecologists, some of the most important wetland areas are dry for much of their year. They are considered wetlands not because of a fixed number of days a year that they are saturated but because of the hydrologic functions and processes they sustain. And ecologists are not necessarily committed to identifying these functions and processes on an *annual* basis, since wetlands may expand and contract in cycles at a much broader timescale.[34] What is more, since ecologists are interested in wetlands functions and processes, they are often as interested in the wetlands interface or margins as they are in the wetlands themselves. Ecologists, then, are likely to have a broader view of wetlands ecosystem boundaries that can include buffer zones that technically lie outside of the wetlands' boundaries. Environmentalists wanted a protocol that was sensitive to the expansive ecological understanding of wetlands.

For the Bush administration and property rights advocates, however, such broad and ambiguous definitions threatened to expand government's regulatory power in unpredictable ways. Justice Oliver Wendell Holmes wrote over 100 years ago that "people want to know under what circumstances and how far they will run the risk of coming against . . . public force through the instrumentality of the courts," and this is precisely what a clear legal definition of wetlands allows.[35] Thus, when agencies write wetlands regulations and search for a precise wetlands definition, one legal scholar explains, they "assess fixed quantifiable endpoints rather than attempting a full assessment of wetland processes or functions as a practical necessity."[36] The agencies hope that "these endpoints are surrogates that indicate a functioning wetland. The difficulty arises because there is an imperfect correlation between these indicators' presence and the existence of wetlands processes or functions."[37]

It is in this sense, then, that the Bush administration privileged politics over science; it privileged the clarity, defensibility, and political palatability of a highly restrictive definition for wetlands delineation over more expansive scientific definitions. It reflected, in other words, an insistence that wetlands were discrete natural objects that should be protected as such rather than hydrologic components of larger ecosystems that should be protected for their ecological function. And the Bush administration took the same approach in dealing with the northern spotted owl, which did drive federal adoption of ecosystem management directly.

The Northern Spotted Owl

The wetlands case illustrated the administration's resistance to ecological science as the basis for regulating private lands. A second case illustrated a similar resistance to an ecological perspective on public land management: the northern spotted owl. The owl was and is considered an obligate of late seral stage (old-growth) forests in the Pacific Northwest. It depends disproportionately on the characteristics of old-growth forests for its nesting, reproduction, and hunting, so the owl's rapid decline was closely correlated with dwindling acreage of these forests. Only about 10 percent of the old-growth forest acreage that existed prior to European contact remained in the Pacific Northwest when President Bush took office, and roughly 90 percent of it was managed by federal agencies. Outside of national parks, wilderness areas, and certain other areas, the Forest Service and Bureau of Land Management (BLM) managed old-growth forest primarily for timber production.[38] The last remaining old-growth forest therefore had two conflicting values—owl habitat and timber—putting the owl and the timber industry on a collision course during the 1980s.

The debate was often framed as a choice between owls and jobs, but it was, in actuality, a much broader tension between two very different ways of seeing forests in the Pacific Northwest.[39] Environmentalists applied a popular ecological lens, seeing the forest as a fragile ecosystem threatened by logging. The owl was an indicator or symptom of the forest's decline. For environmentalists, in other words, the northern spotted owl was a resident Lorax who spoke for old-growth forests of the Pacific Northwest. As the FWS explained in a briefing book for President Bush,

> The controversy surrounding the northern spotted owl is much more than the decline in the owl's population. It represents the merging of two important issues to the environmental community—each of which alone is a bellwether issue, but combined represent a decision that will determine the direction of environmental and preservation movements in the Pacific Northwest for years to come. It is the merger of those forces dedicated to the protection of old-growth forests and those dedicated to the protection of threatened or endangered species.[40]

As environmentalists saw it, the two issues were inseparable, because it was impossible to protect the owl without protecting its old-growth forest habitat. And environmentalists found support from ecologists and conservation biologists both inside and outside of the federal agencies.

For the Bush administration, this view amounted to environmental imperialism, using the northern spotted owl's protection "as a means of solving broader questions of maintenance of biodiversity." The Bush administration was not opposed to owl protection, per se; it simply argued that owl protection needed to be understood and balanced within a multiple-use political framework that disaggregated owl protection and timber production. The owl already enjoyed habitat protection in national parks, wilderness areas, and other acreage closed to logging, and the agencies had already agreed not to harvest timber around known owl nesting sites. For the Bush administration, the owl should not be allowed to claim additional federal old-growth forest acreage at the expense of the region's timber economy.

NEPA and NFMA

As the debate unfolded, though, the courts ruled that Congress had mandated an ecological approach to owl protection through NEPA, NFMA, and ESA, which must have been a particular frustration to those in the Pacific Northwest congressional delegation who had voted for the acts. The Bush administration and Congress made numerous attempts to work around this interpretation, but without significant amendments to at least NFMA and ESA, they could not put the owl back in the Forest Service's and BLM's multiple-use box.

The political and legal battles over the northern spotted owl just before and during the Bush administration were enormously complex. Litigation brought by "the northern spotted owl," environmental groups, and the timber industry named officials representing the Forest Service, the FWS, and the BLM, and it challenged agency and congressional actions under NEPA, NFMA, ESA, and even the Constitution.[41] What is more, many of the cases became entangled in the judicial system as they were joined, appealed, and remanded. Others have followed the owl's entire journey through the courts, and the complete story is beyond the scope of this chapter.[42] What is essential, though, is to see how NEPA, NFMA, and ESA forced an ecological perspective on land-use planning and management in the federal forests of the Pacific Northwest and to see how the Bush administration resisted this perspective and set the stage for the Northwest Forest Plan during the Clinton administration.

Concern for the northern spotted owl began quietly in the early 1970s, when biologists conducting the first serious research on the owl warned that its fate was tied to old-growth forests and that continued logging in these areas would push it to extinction. In response, the FWS requested in 1972 that the Forest Service and the BLM consider the owl's vulnerability in their land-use planning, and the agencies agreed to participate in the Oregon Endangered Species Task Force beginning in 1973.[43]

The first challenge for the task force was that it lacked the forest data necessary to create maps depicting old-growth habitat suitable for northern spotted owls. Faced with these limitations, the task force recommended that the agencies at least prohibit logging within a buffer area around known spotted owl nesting sites. The task force didn't know exactly how large these buffers needed to be, so in recommending 300 acres, it followed the best guess of leading owl biologists Eric Forsman and Chuck Meslow.[44]

If the first challenge was scientific uncertainty, the second challenge was administrative inertia. The Forest Service and BLM rejected the 300-acre buffer for at least two reasons. First, they resisted a guideline that would limit their management discretion. Second, both agencies seemed to view the 300-acre guideline as a slippery slope that would eventually close large acreage to timber harvesting. As agency representatives explained in August 1973, "Rigidly-cast prescriptions tend to become the accepted practice, even though they are intended to be of temporary nature only [and] the 'management by individual animal location' philosophy, when applied to all species which may be identified as requiring old-growth stand habitat, presents a land management specter of considerable magnitude."[45] Still, pressure mounted when

Oregon listed the spotted owl as a threatened species, and the agencies agreed in 1978 to accept interim guidelines for the northern spotted owl, including the 300-acre buffer area. The BLM agreed to protect habitat areas for 100 pairs of owls, and the Forest Service agreed to protect habitat areas for another 200 pairs.[46]

The owl remained an administrative issue through the rest of the Carter administration and most of the Reagan administration. The Forest Service and BLM added the buffer areas to their timber sales, and they both addressed the owl in their forest planning as required by NEPA, NFMA, and FLPMA. But the scientific research on the northern spotted owl conducted concurrently suggested that the agency's steps were not enough. By the late 1980s, both agencies faced litigation under NEPA, and the Forest Service faced additional challenges under NFMA.

The BLM formally incorporated the interim guidelines for the northern spotted owl in its Timber Management Plans and associated EISs prepared in 1979 and 1983. The preferred alternative in each plan set aside land, per the interim guidelines, for a specific number of owls, but timber production still clearly took priority over species protection. For example, the preferred alternative in the Jackson-Klamath Final Timber Management Environmental Statement states that roughly "35 percent of the old growth currently existing in the high intensity lands of JKSYU's would be harvested during the first decade. This could mean a 35 percent reduction of old-growth dependent species such as the northern spotted owl, redback vole and pileated woodpecker on these lands. Old growth would be eliminated on the high intensity lands of the JKSYUs by the year 2018."[47]

In 1986, the Portland Audubon Society requested that the BLM prepare Supplemental EISs (SEIS) for each of the timber management plans in order to incorporate new scientific information about threats to the owl. The BLM completed an Environmental Assessment over the next year and concluded that the information was preliminary in nature and therefore did not warrant any SEISs. Furthermore, it argued, the new information did not indicate any increased risk to the spotted owl.

The Portland Audubon Society sued the BLM in the fall of 1987, alleging violations of NEPA and the Migratory Bird Treaty Act. It argued that the BLM was ignoring important new information, including a status review of the spotted owl by the FWS in 1982, a draft SEIS prepared by the Forest Service in 1986, new analyses by Dr. Russell Lande in 1985 and 1987, and, most importantly, spotted owl analysis prepared by BLM biologists in 1986 and 1987. The court also heard expert testimony in the case from a number of leading scientists working on population modeling and viability assessment.

The information suggested two things. First, an enormous amount of scientific uncertainty remained regarding the northern spotted owl and its habitat needs. Second, while the BLM was focused on protecting specific nesting sites, it was ignoring new scientific studies that addressed broader issues regarding the impact that population size and habitat fragmentation had on the owl's survival. The BLM reports were particularly damning, in that the agency's own staff warned that "if current cutting levels continue for the next four years the BLM will no longer have options of providing: (1) additional spotted owl management areas (SOMAs); (2) habitat corridors between key population segments; and (3) multiple owl SOMAs."[48] The BLM, it seemed, was deliberately closing the door on future mechanisms to protect the owl, while Russell Lande's extensive population modelling indicated that protecting the long-term viability of the spotted owl required preserving virtually all remaining old-growth owl habitat. Other scientists testifying for the Portland Audubon Society concurred that one of the most immediate threats wasn't so much the number of acres of old growth around each nesting site but the fact that the current timber harvesting plans would fragment old-growth habitat in ways that posed an imminent threat to the owl.[49]

The BLM countered that the new information was too preliminary to warrant a full SEIS. The problem, an expert witness for BLM explained, was that predictions that habitat fragmentation would push the northern spotted owl to extinction were "the results of assumptions built into mathematical models. . . . the factors [that led to the predictions] are merely hypothesized in the models as consequences of fragmentation." There was, he said, "no empirical data in existence to establish that any of these factors actually occur."[50] Scientists had used modelling methods in ecology precisely because they could accommodate complexity, uncertainty, and data limitations common in the study of ecosystems. Scientists testifying for the government argued that this might be fine in academic research, but science used in management decisions had to be based primarily on field data and empirical trials. Essentially, they argued that applied science needed to all but eliminate uncertainties before it could justify agency action.

The case, *Portland Audubon Society v. Lujan*, was important because it vetted the science that would later shape ecosystem management surrounding the northern spotted owl. According to Judge Helen Frye, the BLM argued "that the new information submitted by the Portland Audubon Society is speculative in nature and is not accurate. [It argues] that the submission of speculative and inaccurate information about the viability of the spotted owl as a species does not require the BLM to supplement existing Environmental Impact Statements." She ruled that the BLM did not get to decide which scientific studies were correct or incorrect under its NEPA responsibilities. To the extent that the agency had refused categorically to consider "adequate population size or the effects of habitat fragmentation upon the long-term survival of the spotted owl species," it had acted arbitrarily and capriciously.[51]

This points back to the basic problem of agency discretion. Congress had given the BLM enormous discretion in FLPMA, allowing it to prioritize timber production in Oregon. Debate in this case was over the information guiding that discretion. Judge Frye's ruling followed NEPA precedent in saying that while NEPA did not predetermine the BLM's substantive decision on owl protection, it did require the agency to use new science in ecology and conservation biology in its decision-making process. The BLM had resisted because that science would have reframed its decision making and placed new pressure on the agency to change its management practices.

Before Judge Frye's ruling could impact the BLM, Congress intervened in the broader spotted owl situation through a 1989 rider, Section 318, of a continuing budget resolution. The rider instructed the BLM and the Forest Service to continue working on updated forest plans and barred litigation against existing plans for failing to address recent scientific information. Judge Frye dismissed the Audubon Society case as precluded by the rider. But after three more years of legal twists and turns and expiration of the appropriations rider, the Portland Audubon Society returned to Judge Frye's court and renewed the case. Judge Frye once again found that NEPA required the BLM to consider significant new scientific information and ordered the BLM to complete an SEIS accordingly: "There are no defenses to excuse the BLM from complying with NEPA."[52]

The Forest Service ran into NEPA problems as well, although the agency's struggle was further complicated by its NFMA requirement to protect viable populations of vertebrate species. When the Forest Service adopted interim owl guidelines in 1978, it was still finalizing NFMA regulations. In 1984, the agency provided a new Pacific Northwest Regional Guide and companion EIS that included updated owl guidelines. However, to environmental groups, it appeared that the Forest Service had, like the BLM, ignored a growing body of scientific literature on the owl and had instead based these guidelines "upon little more than the collective guess of several agency biologists as to what might be done to save the Northern Spotted Owl within the constraints of timber production goals."[53] In response to public criticism, the Forest Service agreed in 1985 to prepare a SEIS focused exclusively on the northern spotted owl that would incorporate the latest scientific information.

In the SEIS, agency biologists conducted a population viability assessment that included the cumulative impacts of timber harvesting under a worst-case scenario, and it addressed the economic impacts of its various management options. The Forest Service released a draft SEIS (DSEIS) in August 1986, advancing a preferred alternative that proposed 550 habitat protection areas, each 2,200 acres in size. But the Forest Service indicated that it would log within the protected areas, so the net impact of the preferred alternative was the closure of roughly 314,000 to 690,000 acres of land to timber harvesting. This would reduce annual timber harvesting in the owl's habitat by just 5 percent.[54] No one was happy with the DSEIS: only 1 percent of the 41,000 responses to the draft statement indicated support.[55] The timber industry and environmental groups began preparing for political and legal battles over the future of timber harvesting in the Pacific Northwest.

The Forest Service finalized the SEIS over the next two years, releasing the final SEIS (FSEIS) in the middle of the 1988 presidential campaign. The FSEIS made concessions for owl protection, such as enlarging habitat protection areas on the Olympic Peninsula from 2,200 to 2,700 acres, but it still only reduced timber harvesting in owl habitat by about 5 percent annually. The Forest Service was stuck between irreconcilable demands to protect the owl and to maintain timber outputs. As Dick Holthausen, a Forest Service biologist involved in the process, explained, there was "no middle ground that was going to make any elements on either side very happy. So I think the agency was faced with what looked like a lose-lose situation."[56] The Forest Service would face intense opposition no matter what it adopted in the FSEIS.

The Forest Service issued the Record of Decision (ROD) on the FSEIS in December 1988 just as President-Elect Bush was preparing to take office. "The environmental president," who had promised to find win-win solutions to achieve both economic development and environmental protection, took office just as the Forest Service was concluding that a win-win solution was impossible. As if to confirm this, timber groups and environmental organizations quickly appealed the FSEIS and then filed suit one month after President Bush's inauguration.[57] In March 1989, Judge William Dwyer temporarily enjoined 144 timber sales.[58]

As noted above, Congress intervened through an appropriations rider in October 1989, striking its own balance between timber production and owl protection. The rider (Section 318 of the Interior Department's appropriations act for fiscal year 1990) essentially declared that Forest Service and BLM proposed timber sales met the requirements of NEPA and NFMA and were therefore not open to judicial review under those acts. Judge Dwyer dissolved his injunction.[59]

In the meantime, the Forest Service attempted to regain its scientific credibility, which had been damaged by the SEIS. The appropriations rider had also authorized the Interagency Scientific Committee (ISC) to develop a new conservation strategy for the northern spotted owl. The committee was chaired by a Forest Service research biologist, Jack Ward Thomas, and included representatives from the Forest Service, the FWS, and the BLM. Its May 1990 report, "A Conservation Strategy for the Northern Spotted Owl," challenged the existing Forest Service and BLM timber programs: "We have concluded," the committee explained, "that the owl is imperiled over significant portions of its range because of continuing losses of habitat from logging and natural disturbances." The agencies' existing plans, the committee declared, were "a prescription for extinction of spotted owls, at least in a large proportion of the owl's range."[60]

Based on what the committee described as "one of the most complete assessments of a species to date, particularly as applying modern concepts and methods of demographic modeling and conservation biology," it recommended a new strategy of protecting not only owl nesting sites but 8.4 million acres of carefully selected Habitat Conservation Areas (HCAs).[61] The ISC, in other words, reviewed the owl's needs through a broader ecological lens. For the timber industry, which had opposed the SEIS preferred alternative of 314,000 to 690,000 acres, the ISC's recommendations were horrifying.

The Forest Service officially vacated its 1988 SEIS in October 1990, announcing that it would instead manage its Pacific Northwest timber program in a manner "not inconsistent with" the ISC recommendations.

The Seattle Audubon Society sued, arguing that the Forest Service was required by both NEPA and NFMA to address the science itself through the proper forest plans and EISs. In March 1991 Judge William Dwyer once again enjoined Forest Service timber sales in the northern spotted owl's range, ruling that the Forest Service had not demonstrated how its new strategy complied with NFMA. What was particularly significant in this case was Judge Dwyer's interpretation of the Forest Service's NFMA regulations: "The duty to maintain viable populations of existing vertebrate species," he opined, "requires planning for the entire biological community—not for one species alone."[62] Thus, the Forest Service was not only forced to prepare a new EIS, which it issued in September 1991; it was also instructed to take an ecosystem approach to management.

The Endangered Species Act

While the BLM and the Forest Service had taken different approaches to the northern spotted owl dilemma, their paths converged in a set of injunctions. What is more, the courts had affirmed both the validity of population viability analysis and the necessity of an ecosystem approach to species protection. The injunctions suggested that NEPA and NFMA alone might have pushed the BLM and the Forest Service toward ecosystem management. But just as President Bush took office an even more powerful statute, the Endangered Species Act of 1973, became the overarching driver of ecosystem management for both agencies.

Biologists had been arguing that the northern spotted owl was endangered since Congress passed the ESA in 1973. In 1982, the FWS initiated its first intensive status review of the northern spotted owl, which continued for almost six years. Toward the end of the review, and based on research highlighted by the Audubon Society, an environmental organization named Green World petitioned the FWS to list the northern spotted owl as endangered. Green World's primary argument was that the northern spotted owl population had declined because logging had reduced old-growth forest habitat and that continued logging threatened the owl's survival.[63] Green World submitted its petition in January 1987, and in July of that year the FWS announced that the petition had merit and would receive full review.

The FWS Director of Region 1 received the report in November 1987, reviewed it, and decided that the northern spotted owl did not merit listing.[64] The Sierra Club Legal Defense Fund sued in May 1988, arguing that in his decision the director had ignored the agency's scientific findings compiled between 1982 and 1987, and they won the case just days after the November election. The court found that the FWS had acted arbitrarily and capriciously by failing to substantiate its decision with the available scientific data and by failing to consider a lesser listing of threatened for the owl.[65] An investigation by the GAO confirmed problems with the FWS decision, concluding that "FWS officials directed changes to the body of scientific evidence in the report, eliminating an FWS expert's conclusion that the owl was endangered and revising the presentation of other evidence indicating that there were risks to the spotted owl's continued survival if logging of the owl's habitat continued at its current pace."[66]

The court remanded the listing decision back to the FWS for further consideration, and on June 23, 1989, just five months into the Bush administration, the agency reversed course and proposed listing the northern spotted owl as threatened. Top officials in the Bush administration were open about their disagreement with the decision. Assistant Secretary of Agriculture James Cason, who had buried northern spotted owl reports by BLM biologists several years earlier because he argued that they lacked scientific credibility, rejected the scientific basis for the FWS decision.[67] Interior Secretary Lujan also rejected the validity of the FWS's scientific conclusions, arguing that the agency had been manipulated by environmentalists who really just wanted to protect old-growth forests.[68] The fact that the FWS moved forward with its decision is a testimony to the power of the ESA as enforced by the courts.

By law, the FWS had one year to make a final determination about the listing. The Forest Service and BLM tried to convince the agency that they could protect the owl without ESA listing by agreeing to "provide for a number and distribution of spotted owls . . . so they may interact with spotted owls throughout the geographic range of the species," but it was not enough.[69] The FWS officially listed the northern spotted owl as threatened in June 1990: "The biological evidence," the agency's director John Turner explained, "says the northern spotted owl is in trouble. We will not and, by law, cannot, ignore that evidence. . . . But I strongly believe there is room in the world to protect both owls and loggers. Our intent now . . . is to find ways to protect the owl with the least possible disruption to the timber economy of the Northwest."[70]

That summer and fall, the Forest Service and the BLM responded differently to the listing, but both ran very quickly into legal trouble.

As explained above, the Forest Service vacated its 1988 SEIS and announced that it would follow the ISC's recommendation for HCAs and its stipulation that logging in other areas of owl habitat would be prohibited unless 50 percent of each quarter township had forest stands with a mean diameter of 11 inches and a 40 percent canopy closure (50-11-40 rule). The BLM took a different approach because, BLM Director Cy Jamison explained, the ISC recommendations were just too tough economically. The BLM estimated that full compliance would eliminate 7,600 timber-related jobs. Instead, in September 1990, the BLM released its own "Management Guidelines for the Conservation of the Northern Spotted Owl," often referred to as the Jamison Strategy. It included significant owl protection measures, such as an agreement not to offer green timber sales in the ISC-proposed HCAs, but it did not follow the 50-11-40 rule. In other words, it agreed to additional, discrete protection measures, but it refused to apply something like the 50-11-40 rule to all of its forested land. The BLM estimated that this difference would lead to the loss of only 1,000 timber-related jobs.[71]

The Lane County Audubon Society sued BLM, arguing that the Jamison Strategy was itself a major federal action that required the agency to consult with the FWS pursuant to Section 7 of the ESA. The district court ruled in their favor, and the appellate court affirmed the decision, enjoining BLM timber sales until the FWS reviewed the Jamison Strategy. This brought the BLM's timber program to a temporary halt.[72]

The crisis continued to escalate politically as the FWS first avoided and then fulfilled its other ESA obligations. The agency initially argued that it did not have adequate scientific information to make a critical habitat determination. In 1991, Judge Thomas Zilly, who had ruled against the FWS in 1988 for its decision not to list the owl, once again ruled that the FWS had violated the ESA and ordered the agency to designate critical habitat by May 1991.[73]

Under court order, the FWS turned to what at the time was the best available science, including the ISC report, and it designated critical habitat on schedule. If the Bush administration had been alarmed by the ISC recommendations, the FWS's critical habitat designation was politically terrifying. Whereas the ISC had recommended protection for 8.4 million acres, the FWS identified 11.6 million acres of land as critical habitat, including 3 million acres of private land, and this was in addition to the nearly 2 million acres already closed to logging in national parks and wilderness areas.[74]

Three months later, after significant public comment, the FWS released a revised final designation for critical habitat. The revised designation reduced the listed habitat from 11.6 million acres to 8.2 million acres and eliminated most of the critical habitat proposed on private lands.[75] Although this was a significant reduction, it did little to staunch criticism, since the designation would invariably mean dramatic reductions in timber harvesting.

In May 1991, with the crisis spiraling out of control, Congress intervened once again by creating its own "Scientific Panel on Late-Successional Forest Ecosystems." The committee, often called "The Gang of Four," included the ISC chair Jack Ward Thomas, along with two other forest ecologists and an economist. The Gang of Four had a broader mandate than the ISC, which was to assess the viability of *all* vertebrate species associated with old-growth forests and all vulnerable fish stocks in the owl's habitat range as well as the region's ecological integrity. In its October 1991 report, the panel concluded that the remaining old-growth forest in the owl's habitat range was insufficient to protect both species viability and current timber harvest levels. It recommended an even more expansive acreage of late seral-stage and old-growth reserves. It also recommended additional owl protection areas and protection of key watersheds necessary for vulnerable fish species.[76] In other words, the Gang of Four's recommendation expanded, rather than reduced, conservation measures that would constrain timber harvesting.

The White House

Rather than resolving the issues, the Bush administration's response helped escalate the crisis to a breaking point. What the administration wanted in the Pacific Northwest was traditional multiple-use management, which it believed was under attack by the environmental movement:

> At the heart of this movement, and the greatest risk posed by it, is the overarching purpose to attack the policy of "multiple-use management" on public lands. Similar to the wilderness battles of the 1970s and 1980s, the environmental community has chosen old growth and spotted owls as the latest tools in their attempt to constrain use of Federal natural resources and move the multiple-use management of the national forests toward the

singular, preservation philosophy most consistent with the National Park Service (NPS).[77]

As the FWS had warned the president just before listing the northern spotted owl as threatened, the HCAs recommended by the ISC were "*quasi-wilderness areas*" that would slash federal timber harvests by 50 percent.[78] This was not the kind of balanced environmental protection that the president had envisioned: "I'm interested in the owl," the president said, "but I'm also interested in jobs for the American family. . . . I want to be known as the environmental President, but I also want to be concerned about a person's ability to hold a job."[79]

As the spotted owl controversy moved through the courts and through Congress, then, the Bush administration formed its own spotted owl task force to find an alternate strategy for balancing owl protection and timber harvesting.[80] The new task force promised to study the federal timber program and recommend actions to limit the economic impact of owl protection. It also promised to recommend ESA amendments to ensure that economic needs would not be trumped by the act in the future. Both sets of recommendations were promised by September 1, 1990.[81]

As the summer of 1990 wore on, tension among task force members grew. Scientists who consulted with the task force supported the ISC report, arguing that it was the only plan that ensured adequate protection for the northern spotted owl.[82] Secretary Lujan, though, signaled that the task force would propose a plan with far more modest protection measures: "Environmentalists are not going to like it, but you have to have balance. . . . You can't throw 21,000 people out of work and not give it consideration."[83] Divided, the task force missed its September 1 deadline and released a scant 2½-page plan on September 21. The plan called for significant cuts in Forest Service timber harvesting but nothing close to the 50 percent reduction that the ISC recommended. Furthermore, the plan did not explain where the timber would be harvested, and it essentially asked Congress to exempt the timber sales from NFMA and ESA requirements. Frustrated members of Congress complained that the administration had simply kicked the problem over to the Capitol building.[84]

With hope in the task force lost and no hope of swift congressional action, the Bush administration turned to one of the few remaining administrative tools available: the Endangered Species Act Committee, or as it is usually called, the God Squad. Congress had authorized the God Squad in an early ESA amendment precisely so that economic and other considerations of national or regional significance could, in exceptional circumstances, trump the protection of endangered and threatened species. Invoking the God Squad had political risks, but the administration was running out of options. Lobbyists from the timber industry were cautiously optimistic that the God Squad would decide that the region's timber economy was more important than the northern spotted owl. "If the Act's mechanisms are allowed to work (ie: a jeopardy opinion is rendered and the God Squad called into service) the desired results will follow. . . . The problem therefore is how to get the Endangered Species Act Committee to function and function appropriately."[85]

President Bush's chief of staff, John Sununu, and other top administration officials were exploring this option throughout 1990 as they grappled with the FWS's listing decision. According to ESA regulations, the God Squad is permitted to grant exemptions for specific actions such as timber sales if there are no reasonable alternatives, the benefits outweigh the benefits of species conservation, the actions are of regional or national significance, and the actions include mitigation measures to minimize the negative impact on the listed species.[86] For specific projects, the God Squad could essentially remove the ecological lens imposed by the ESA and advocated by the ISC and the Gang of Four, adopting instead an economic and political lens.

The administration had one last opportunity to soften the ESA's economic blow before turning to the God Squad: recovery planning for the owl. In December 1990, the FWS formed a recovery team, which included not just staff biologists but economists and political appointees as well. A vice president of the Wilderness Society predicted, "The wolf politicians and the lamb biologists may lie down together, but the lamb isn't going to get much sleep."[87] The stakes were incredibly high. According to Secretary Lujan, "The development of a recovery plan for the Northern Spotted Owl may be the most important effort of its kind since the passage of the Endangered Species Act."[88]

As the recovery team worked in 1991, however, the administration concluded that it couldn't postpone the God Squad strategy any longer. The administration faced charges from a wide range of critics that it was creating a "train wreck" out of the northern spotted owl case intentionally so that Congress would be forced to amend the ESA when it came up for reauthorization in 1992.[89] While the administration might have been able to ignore this accusation from environmentalists on the left, members of its own party were cautioning that this charge was gaining traction. Senator Slade Gorton (R-WA) wrote to John Sununu, warning, "Many members of Congress are [pointing] the finger of blame at President Bush and accuse the Administration of a full-scale conspiracy to cause a 'train wreck' within the timber industry. . . . The Administration must act now to reverse this trend of blaming the Administration and must act now to save the working communities and families of the Northwest."[90] With no remaining options, the administration turned to the God Squad.

In January 1991, BLM consulted with the FWS over 174 specific timber sales. The FWS cleared 122 but determined that the remaining 52 were likely to jeopardize the owl in large part because the Jamison Strategy did not include the 50-11-40 rule.[91] In September, Jamison petitioned the God Squad to exempt 44 of the 52 timber sales for which the FWS had issued a jeopardy finding. The God Squad began its review divided. Administrators of the EPA and the National Oceanic and Atmospheric Administration (NOAA) suggested that the God Squad had been assembled prematurely and that the BLM had not done due diligence in its consultation before appealing to the secretary. As John Knauss, head of the NOAA, complained, "I am not satisfied that BLM did an industrious enough effort in finding or attempting to find reasonable and prudent alternatives."[92] They were also concerned that the BLM did not have a long-term solution to the northern spotted owl problem and might therefore return year after year seeking additional exemptions. Other members of the committee, particularly Secretary Lujan, were adamant that exemptions were justified.

Of the forty-four timber sales for which the BLM sought exemptions, the committee agreed that thirty-one sales did not merit exemption because reasonable alternatives existed, because they were located within the owl's critical habitat areas, or because they were not of regional economic significance.[93] But on May 14, 1992, the committee voted five to two to exempt thirteen timber sales that the FWS had determined might jeopardize the northern spotted owl. To avoid regular God Squad petitions in the future, however, the committee added a specific caveat that the BLM was required to abide by the FWS's northern spotted owl recovery plan in all future timber sales.[94]

The God Squad decision settled the conflict over thirteen specific timber sales, but it did nothing to address the larger crisis. The FWS's recovery plan, released a month before the God Squad decision, called for preserving 5.5 million of the spotted owl's 8.2-million-acre critical habitat. Although this was less than environmental groups wanted, it was still projected to cut timber harvesting in half compared to harvesting in the 1980s.

Having seen an advance copy of the recovery plan in January and knowing that the God Squad could provide only limited exemptions, Secretary Lujan had created a separate working group in February 1992 to find alternatives that would protect the valuable timber harvesting.[95] "I am specifically interested," the secretary had written to the group, "in formulating and evaluating options whose implementation would result in significantly low economic impacts and in a high probability of owl preservation and persistence, even though they might not achieve recovery of the species throughout its range or even in some physiographic provinces as required by the ESA. This goal clearly distinguishes your effort from that of the recovery team."[96] Secretary Lujan had requested that the working group provide its recommendations on April 15, the date scheduled for public release of the draft recovery plan, hoping that when Congress saw the projected economic impact of the plan, it would be willing to amend the ESA.[97]

The working group essentially went through a recovery planning process that was not constrained by the ESA, completing its work contemporaneously with the God Squad and recovery team decisions. Whereas the official draft recovery plan called for protecting 2,300 pairs of owls on 5.5 million acres of land, the working group recommended protecting 1,300 pairs of owls on 2.8 million acres of land.[98] The plan provided the northern spotted owl with at least a 50 percent chance of survival, essentially the same odds as the Forest Service's 1988 SEIS, and reflected Secretary Lujan's argument that species protection should be *a* goal but not *the* goal of land management. Secretary Lujan announced the plan on the same day as the God Squad decision, encouraging Congress to act.

The Presidential Campaign

The northern spotted owl crisis—court injunctions halting most federal timber sales in the owls' habitat range, the God Squad decision to exempt thirteen of forty-four BLM timber sales from the ESA requirements, the FWS's draft recovery plan, and the administration's alternative recovery plan—peaked in the late spring of 1992 in the middle of a hard-fought presidential campaign. Perhaps no other individual species has enjoyed as much campaign attention as the owl did that year.

During the 1992 campaign, President Bush once again campaigned on a platform of *balanced* environmental policy. He emphasized his achievements in improving air quality and reducing off-shore oil drilling as evidence of his commitment to environmental protection, but he pointed to the ESA as an example of an environmental law completely out of balance with voters' needs, particularly in the Pacific Northwest. This message, Republican strategists believed, would win Oregon and Washington in the election. As one Republican strategist explained to John Sununu, "If the President is perceived as stepping in to save timber jobs, he could put together a coalition of the traditional Republicans with non-Republican blue collar voters which would carry [Oregon and Washington]. If he allows the U.S. Fish and Wildlife proposal to protect an additional 12 million acres to stand or to be modified only a little, he can kiss Washington and Oregon good-bye."[99] This sentiment was echoed in the letters that Secretary Lujan received from Republicans and Democrats alike in Oregon. As the Democratic Oregon state senate majority leader Bill Bradbury complained in early 1992, "We are tired of having Northwest timber workers 'jerked' around. . . . Your Bureau of Land Management has failed Oregon's workers and timber industry miserably by refusing to do the necessary staff work supporting an exemption of their 44 sales from the Endangered Species Act."[100]

So on the campaign trail in the Pacific Northwest, the president once again emphasized his environmental commitments, and he once again identified himself as a conservationist in the tradition of Teddy Roosevelt. This time, however, President Bush was careful to choose sides in the struggle between preservationists like John Muir and conservationists like Gifford Pinchot: "Teddy Roosevelt once said," he told an audience during a campaign stop in Washington, "'Wise forest protection does not mean the withdrawal of forest resources from contributing their full share to the welfare of people.' What President Roosevelt had in mind, and what the American people have always wanted, is balance."[101] The Endangered Species Act, the president insisted, had been perverted and knocked out of balance: "Today, the act and other laws are being used by people with extreme views . . . to achieve in the courts what no sane official would ever have voted for, the complete lockup of the most productive forests in the entire United States."[102]

The president pledged to restore balance to endangered species protection if reelected: "The law is broken, and it must be fixed." Specifically, he pledged not to sign an extension of the ESA "unless it gives greater consideration to jobs and to families and to communities. I will not sign it without a specific plan in place to harvest enough timber to keep timber families working in 1993 and beyond. It is time to make people more important than owls."[103] Other Republicans, such as Senator Mark O. Hatfield (R-OR), echoed this message that it was time to put people above owls: "The fact is that Congress always considered the human element as central to the success of the Endangered Species Act. But today, with the act being applied across entire regions, the human factor has been lost."[104] President Bush's campaign message, then, was clear. He supported environmental protection, including protection for endangered species. However, he rejected, with Senator Hatfield, a broad, ecological approach, which he argued was simply a ploy by extreme environmentalists to lock up land and resources.

The Surprising Twist: Ecosystem Management Becomes Official Forest Service Policy

Given the president's campaign message and environmental record, it is perhaps surprising that the Forest Service adopted ecosystem management as official policy during the 1992 presidential campaign. After all, the administration had resisted framing wetlands delineation in ecological terms and absolutely rejected the idea that protection for the spotted owl should entail region-wide protection of old-growth forests. Nonetheless, on June 4, 1992, Forest Service Chief Dale Robertson announced that the Forest Service was committing itself to ecosystem management. The agency, he explained, had "been courting the ecosystem approach. . . . Today, I am announcing the marriage and that the Forest Service is committed to using an ecological approach."[105] What was it that could make the Forest Service shift rhetorical course so quickly and embrace the very language associated with advocates of old-growth forest protection?

For the White House, the decision to have the Forest Service adopt ecosystem management was made in great haste on the basis of a fairly clear political calculation. In June 1992, President Bush was scheduled to appear at the Rio Earth Summit, which the United Nations had organized to address issues including biodiversity protection and forestry. The United States had, for a decade, pressured other countries to stop wholesale deforestation and to protect biodiversity, making its own forest management in places like the Pacific Northwest look hypocritical. Furthermore, the president's approval ratings on environmental issues were abysmally low; the president had already signaled that he would not sign environmental treaties that threatened economic growth; and environmental critics such as Senator Al Gore were arriving at the conference early in order to attack the president's environmental record.[106] The president might have been willing to accept domestic criticism from environmental organizations for his handling of the northern spotted owl crisis, but he was not willing to expose himself to international disapprobation in Rio.

William Reilly, the administration's lead in Rio, urged Clayton Yeutter, the president's counselor on domestic policy, to prepare a significant environmental policy statement for the president to give in Rio that would diffuse international criticism. Reilly knew the Forest Service chief well and had talked with him extensively about the environmental problems with clearcutting. Yeutter also knew Robertson, having been Robertson's boss from 1989 to 1991 when he was agriculture secretary. With only about a week to go before the Rio Summit, Yeutter bypassed the chain of command in the Department of Agriculture and asked Robertson to write a statement eliminating Forest Service clearcutting. According to Roberston, he recommended situating the end of clearcutting within the broader framework of ecosystem management, and though Yeutter said that he and the president didn't understand ecosystem management, they liked the sound of it.[107]

In less than twenty-four hours, Robertson himself had crafted a statement adopting ecosystem management as official Forest Service policy, and within a few days he had announced it publically. According to Robertson, the secretary of agriculture had not even weighed in on the decision. The political urgency kept the circle of decision makers small and provided no time to discuss what ecosystem management would mean for the Forest Service beyond its specific commitment to end clearcutting, at least as a standard practice. In a matter of days, the Bush administration had thrown the mantle of ecosystem management over the Forest Service.

This explains the adoption of ecosystem management from the White House's perspective, but for the Forest Service, it was a more complicated story. Robertson might have decided to adopt ecosystem management in less than twenty-four hours to help the president, but it had much deeper roots in the agency. Indeed, Forest Service researchers and some leaders had been experimenting with ecosystem management concepts for some time in an effort to rehabilitate the agency's beleaguered multiple-use management framework that Robertson later explained had hit a wall and to break the cycle of litigation that had all but eliminated the agency's discretion in the Pacific Northwest.[108]

The Forest Service's multiple-use crisis was epitomized by the battle over old-growth forests, but it was caused by broader criticism from policymakers and scholars as well. One of the most concise records of this criticism came out of a March 1992 workshop on multiple use and sustained yield (MUSY), hosted by the Congressional Research Service. For some, the Forest Service simply hadn't lived up to its MUSY promise. Richard Behan, from the School of Forestry at Northern Arizona State University, complained, "If the concepts—multiple use and sustained yield—were intended, as I believe they were, to protect and maintain the health, beauty, and productivity of federal land biophysical systems, primarily to serve the general public welfare, they have not worked. They have *been worked*, on the contrary, to provide what Marion Clawson once described as 'the private use of public lands', the *de facto* land use policy for the federal estate."[109]

For many critics, MUSY itself was part of the problem because it was built on outdated science and an outdated understanding of forest values. Multiple use, Behan also argued,

> focuses, by definition, on uses, not resources. . . . Instead of focusing on uses, I believe we must look instead at the resources, or more accurately at the resource systems found on the federal forest and rangelands. . . . Instead of sustaining yields, we should have been sustaining, it is becoming clear, the biophysical systems in which they originated. We should indeed have focused on resources, not uses.[110]

Or as ecologist Chris Maser offered, "Long-term, ecological wholeness and biological richness of the landscape must become the measure of economic health."[111] This echoed Judge Dwyer's recent ruling: "The duty to maintain viable populations of existing vertebrate species requires planning for the entire biological community—not for one species alone."[112]

The Forest Service was chastened by the growing chorus of criticism and, more importantly, by the growing case history of timber sale injunctions. So while simultaneously trying to protect the preeminence of its timber program, it cast about for ways of greening MUSY by reinterpreting it through an ecological lens.[113] Forest Service leaders turned to the agency's research ecologists, who had been developing concepts of ecological forestry for twenty years. Beginning in the early 1970s, for example, Jerry Franklin (Forest Service), Chris Maser (BLM), and others had begun publishing reports on the importance of wildlife research in Oregon natural areas and of ecological communities in the national forests.[114] By the 1980s, they had begun to advocate for a new, ecological forestry.

The "new forestry" approach, as they called it in the late 1980s, emphasized the need to maintain "ecosystem resilience—an ability to absorb stress or change without significant loss of function." This means, according to Franklin, "Foresters must manage to retain greater ecological margins in order to sustain long-term productivity and buffer against uncertainties. . . . In new forestry, what's left behind on the site is more important than what is taken out."[115] Franklin and his colleagues emphasized changes in timber sale designs, such as leaving snags (dead standing trees), large logs, and shaded riparian areas, and they tested these changes in the Willamette National Forest in Oregon.[116]

Forest Service research ecologists initially had a tough time selling their work. Part of the problem was that their approach seemed so much less efficient than traditional clearcutting and replanting practices. After seeing some of the tests, for example, Chief Robertson noted that they "looked like a sloppy, unfinished logging job."[117] Researchers were also worried that their work would be misunderstood in an agency that emphasized resource development, and they echoed Pinchot's assurances that new forestry was not a preservationist agenda. "This perspective," Franklin and colleague F. J. Swanson assured their readers, "does not constitute a 'naturalistic ideology' in the sense of managing ecosystems for the sake of naturalness." Rather, they saw new forestry as a way to avoid stark trade-offs between economic and ecological values. They argued that their perspective was a new form of multiple use that could provide timber and other resources without foreclosing on "future opportunities to manage for ecological objectives."[118]

Given new forestry's tentative foothold in Forest Service research stations and a few ranger districts, agency leaders began a cautious experiment with the approach. When Robertson became Forest Service chief, he supported the effort but changed the name to "New Perspectives," which "was a good, neutral term and it would be hard for anyone to disagree with."[119] In 1989, Robertson appointed Hal Salwasser, then head of the agency's wildlife management staff in Washington, to lead the New Perspectives initiative. This was not intended to be a comprehensive reform of Forest Service policy but a concrete exercise in improving forest practices. Thus, the goals were "stimulating innovation in the field, capturing ideas, and disseminating them around the National Forest System."[120] For the next two years, the Forest Service ran the initiative as a pilot program on forests around the country, waiting until 1991 to expand its work throughout the agency. The goals in 1991 were to "1) strengthen the ecological basis of land management; 2) sustain the diversity and productivity of the land for multiple-resource values and uses through ecosystem management; and 3) improve the responsiveness of land management to public concerns."[121] Like multiple-use management generally, then, New Perspectives was a promise to please everyone.

For the Forest Service, the New Perspectives initiative was an answer to the northern spotted owl crisis, because it suggested that the agency had a new, broader management approach that could find the elusive balance between timber harvesting and species protection. As with any compromise approach, New Perspectives drew its share of criticism. After Robertson designated the entire Ouachita National Forest a New Perspectives project in 1990, the timber industry worried that this new approach threatened intensive timber management more broadly. At the same time, conservation biologists and environmentalists were also suspicious of the New Perspectives language. Articles appeared in *Conservation Biology* with titles such as "Is There Any Conservation Biology in 'New Perspectives?'" and "New Perspectives or Old Priorities."[122] For conservation biologists and environmentalists, "Conserving biological diversity is a necessity," or at least "achieving it ought to be the preeminent goal of the United States Forest Service."[123] They remained skeptical that the Forest Service was prepared to shift its focus and culture enough to privilege ecosystem resilience over resource outputs. In other words, both sides wanted the agency's management framework to constrain its discretion in ways that favored their position, so they were suspicious of a new, even broader framework that dripped discretion from every pore.

In retrospect, both sides had reason for suspicion. Salwasser later reflected, "it was real clear in the discussions we had that [Forest Service leadership] was simultaneously managing the agency along the traditional budget lines and priorities and supporting and encouraging the New Perspectives that . . . was changing the direction the agency was going."[124] Robertson was reluctant to push the agency in one clear direction or the other. He had chosen not to use the phrase "new forestry" because he was afraid that environmentalists would hear "forestry," and he had resisted the language of ecosystem management because he was afraid that the timber industry and Republicans would hear "ecosystem."

When the White House asked Robertson for a strong new policy statement about American forestry, however, it essentially asked him to choose a phrase that clearly emphasized ecological protection. New Forestry and New Perspectives had worked in the Forest Service context, but they would not have communicated anything in Rio. The phrase "ecosystem management," by contrast, was associated with ecology and conservation biology, so it was likely to be interpreted as favoring ecosystem protection.

This ambiguity was precisely what troubled the timber industry, environmentalists, and conservation biologists in the United States after the Forest Service adopted ecosystem management as official policy. The timber industry wanted to make sure that it meant only a process of large-scale planning that continued to favor timber production; the environmentalists and conservation biologists wanted to make sure that ecosystem management meant a fundamentally new paradigm for the Forest Service that prioritized biodiversity and ecosystem resilience over timber harvesting. For the Forest Service, ecosystem management meant something in between. It needed to lead to some new levels of ecosystem protection if the agency was to fulfill its NFMA and ESA obligations and escape continual court injunctions; at the same time, it needed to achieve timber production targets that the agency promised and that Congress all but ordered in annual appropriations bills.

Given ecosystem management's roots in New Forestry and New Perspectives, it would be incorrect to see ecosystem management as a greenwashing effort on the part of the agency. It was rooted in substantive scientific research and experimentation and reflected some things that were genuinely new and designed to address NFMA and ESA requirements. It would also be wrong, though, to see ecosystem management as a new paradigm that constrained or challenged multiple-use management. Indeed, ecosystem management was, for the Forest Service, an even more expansive interpretation of multiple-use management. The Forest Service had expanded its definition of multiple-use management before to include not only various economic values of the national forests but also recreational and aesthetic values. In ecosystem management, the Forest Service was not renouncing any of these established values; it was simply promising to expand the multiple-use paradigm even further to encompass biodiversity, ecosystems, ecological processes and functions, etc. Ecosystem management was a promise that the agency would now use its broad discretionary authority, which was currently locked up in the courts, to protect all of the values that Americans saw in the national forests.

For Forest Service leaders, ecosystem management was not a substantive "system of management" with predetermined priorities but a "management approach" to multiple use that incorporated ecological concerns.[125] As Deputy Chief James Overbay explained in a workshop entitled "Taking an Ecological Approach to Management," "it is time to embrace the concept of managing ecosystems to sustain both their diversity and productivity and to chart a course for making this concept the foundation for sound multiple-use and sustained yield management."[126] Or, as Robertson explained to the agency in 1992, "an ecological approach will be used to achieve the multiple-use management of the National Forests and Grasslands. It means that we must blend the needs of people and environmental values in such a way that the National Forests and Grasslands represent diverse, healthy, productive, and sustainable ecosystems."[127]

Conclusion

The four years of George H. W. Bush's presidency were a time of remarkable developments in federal land and resource management, particularly in the Pacific Northwest. Bush had taken office promising to be the "environmental president" and had followed through on that promise through achievements like the Clean Air Act amendments of 1990. What the Bush administration demonstrated, however, was that the president was committed to environmental protection where this could be framed in clear political terms that did not threaten economic development. The administration was particularly hostile to developments—e.g., the Vision document in the Greater Yellowstone Ecosystem, wetlands delineation, and spotted owl protection—in which the issues were reframed in ecological terms that emphasized connections across political and legal boundaries, complexity, ambiguity, and precaution. This, for the administration, was the stuff of regulatory mischief that hurt the American economy.

Ironically, perhaps, despite the administration's efforts to hinder an ecological framing of land and resource management issues, its escalation of the northern spotted owl controversy led to judicial rulings that demanded an ecosystem approach, and the president's trip to Rio occasioned official Forest Service adoption of an ecosystem management policy. When President Clinton took office in 1992, then, many of the conceptual resources were in place to rethink ecosystem management, and the northern spotted owl crisis provided a laboratory in which to test it.

5 | Putting Ecosystem Management to the Test in the Clinton Administration

Introduction

President Clinton took office in January 1993 just seven months after Chief Robertson committed the US Forest Service to an ecosystem management approach, and the agency had barely begun to think through what this would mean for its various resource programs. And just months earlier, President Clinton had campaigned in the Pacific Northwest, where the previous administration had fought an ecosystem approach to managing the northern spotted owl and other old-growth-dependent species. When President Clinton took office, in other words, the meaning of ecosystem management was unclear and its future was uncertain.

What is more, the Clinton administration's first actions hardened opposition to reforming federal land and resource management. The president's first budget proposal waded naively into contentious and divisive issues, proposing to create the first-ever royalty program for federal hard-rock minerals and to raise grazing fees to fair market value. Environmentalists were ecstatic; the mining industry, the livestock industry, and virtually the entire western congressional delegation were furious. After western Democrats threatened to block the budget, the administration withdrew the royalty and grazing fee proposals. It hardly seemed like the best political climate in which to advance an overarching framework for federal lands and resources that had the word "ecosystem" in its title, and the climate became far, far worse after Republicans took control of Congress in the 1994 election.

Yet by 1995, all federal land and resource agencies had joined the Forest Service in adopting ecosystem management as official policy—its definition broad, its importance symbolic—and they had initiated ambitious projects in the Pacific Northwest, the intermountain West, and the

Florida Everglades, all under the ecosystem management banner. As the agencies worked to implement ecosystem management, two things became increasingly clear. First, for the Clinton administration, ecosystem management meant elevating the role of science and scientists in land-use planning and management decisions. Officials emphasized this in congressional testimony, published essays, and popular speeches, and the administration demonstrated this through a variety of high-profile policy decisions. Second, it meant leveraging regulatory statutes or the threat of those statutes to get ecosystem management projects off the ground. The administration naturally did not advertise this aspect of ecosystem management, but it was a pragmatic reality nonetheless. As Interior Secretary Bruce Babbitt explained after leaving office, "you have to have a legislative stick to get anywhere. You have to come from a position with some power to cause a result that will inflict pain."[1]

What remained unclear at the end of the Clinton administration was the underlying legal and political meaning of ecosystem management. It was clearly an approach that emphasized flexibility and discretion, which the administration promised to use to find win-win solutions to land-use conflicts. But this left open the more important question, how would federal agencies use this discretion? Or asked another way, when the administration could not find a win-win solution, which way would ecosystem management cut? Some of the administration's rhetoric and concrete projects suggested that ecosystem management was a substantive new paradigm that prioritized ecosystem integrity and biodiversity protection over other established, multiple-use goals. At other times, its rhetoric and projects suggested that ecosystem management was a procedural paradigm that promised to ensure that ecosystem planning incorporated the best available science and open, public deliberation. The former suggested ecosystem management built on the model of the Endangered Species Act (ESA); the latter suggested a model built on the model of NEPA.

Ecosystem Management in the Forest Service

When President Clinton took office, the Forest Service was working on defining its new ecosystem management approach. At workshops throughout 1993—e.g., "Sustainable Ecological Systems: Implementing an Ecological Approach to Land Management" and "Silviculture: From the Cradle of Forestry to Ecosystem Management"—Forest Service scientists and officials struggled to reinterpret multiple-use management through an ecological lens.[2] Hal Salwasser and Robert D. Pfister explained at one of the workshops that "an ecosystem perspective enlarges the focus of land management and resource conservation to whole ecosystems rather than selected parts or processes . . . it focuses on sustaining desired ecosystem conditions of diversity, long-term productivity, and resilience, with yields of desired resources and uses being commensurate with the larger goal of sustaining those conditions."[3] And Ann Bartuska, a leader in the Forest Service's ecosystem management office, described it as "a shift from focusing on a particular unit of production on a specific piece of land to a broader view. The broader view first identifies the overall condition of the ecosystem at issue, and then attempts to identify the multiple products which are available from that larger landscape, where long-term sustainability of the ecosystem and its health is a priority."[4]

Yet participants recognized that these ideas were not self-executing. The goals of ecosystem management were relatively abstract compared to timber production targets; they were articulated in phrases such as "maintain a statistically stationary pattern over time"; and the tools for achieving them were complex combinations of "geographic information systems (GIS), ecological simulators, and decision support models."[5] This was hardly a detailed blueprint. Somehow, these broad goals and methods needed to be translated into detailed prescriptions for activities as specific as individual timber sales.[6]

In order to move further toward concrete applications of ecosystem management, each region of the Forest Service developed its own set of guidelines. As an example, the Pacific Southwest Region worked on its draft guidelines from March 1993 to February 1994, with three objectives: "1. Defining the major ecosystems in the Region and at what scale they operate, defining their basic components and functions, and developing measures and thresholds of functionality and integrity. 2. Developing clear objectives for ecosystem management in the Region. 3. Developing a process by which these objectives will be incorporated into planning."[7]

The first goal could be met incompletely by the agency scientists working in and through the Pacific Southwest Research Station. They could certainly categorize the major ecosystems of the region and identify some of their components and functions. Developing measures of functionality and integrity, however, was an ongoing research problem for ecology as a science, and it posed a significant challenge for the federal scientists.

The second goal was significantly more difficult, because as the draft guidelines explained, this was a process of merging scientific data with public demands and interests to formulate "desired conditions" (DC) for each ecosystem: "Typically, DC is an integrated and pragmatic expression of the state or condition in which we prefer our ecosystems to exist, now and into the future. Ideally, it reflects a range of conditions, recognizing that landscapes change and develop over time. Furthermore, DC must be based on the natural capabilities of the land."[8] This, of course, went to the heart of federal land and resource politics, since various constituencies had very different "desired conditions" in mind, and scientific information was as likely to drive the agencies' constituencies further apart as it was to bring them together with a common vision.

The third and final goal pointed to another enormous challenge given the agency's existing legal, political, and administrative infrastructure. Achieving desired conditions for ecosystems, the authors noted, would be difficult within the existing process of forest planning:

> The more [ecosystem management (EM)] concepts are implemented, the more obvious it becomes that existing Forest Plans did not carefully consider sufficiently large areas. Thus, the emerging need for ecoregion/subregion level analyses becomes quite evident. To effectively implement EM objectives, the Forest Plans of the future will most likely encompass multiple National Forests. Additionally, these plans will need to assess land management practices on multiple ownerships both intermingled and extending beyond National Forest boundaries, to the extent possible. Cooperative work with other ownerships may be crucial to the maintenance or restoration of some ecosystems. The planning process of the future may well be labeled "ecoregional multi-agency ecosystem based planning."[9]

In essence, the authors recognized that individual forests could not implement ecosystem management in isolation. They would need to make decisions in the broader context of multi-jurisdictional landscapes. While federal law did not prohibit this, it certainly made it difficult to do in any coherent fashion.

The three goals of Region 5's guidelines captured succinctly the challenges that the Clinton administration faced in advancing ecosystem management as an overarching framework. The first challenge was at some level one of scientific research. In many cases, federal agencies simply lacked basic scientific data needed to map and define ecological systems. But Region 5's first goal captured a more complicated challenge for scientists. Identifying precise thresholds of ecological resilience—predicting the point past which an ecosystem's processes and functions would be altered in fundamental ways—was certainly a scientific challenge, but it was also inextricably linked to a normative challenge. To describe ecosystem integrity, or ecosystem health, is already to imply that there is a way that ecosystems *ought* to function, so the science itself would be advancing management goals.[10]

The second challenge for ecosystem management was deciding how to determine goals or "desired conditions" for an ecosystem beyond a scientific perspective. What kind of relationship did proponents of ecosystem management envision between science and politics? More specifically, how normative were scientific measures of ecosystem integrity in federal planning, and what kinds of limits could or should be placed on public desires? What is more, the federal land agencies already had certain management goals in their authorizing statutes, so how would new goals implied by ecosystem management integrate those diverse mandates?

The third and closely related challenge was establishing a process for ecosystem management. The agencies already had extensive planning processes focused on lands and resources within their jurisdiction. They had different methods of gathering data and different formats for data storage. The Clinton administration had to develop a new infrastructure for integrating agency planning around the new goals of ecosystem management. These three challenges—science, substance, and process—were all significant in shaping ecosystem management during the Clinton administration.

The Political Authority of Science

When President Clinton took office, environmentalists were hopeful that he would be a greener "environmental president" than his predecessor, though their hope rested less on Clinton himself than on other key members of the administration. They were particularly encouraged when the president appointed Bruce Babbitt Secretary of the Interior, giving him authority over all federal land and resource agencies other than the Forest Service.[11] Babbitt had served nearly three terms as governor of Arizona, which had vetted his skills in dealing with contentious land and resource issues, and he had most recently served as president of the League of Conservation Voters, validating his concern for biodiversity and other environmental issues.

Secretary Babbitt took office critical of the Interior Department and of traditional multiple-use management. One *Washington Post* article in January 1993 quoted Babbitt as saying that the BLM was an agency "tainted by politics and incompetence in upper management and heavily influenced by mining and livestock constituencies" and that the Bureau of Reclamation's work was "the most environmentally destructive of all the public land agencies. . . . It seems to know the subsidized price of everything and the long-term value of nothing."[12] Multiple use, Babbitt argued, had been interpreted in a way that favored extractive industries and led to environmentally destructive practices: "The next step in the evolution of public land use policy is to replace multiple use management with a new concept . . . that gives priority to recreation, wildlife and watershed uses. [It] would be a mandate to reconsider destructive resource exploitation that is of marginal economic importance."[13] After taking office, he advocated ecosystem management as this next step. Already, the difference in interpretation of ecosystem management between the Bush administration and the Clinton administration was clear.

For Babbitt, reform began by reframing federal land and resource issues in comprehensive, ecological terms that exposed negative environmental externalities and that forced broader deliberation about environmental trade-offs. This involved two significant commitments. The first commitment, he emphasized repeatedly, was looking at land-use problems not simply within the traditional metes and bounds of land surveys but within broader ecological boundaries as well: "Congress and my predecessors handled resource conflicts by drawing enclosures," Babbitt lamented in an essay defending the Clinton administration's plan for the northern spotted owl, "but what we're learning, thanks to modern biology and a more reasoned and thoughtful understanding of the landscape, is that we cannot protect the splendor and biological diversity of the natural world by simply fencing off a few protected areas within an overall landscape of exploitation."[14] The corrective, he explained, was "looking across fences, using good science, integrating the land management agencies together."[15]

The second commitment, which he also emphasized repeatedly, was integrated, proactive land-use planning: "If we could identify the problems with potentially endangered species before they reach the crisis point," he wrote after leaving office, "scientific understanding would give us more time and a lot more flexibility in working out solutions that could accommodate both landowner expectations and the requirements of the Endangered Species Act."[16] The call for more comprehensive land-use planning struck a nerve among Republicans because it sounded like code for regulation, but Babbitt dismissed this as disingenuous and hypocritical: "Land use planning has . . . been a federal function since the nation's founding. And so long as the planning is intended to facilitate development, hardly a discouraging word is ever heard. Yet parallel planning for protection of our remaining open space and conservation of our natural resources evokes strident opposition."[17]

Babbitt's efforts to apply an ecological lens also raised an important question politically, namely the role of ecology, or science more broadly, in ecosystem management. At times, Babbitt emphasized the neutral, descriptive authority of science and the importance of using science in a democratic, planning *process*. This process would be truly democratic precisely because it would start with the assumption that Congress and the agencies had not predetermined the goals of planning but expected to establish them deliberatively. At other times, Babbitt emphasized the normative implications of concepts like ecosystem integrity, suggesting that public deliberation and choices needed to be tightly constrained.[18]

The tension here was not isolated in political debate. As explained in chapter 1, it was embedded in the field of conservation biology and ecology. For example, when the Ecological Society of America issued a report on "the scientific basis of ecosystem management," it provided the following working definition: "*Ecosystem management is management driven by explicit goals, executed by policies, protocols, and practices, and made adaptable by monitoring and research based on our best understanding of the ecological interactions and processes necessary to sustain ecosystem structure and function.*"[19] Read carefully, this definition does not delineate the explicit goals that should drive ecosystem management, suggesting that ecology simply explains what managers would need to do *if* they wanted "to sustain ecosystem structure and function." Immediately after this definition, however, the report lists eight principles that should guide ecosystem management. The first principle is clearly normative: "Ecosystem management assumes intergenerational sustainability as a precondition for management."[20] Ambiguity about the role of science—a neutral resource or normative guide—made the Clinton administration's efforts to elevate science and scientists in federal land and resource decision making sometimes surprisingly controversial. Two examples will suffice in illustrating the contours of this debate: the National Biological Survey (NBS) and Rangeland Reform '94.

National Biological Survey

The Clinton administration proposed a complete reorganization of federal research in the biological sciences to increase its credibility and, indirectly, its power in decision making. Almost as soon as he took office, Secretary Babbitt announced plans to create a new agency in the Interior Department called the National Biological Survey, which he likened to the US Geological Survey (USGS): "Just as the US Geological Survey gave us an understanding of America's geography in 1879, the National Biological Survey will unlock information about how we protect ecosystems and plan for the future."[21] The new agency was to be a true scientific research agency rather than a regulatory agency, and it would take research scientists out of the political fray of policy and management decisions. One of the NBS's key tasks would be to conduct "a systematic biological inventory of the entire nation at an appropriate scale and feasible level of detail."[22] This would serve as the foundation for truly ecological management.

Creating the National Biological Survey, he told Congress, was a necessary step for two reasons. First, existing scientific research was scattered throughout the department but concentrated in the US Fish and Wildlife Service (FWS). Agencies like the National Park Service and the BLM did not have adequate research capabilities, and it was difficult to collate and synthesize the research results across jurisdictional lines.[23] The new agency would ensure that all agencies had equal access to the research information they needed. Second, because it would be somewhat insulated from political and regulatory pressures, the survey would provide a credible scientific foundation for proactive management and regulatory decisions: the NBS would "provide a road map to enable us to get ahead of the endangered species listing process and constructively solve environmental and economic conflicts."[24] In other words, a more complete scientific assessment of land and resources would allow management agencies to find new opportunities for win-win management decisions.

To build support for the NBS as a legitimate research science agency, Secretary Babbitt asked the National Research Council in February 1993 to convene a special committee of leading scientists to outline the NBS's appropriate role. While much of the report dealt with administrative matters related to organizational structures and research partnerships, the final recommendations outlined top research priorities for the new agency. Within three years, the committee urged that "the NBS should establish or expand its research programs in environmental indicators and inventories of areas rich in biological diversity, of unique ecosystems, and of potential candidate areas for restoration." Within five years, it "should broaden its scientific priorities to include research efforts in restoration biology and expand inventories, should develop models to predict the status and trends of ecological systems, and should work to fill remaining gaps."[25] Thus, both Babbitt and the committee recommended a research agenda that was explicitly oriented to the protection of biodiversity.

Democrats in the House and Senate sponsored bills to create the National Biological Survey in the early summer of 1993.[26] Appearing before the House Merchant Marine and Fisheries Committee on July 15, Babbitt reiterated his case that the NBS was crucial to avoiding future ESA "train wrecks":

> The train wrecks in my judgment are unnecessary. They have arisen because we have been passive, awaited until the 11th hour when the species is at the brink of extinction tumbling toward oblivion in which the management options on both Federal and private land have been reduced to a lot of very unhappy alternatives, usually being administered not by the executive branch of government but by Federal judges.[27]

Secretary Babbitt was adamant that the new NBS would serve not only environmental interests but the interests of economic development and private property rights as well: "One simple fact. . . is that the use of good science in almost every circumstance clears up problems, creates stable markets, enhances property values, and makes it vastly easier for everyone in the private sector to do their business."[28]

Republicans and some western Democrats were particularly suspicious of Babbitt's claim that the NBS's work was neutral science and condemned it as a ploy to expand government control of private property and stymie markets. Their accusations ranged from more charitable assertions that the data collected would eventually lead to more ESA listings and more train wrecks to charges of an environmental conspiracy.[29] More scientific information, they feared, would provide support for new land-use restrictions.

House subcommittees approved two different versions of the bill, but only after a series of amendments designed to safeguard against environmental abuse. One amendment written by a Minnesota Democrat stated that employees of the National Biological Survey would comply with existing law, "including laws relating to private property."[30] Other amendments were designed to ensure that the science was duly refereed to avoid irresponsible advocacy from the NBS. In the end, a cloud of suspicion hung over the bills. As another Democrat warned, "This scientific survey, you mark my words, will become the basis for more encroachment on people's lives and more regulations."[31] And when the bill finally reached the House floor in October, a heated six-hour debate ensued. Don Young (R-AK) said that the bill was part of "the socialist agenda to make sure that Big Brother, big government controls all and everyone." And Jack Fields (R-TX) said that volunteers collecting information for the survey would be "an environmental Gestapo."[32]

With bills in Congress faltering, Secretary Babbitt created the NBS by secretarial order on September 29, 1993. At this point, agencies themselves had misgivings, since they would be required to give many of their research scientists to the NBS.[33] And Congress was not about to lavish funds upon the NBS, so money in the FY 1994 budget was mostly just consolidated from agencies' budgets. Roughly 1,300 of the 1,800 scientists and two-thirds of the budget were taken from the FWS.[34] To head the new agency, Babbitt recruited ecologist Ron Pulliam from the University of Georgia. Initially hesitant to leave the university context for Washington, Pulliam later remembered Secretary Babbitt saying that "he would handle the politics, and I would tend to the science." But, he continued, "This was a deal too good to be true."[35]

The National Biological Survey never really got off the ground. For the next three years, Congress appropriated less than requested, and Pulliam recalls regular, hostile conversations with members of Congress:

> What I heard was always the same. People were fed up with government regulations and Washington bureaucracy. They didn't want any new government agency and, furthermore, many were convinced that NBS was part of a land-grab conspiracy. I was told that NBS scientists would soon be cutting fences in the middle of the night and trespassing on private property with

> the aid of hoards [*sic*] of untrained volunteers, most of whom were probably card-carrying members of radical groups like Green Peace and Earth First; they would find endangered species so that government regulators had an excuse to seize land from private citizens.[36]

In January 1995, Secretary Babbitt changed the survey's name to the National Biological Service to avoid the assumption that the agency's main job was surveying every square inch of the country for endangered species, but this did nothing to mollify its critics. In 1995, the 104th Congress convened with Republican majorities, which called for outright abolition of the NBS. Pulliam remembers this as a particularly difficult time for the agency. In a meeting with one of the Republican leaders in Congress, he suggested that they rename it the Biological Research Service to emphasize the agency's policy-neutral role. As he remembers it, the leader retorted, "Biology is no longer an acceptable word in this Congress."[37] After only three years, Congress cut funding for the NBS altogether, and the agency's scientists were absorbed into the USGS's new Biological Resources Division. As the division head explains, this was a political compromise in which Republicans could claim victory in abolishing the NBS without eliminating the Interior Department's research capabilities.[38]

The NBS story is revealing of the politics of federal ecosystem management. Assuming that claims of a Nazi-inspired conspiracy theory were unfounded, the agency's role was indeed that of research science formally separated from policy and management decisions. But framing land and resource management issues in ecological terms undoubtedly challenged the existing political dynamics in each land and resource arena. This was precisely what the Bush administration wrestled with in the case of wetlands regulation and northern spotted owl protection, since framing the issues ecologically would have expanded federal land and resource protection and restricted development. Separate balances of power at a variety of scales existed for timber, minerals, grazing, and other resource arenas, and agreeing in advance to base decisions on ecological science, no matter how reputable, would shift the emphasis of management. And to the extent that fields like conservation biology had an open goal of preserving biological diversity, framing land and resource issues in ecological terms threatened to dethrone resource development interests.[39]

Rangeland Reform '94

In addition to general support for scientific research, the Clinton administration worked to place specific land and resource programs on an ecological foundation. Perhaps the best example outside of forestry was the Interior Department's work revising the BLM's range regulations. Just seven months into the Clinton administration, Secretary Babbitt announced Rangeland Reform '94, which proposed new grazing regulations for the BLM. These regulations included a second attempt to raise grazing fees on public lands, but the revision was more modest than the first. Rangeland Reform '94 tackled other significant controversies involved in public lands grazing, but the one most relevant to ecosystem management was the proposal to establish a national rangeland standard that emphasized certain ecological conditions and protection of riparian zones.

Establishing rangeland standards might seem innocuous, but it addressed a long and bitter struggle over the BLM's functional power to regulate livestock grazing on public lands. Congress had given the newly formed BLM a meager range budget in 1946 to ensure that the agency did not have enough range conservationists to conduct scientific surveys of range conditions. And for the next forty years, the BLM's efforts to set clear, scientific guidelines governing the number of livestock a permit holder could graze on public lands remained a source of bitter controversy. For example, after the BLM's first extensive EIS analysis for grazing began to show that the public lands were overgrazed, Interior Secretary Watt and BLM Director Robert Burford blocked all grazing reductions, arguing that the science behind the range surveys was inaccurate and that the only way that the BLM could demonstrate range deterioration from overgrazing was through five years of range monitoring.[40]

For this reason, pledging to establish ecological rangeland standards caused a political firestorm. Soon after the announcement, Senator Pete Domenici (R-NM) attempted to block the proposed regulations in the appropriations process. After a three-month skirmish that the Democrats were likely to lose, Secretary Babbitt withdrew the proposed regulations and promised to start over. In March 1994, Babbitt announced a revised Rangeland Reform '94 that worked toward the same goals but with significant concessions. In a Senate hearing, Babbitt essentially apologized for the first round of draft regulations, which he said were "way, way too prescriptive."[41] Rather than a single national rangeland standard, the new regulations proposed separate standards for each state developed through a collaborative process, with the negative incentive that the BLM would impose a national standard for any state that hadn't completed its standards within a year and a half. This was not enough to sway Congress, which relented only after Babbitt agreed to drop grazing fee increases.[42]

The battle over Rangeland Reform '94 can be understood as a struggle not so much over science itself as over the means of controlling the significant discretion that Congress had granted the BLM in the Taylor Grazing Act, FLPMA, and other authorizing statutes. Congress generally controlled that discretion through appropriations and oversight hearings that largely protected the status quo. To agree in advance to ecological standards and benchmarks for range conditions was to turn that discretion over to scientific judgments, ultimately constraining established tools of political control over range management decisions.

Defining, or Not Defining, Ecosystem Management

Clearly, then, ecosystem management for Babbitt and others in the Clinton administration meant a heavy emphasis on science, particularly ecology, but in their effort to sell the NBS they shied away from articulating the ends that new science would serve. The Forest Service had adopted ecosystem management as an approach to multiple-use management, which seemed to mean primarily that the agency would employ ecological analysis across broader landscapes in determining multiple-land uses. The FWS, by contrast, embraced it as an approach to protecting biodiversity, broadening the ESA's habitat protection mandate.

The Clinton administration embraced this ambiguity, offering ecosystem management as something "implicit, broad, and inclusive."[43] At a symposium hosted by the Congressional Research Service in June 1994, Will Stelle from the Office on Environmental Policy explained that the administration didn't want to impose a definition. "We are in the very early stages," Stelle said, "If we rush too quickly to give it too much definition, we may destroy it. . . . Our challenge is to take the basic philosophical foundation of ecosystem management, and give it substance, give it meaning, and give it heft."[44] In other words, the administration chose to leave ecosystem management ambiguous so that politically it would be a difficult target to hit.

Papers presented at the 1994 symposium illustrated the two very different emphases within ecosystem management. One interpretation focused on "changes in the public policy process based on improved interaction among stakeholders. Ecosystem management, in this view, meant expanded participation and greater cooperation. Some participants appeared to believe that changes in process that allowed all stakeholders to collaborate were sufficient to qualify as ecosystem management."[45] This, like NEPA, promised that agencies would *consider* biodiversity and ecological integrity in their planning, allowing it and public participation to determine their management decisions. This was a politically palatable interpretation that did not address potential trade-offs in land-use planning and management. The other interpretation centered "on changes in resource conditions—making ecosystem management another phrase for conservation of biological diversity and of ecological processes and services."[46] This, like the ESA, began with predetermined protection goals and used public participation to explore ways of achieving them. As with the ESA, this was a politically contentious interpretation of ecosystem management precisely because it promised to accomplish specific goals.

One of the most revealing things about the papers presented was the fairly clear divide between policymakers and scientists. Policymakers equated successful ecosystem management with collaborative problem solving or successful political process, while scientists at the symposium tended to equate it with increased scientific understanding of ecological systems and improved ecosystem health, as defined by various indicators. Policymakers tended to argue "that improved decision-making process was central to ecosystem management and would benefit resources as well," while scientists tended to be more ambivalent about political deliberation and stakeholder collaboration.[47] The Clinton administration's embrace of a broad and amorphous meaning of ecosystem management held these two emphases together in principle, but they clashed in various projects when the two led in different directions. As editors of the CRS symposium proceedings concluded, "Perhaps the best measure of how ecosystem management succeeds or fails in the future will be whether it can continue to attract and accommodate all, or at least what may be defined as the most critical, interests to work together, and whether it can continue to accommodate diverse resource goals that various interests can work toward rather than fight against."[48]

The Clinton administration advanced the loosely defined ecosystem management paradigm in a number of important ways, capitalizing on its ambiguity. On the one hand, administration officials offered ecosystem management as a guide to procedural reform that would increase government efficiency and greater public satisfaction. On the other hand, they offered it as a way to increase the government's effectiveness in achieving environmental goals such as biodiversity protection and maintaining ecosystem integrity.

Ecosystem Management, Procedural Reform, and NEPA

The first category of arguments was spearheaded by the Clinton administration's National Performance Review, led by Vice President Al Gore. During the 1992 campaign, Governor Clinton and Senator Gore had published a book that sounded the well-worn theme of reinventing government: "We can no longer afford to pay more for—and get less from—our government," they argued, "It is time to radically change the way government operates—to shift from top-down bureaucracy to entrepreneurial government that empowers citizens and communities to change our country from the bottom up."[49] It was time, they argued, for the federal government to take customer service seriously in how it interacts with citizens. This was the popular movement discussed in chapter 1 to increase governance while decreasing the government's coercive role. Just over a month after taking office, President Clinton announced that Vice President Gore would lead an initiative to reinvent government in order "to make the entire Federal government both less expensive and more efficient. . . . We intend to redesign, to reinvent, to reinvigorate the entire national government."[50]

After six months, the National Performance Review released a report entitled "From Red Tape to Results: Creating a Government That Works Better and Costs Less." The review, the vice president explained, reflected the administration's conviction that "it is time for a new customer service contract with the American people, a new guarantee of effective, efficient, and responsive government."[51] Recommendations offered in the report would reduce FY 1995 spending by $108 billion and reorient federal agencies to customer service.

It was in this context that the vice president recommended establishing ecosystem management as official policy across all areas of government.[52] More often associated with the goals of ecosystem protection, ecosystem management in this context was recommended as a way to reduce redundancy and conflict among the many agencies involved in land and resource management. The report laments, "Different agencies, with jurisdictions over the same ecosystem, do not work well together. Even within the same agency, bureaus fight one another. At the local level, a hodgepodge of government agencies control activities that affect the environment."[53] The results, it indicated, were gridlock, inefficiency, public frustration, and environmental degradation, which a collaborative, cross-jurisdictional ecosystem management approach would hopefully change.

In response to the National Performance Review's preliminary findings, Council on Environmental Quality chair Katie McGinty created and led the Interagency Ecosystem Management Task Force (IAEMTF) established in August 1993. It was made up of assistant secretaries and other officials from fourteen different departments and offices. After two years, the task force issued a three-volume report entitled "The Ecosystem Approach: Healthy Ecosystems *and* Sustainable Economies."[54] Given the task force's mandate from the National Performance Review, it is not surprising that the emphasis was on ecosystem management's potential to break gridlock, increase collaboration and cooperation, and decrease costs. "It is goal driven," the task force wrote, but "it is based on a collaboratively developed vision of desired future conditions that integrates ecological, economic, and social factors."[55] Indeed, the very first principle of ecosystem management that the task force advocated was to "Develop a shared vision of the desired ecosystem condition that takes into account existing social and economic conditions of the ecosystem, and identify ways in which all parties can contribute to, and benefit from, achieving ecosystem goals."[56]

In the end, the task force outlined a procedural model that fits very well with agencies' mandated NEPA analysis. The framework was not intended to predetermine ecological or economic goals. Rather, it was designed to ensure that agencies follow a particular process, emphasizing issues of scale and ecological dynamism that had received less attention in the past. The framework had eight steps, which are worth listing in abbreviated form:

- Define the Area of Concern/Interest
- Involve Stakeholders
- Develop a Shared Vision of the Ecosystem's Desired Future Conditions
- Characterize the Historical Ecosystem and the Present Economic, Environmental, and Social Conditions and Trends for the Ecosystem
- Establish Ecosystem Goals
- Develop and Implement an Action Plan for Achieving the Goals
- Monitor Conditions and Evaluate Results
- Adapt Management According to New Information[57]

This sounds very much like the rational, public decision-making process required in an Environmental Impact Statement. The main difference is that the area of concern is open-ended, and there is a stronger emphasis on uncertainty and the need to manage adaptively after the decision is made. Of course, these guidelines offered no greater guarantee that the agencies would be able to develop a shared vision with each other or with their constituencies.

In December 1995, fourteen federal departments and lead agencies signed a memorandum of understanding, as recommended in the task force report, to "Foster the Ecosystem Approach." This was certainly not a reinvention of federal land policy or even of the operations of federal land agencies. As with the task force recommendations, the emphasis of the memorandum was more on political process than it was on ecological objectives: "Federal agencies should ensure that they utilize their authorities in a way that facilitates, and does not pose barriers to, the ecosystem approach. Consistent with their assigned missions, federal agencies should administer their programs in a manner that is sensitive to the needs and rights of landowners, local communities, and the public, and should work with them to achieve common goals."[58] The key emphases included:

- Striving to consider all relevant and identifiable ecological and economic consequences.
- Improving coordination among federal agencies.
- Forming partnerships.
- Improving communication with the general public.
- Carrying out federal responsibilities more efficiently and cost-effectively.
- Basing decisions on the best science.
- Improving information and data management.

- Adjusting management direction as new information becomes available.[59]

Thus, the emphases to which the departmental representatives agreed were fairly innocuous commitments to improve the quality of information they used and communication of that information to the public, finding more cost-efficient ways to operate, and remaining flexible in the face of changing conditions and goals.

On its face it is difficult to find anything politically volatile in this memorandum, since few people oppose efficiency, good communication, and improved data management. Yet the very fact that department leaders adopted something called *ecosystem* management that would frame federal land and resource questions with ecological science caused considerable agitation among property rights groups, libertarians, and many Republicans. While the memorandum emphasized process goals over substantive outcomes, critics feared that these recommendations disguised more substantial ecological commitments. The MOU, after all, was focused on processes in which the agencies had enormous administrative discretion, and ecosystem management critics were convinced that the Clinton administration intended to redirect discretion in ways that would privilege environmental over economic interests. The author of one Cato Institute report warned that the ecosystem management initiative was a "radical departure from past environmental policy" that would "significantly expand federal control of the use of privately owned land and lead to increased restrictions on the use of the nation's public lands for economic purposes. Economic activity and private property rights would be subordinated to ecosystem protection."[60] The author also sought to expose an environmentalist conspiracy behind ecosystem management: "'Cardinal Green' of the eco-church and his minions threaten freedom and improvement in human well-being at this moment in our time, just as King George and his officials threatened them at our nation's founding. As we once rejected the king, I suggest that we rebuff the 'cardinal.'"[61]

Other critics echoed this fear. In his 1995 book, *In a Dark Wood: The Fight over Forests and the Rising Tyranny of Ecology*, philosopher Alston Chase outlines a similar argument: "Unknown to the public," Chase asserts, "the administration, under the rubric of 'reinventing government,' had adopted biocentrism as the guiding philosophy of all federal land management."[62] He goes on to explain that Vice President Gore's

National Performance Review had indeed reinvented government by shifting "its goal from protecting people to that of safeguarding ecosystems from them."[63] And less sensational critiques abounded. Economist Ronald Johnson argued that ecosystem management was based on alarmist claims about the fragility of ecosystems and the imminent threat posed by rising population and resource use. The ecosystem management task force, he continued, makes the false assumption that healthy ecosystems are the source and cause of healthy economies, thus prioritizing ecosystem protection above other goals. Ecosystem management would therefore be used to achieve "less harvesting of timber, no clear-cutting, bans on the use of chemicals, strict controls of population growth, and a whole new world economic order."[64] In Johnson's view, the true science necessary for environmental management was neoclassical economics, through which environmental decisions could be understood appropriately in marginal terms.

So what, on its face, appeared to be general procedural reforms of federal planning and management nonetheless became a political lightning rod, particularly after Republicans took control of Congress in the 1994 elections. As all politicians understand, controlling discretionary processes always allows for greater control of substance and outcomes, and permitting federal agencies to operate under the procedural guidelines of ecosystem management was seen as a tool to privilege ecosystem protection over human interests.

Ecosystem Management, Substantive Reform, and the ESA

The second category of arguments focused on substantive ecological goals of ecosystem management, particularly goals of protecting biodiversity and ecosystem integrity, though they also promised procedural benefits. As explained in chapter 1, a central driver for ecosystem management had been mounting concern over the loss of biodiversity. Biodiversity issues emerged in connection with a number of federal laws. During the final year of the Bush administration, for example, the Council on Environmental Quality (CEQ) and the Environmental Protection Agency (EPA) had hosted a series of meetings on how to address biodiversity protection in NEPA analysis, and in January 1993 the CEQ released a report entitled "Incorporating Biodiversity Considerations into Environmental Impact Analysis under the National Environmental Policy Act." The authors explained that biodiversity has often been understood primarily as species diversity but that this is only one element of a far more complex idea of "the variation between and among major ecological elements." Thus, those interested in protecting biodiversity need to "maintain naturally occurring ecosystems, communities, and native species."[65] The authors provided a list of eleven principles necessary for biodiversity protection, most of which resemble the ecological principles of ecosystem management: "Take a 'big picture' or ecosystem view"; "Protect communities and ecosystems"; "Monitor for biodiversity impacts. Acknowledge uncertainty; Be flexible."[66]

Building on these principles, the authors made a number of recommendations for reform, two of which are particularly important for ecosystem management. First, they recommended that agency staff "acknowledge the conservation of biodiversity as national policy and incorporate its consideration into the NEPA process." As noted earlier, this is not simply the protection of species diversity but the whole complex of genetic, species, and ecosystem diversity. Second, they recommended that agency staff "encourage and seek out opportunities to participate in efforts to develop regional ecosystem plans."[67] In other words, they recommended that agencies acknowledge biodiversity protection as a core goal and plan at the broad scales necessary to achieve it.

Yet clearly NEPA reform would only mean that federal agencies had to *consider* biodiversity in their planning; it would not mean a new federal mandate to choose biodiversity protection over other land and resource options. This is why it is difficult to overstate the importance of the ESA as the primary legal driver of biodiversity protection, along with the National Forest Management Act (NFMA) in the case of the Forest Service. It was the legal muscle that could elevate biodiversity protection to the highest management priority.

The ESA, though, was also a blunt tool that caused collateral political damage, as it had in the Pacific Northwest. During the Clinton administration, the ESA was up for reauthorization, and in the midst of the northern spotted owl crisis and the start of a Republican Congress, the mood did not favor strengthening or even maintaining the act. The Clinton administration therefore offered ecosystem management as an approach to ESA implementation that it promised would be less disruptive of private property rights and economic development than the traditional species-focused approach.

During President Clinton's first term, debate over the ESA was particularly intense. The act's authorization had expired in 1992, and the

102nd Congress had postponed reauthorization because it was an election year. Environmentalists and biological scientists fought to protect the act, but they also criticized its implementation for being too narrow and ineffective. The act's detractors fought to amend or repeal the act because of its perceived impacts on economic development and private property.[68] Both sides had reason to complain. For environmentalists and scientists, the act was ineffective because it was underfunded and underimplemented. What is more, where it was implemented, the regulatory agencies tended to follow the act's more narrow species orientation rather than interpreting its habitat protection mandate in broad, ecological terms. The Clinton administration was therefore under political pressure from many Democrats to provide more robust administration of the act.

Furthermore, the Clinton administration had inherited a legal agreement made in the final months of the Bush administration that obligated the FWS to improve ESA implementation. In the previous decade, the FWS had fallen woefully behind in its review of candidate species under the ESA. Of particular frustration to environmentalists, the agency had a backlog of roughly 400 "Candidate 1" species—species for which there was sufficient information to merit a threatened or endangered listing but for which the agency still had not made a final determination. In 1992, the Fund for Animals had sued Secretary Lujan and FWS director John Turner,[69] claiming that the agency's delay expressly violated the ESA. On December 15, 1992, the Interior Department reached a settlement agreement under which it pledged to make final determinations for all 400 species by 1996, essentially doubling its current listing rate. This was certain to increase the number of ESA conflicts at the very moment that Congress discussed the act's reauthorization.[70]

Property rights groups and Republicans, by contrast, complained that existing implementation of the act was already too onerous. While they lacked quantitative data to support a broad argument about the act's negative impacts, the ESA's detractors could point to a number of celebrity cases to win support.[71] In the Pacific Northwest, for example, ESA protection for the northern spotted owl led to a landmark Supreme Court case in 1995 affirming the act's power over private property.[72] And the Clinton administration took office just as the FWS was finishing what remains one of the most notorious ESA listings: the Delhi Sands Flower-loving Fly. There was no controversy about whether or not the fly was endangered. Only 2 percent of its original habitat of Delhi Sands remained in San Bernardino County, California, and only a handful of colonies were known to exist. The obvious controversy, which played well for property rights groups, was that the FWS listed the fly on September 23, 1993, just one day before San Bernardino County started construction of a $473 million hospital complex in the fly's habitat. The listing stopped the project dead in its tracks and blocked development on private property as well. The idea that a fly, which most people considered a pest, could stop a project of clear public value provided more than adequate support for those who feared the ESA's regulatory reach.[73]

The Clinton administration, then, had two ESA priorities. On the one hand, the administration needed to prove that the ESA could be administered more effectively to protect biodiversity. On the other hand, it had to prove that it could carry out more rigorous administration of the act while simultaneously reducing its impacts on economic development and private property. Secretary Babbitt embraced the challenge of pursuing these twin priorities. "I have focused on the ESA," he explained to Congress, "because it can be the most useful tool that has ever been fashioned for the land and resource planners of this country."[74] And the twin goals of biodiversity and property rights protection did not need to be mutually exclusive. The reason that they had conflicted in the past, he insisted, was that "no one had ever really tried to make the Act work," and he promised to "dust off the Act and bring it to life" by using its "flexibility and potential for innovation."[75]

Secretary Babbitt argued that many complaints about the ESA, from both environmentalists and property-rights groups, could be resolved by exercising the act's discretionary provisions within a more comprehensive ecosystem approach. Given the limited resources available, the regulatory agencies—the FWS and the National Marine Fisheries Service (NMFS)—could not fulfill their listing obligations let alone ensure that listed species were recovered. In 1991, for example, federal agencies spent roughly $175 million on species protection, and over half of that went to 7 of the 639 listed species.[76] Waiting for Congress to appropriate enough money to fund full recovery programs for each and every listed species was essentially condemning them to continued decline. What is more, in following the letter of the law, the regulatory agencies implementing the ESA addressed one species at a time, generally in isolation from other candidate and listed species. This ensured that ESA implementation was reactive, since the agencies only addressed species already in danger of extinction, making it difficult for ESA implementation to address common, overarching causes of species decline.

The answer, Babbitt argued, was an ecosystem management approach. This, he explained, would allow agencies to identify and address the foundational causes of species extinction rather than just trying to treat the symptoms one species at a time. And as a result, the ecosystem approach would be less expensive and more cost-effective. This approach was also in keeping with a number of judicial decisions and the FWS's position outlined in the December settlement agreement:

> Defendants [FWS] recognize that a multi-species, ecosystem approach to their listing responsibilities under the ESA will assist them in better analyzing the common nature and magnitude of threats facing ecosystems, help them in understanding the relationships among imperiled species in ecosystems, and be more cost-effective than a species-by-species approach to listing responsibilities.[77]

As the Director of Cooperative Research Units for the FWS explained in the first months of the Clinton administration, "the single greatest change in resource management should be the shift to an integrated, holistic approach to the management of large ecosystems as opposed to single, individual species."[78] The assumption here was that protecting ecosystems would protect the species that depended on them.

The FWS announced in July 1994 that it was adopting "An Ecosystem Approach to Fish and Wildlife Conservation: An Approach to More Effectively Conserve the Nation's Biodiversity" as a framework for ESA implementation.[79] The ecosystem approach, the agency explained, meant "protecting or restoring the function, structure, and species composition of an ecosystem, recognizing that all components are interrelated."[80] To ensure a truly comprehensive and holistic perspective, the agency first divided the country into fifty-three broad ecosystem units, following hydrologic mapping by the USGS, and they established a process to prioritize them for conservation purposes. Given the agency's legal responsibilities for threatened and endangered species on all federal lands as well as state and private lands, the FWS announced at the same time that it would develop new cooperative mechanisms to achieve the ESA's ecosystem-based goals: "Species will be conserved best," the announcement read, "not by a species-by-species approach but by an ecosystem conservation strategy that transcends individual species, [which] will depend on the cooperation of partners, (federal, state, and private)."[81] In listing decisions and recovery planning, the FWS pledged to work cooperatively with other federal agencies, state agencies, and private parties so that its work could focus more effectively on the ecosystem scale and better protect the ecosystems necessary for listed species. In keeping with the Clinton administration's reinventing government initiative, the agency was promising better ESA governance with less federal coercion.

Several months later, fourteen federal agencies with ESA responsibilities signed a memorandum of understanding (MOU) reaffirming "the common goal of conserving species listed as threatened or endangered under the [ESA] by protecting and managing their populations and the ecosystems upon which those populations depend."[82] Among other things, the MOU committed the signatory agencies to a working group to identify the most important geographic areas for coordinating an ecosystem approach to ESA implementation, suggesting a fundamentally different scale of implementation that would yield more flexible and effective ESA implementation. Mollie Beattie, director of the FWS, explained, "By looking at the big picture, by focusing our resources and efforts on ecosystems rather than individual species, we can get away from crisis management where our choices are limited and our costs are high."[83] Again, the working assumption was that protecting ecosystems would itself provide protection for individual species.

Property Rights and the ESA

The ecosystem-based approach, the FWS argued, was necessary first and foremost because it would yield more effective and efficient ESA implementation. The sticking point was how this broader ecosystem approach would affect state and private lands. The FWS and the NMFS needed to integrate private lands into their ecosystem management to effectively administer the ESA and achieve their goals, since "Ninety percent of the listed endangered and threatened species [had] habitat on private land; two-thirds of listed species [had] more than 60 percent of their total habitat on private land."[84] But federal regulation of private property was an incredible political liability that threatened the ESA's very existence. This was true when President Clinton took office, and the 104th Congress's Republican majorities ensured that these criticisms received a sympathetic ear in debates over ESA reauthorization. What is more, legislatures in every state in the country were considering legislation to protect private property rights during President Clinton's first term, with over 120 bills introduced in 1995 alone. By 1997, over half of the states had enacted some type of property rights legislation.[85] If the Clinton administration wished to keep the ESA as a tool for conservation, and if the administration wanted to continue protecting threatened and endangered species beyond wildlife refuges and national parks, it needed to show that ecosystem management would make the act less onerous to economic development and private property owners.

The dilemma was hardly new to the ESA or ecosystem management. For the past century, popular land and resource regulation on federal lands had hit a solid wall when policymakers tried to extend it to private lands. One has only to recall Gifford Pinchot's desire to gain authority over private forestry practices in the early twentieth century. Even here, at the height of the Progressive Era and with the ear of a powerful president, Pinchot could not reach beyond federal lands.[86] Or one might consider the ill-fated bill introduced by Senator Henry Jackson in 1970 entitled the National Land Use Policy Act. This act did not authorize any federal regulation; it authorized only funding and a national data system to assist states and municipalities in creating land-use plans. Jackson's vision was that states would develop comprehensive plans for watersheds and other ecologically bounded areas in order to avoid the redundancy and conflict associated with so many disparate sources of authority and planning.[87] This bill died in the same Congress that passed some of the most significant and far-reaching environmental statutes the United States has ever had. Suggesting greater federal control or even systematic and comprehensive land-use planning that includes private property has always been a political dead end.

Secretary Babbitt and other leaders in the administration therefore set out to both increase coordinated biodiversity protection across land-ownership boundaries *and* decrease federal regulatory restrictions and uncertainty on private property. There was one way around this paradox, they argued: ecosystem management. First, by adopting an ecosystem approach to fulfilling their other statutory mandates, federal agencies could improve ecosystem protection, thereby avoiding the need to list as many species under the ESA in the future. Ecosystem management offered a prophylactic against species decline and regulatory "train wrecks," thereby avoiding potential economic impacts associated with species listing. The administration emphasized this in some of its major ecosystem management projects, from the Northwest Forest Plan to the Sierra Nevada Ecosystem Project and the Comprehensive Everglades Restoration Plan. Second, and more important politically, Secretary Babbitt insisted that a more holistic ecosystem approach would give the FWS and the NMFS greater flexibility, which he pledged the agencies would use to protect both listed species *and* private property rights. Specifically, the flexibility would allow the agencies to shift as much burden as possible to federal lands and management activities, lightening the burden for other property owners: "The first step in the process of any kind of endangered species situation is to ask a simple question," Babbitt explained, "Are there public lands available to use as the core of the protection scheme? Public land is owned by all of us so there is no 'takings' question."[88]

Clearly, though, this broad promise of more effective management and administrative discretion would do little to assuage the anger and frustration on the part of property rights advocates or to boost the confidence of environmental groups, so Secretary Babbitt pursued concrete administrative ESA reform to demonstrate his win-win strategy. In doing so, he led the most significant reform of the ESA since the congressional amendments of 1982.[89]

Much of the political furor over the ESA, Secretary Babbitt argued, was unnecessary because it simply ignored the flexible provisions of the ESA. It was part of what he called the kabuki drama of land and resource politics: "We tend to argue the future by simply adopting the costumes, the masks and the rhetoric of the past."[90] What the rhetoric amounted to was a petty complaint that "We do not like environmental laws, and if they inconvenience us, we will send the government the bill and ask the public to pay."[91] Critics, he suggested, were holding on to an outdated frontier mentality.

Nonetheless, he agreed that "the government should administer the ESA in a way that is sensitive to private property, and demonstrate that the administration of the ESA . . . does not inflict unnecessary inconvenience and hardship on citizens."[92] The problem with ESA implementation to date was not that it violated private property rights but that it caused uncertainty among landowners: What would happen if a species were listed on their property? If they had permission to develop property that was habitat for a listed species, what would happen if biologists later decided that the species needed more stringent protection? These questions, Babbitt conceded, were critically important given the scope of the ESA's regulatory power. What federal agencies needed, then, was the ability to negotiate firm agreements with property owners, enabling them to accurately predict the ESA's future impacts on their property. What they needed were tools that would shift some of the ESA-related risk from property owners to the federal government.[93]

All of this could be done, Secretary Babbitt argued, without waiting on Congress to break the ESA gridlock; it simply required more effective use of the act's 1982 amendments. Here, Congress had provided an exemption process through which the secretary could issue "incidental take" permits to private property owners, allowing them to use their property in ways that would "take" listed species "if such taking is incidental to, and not the purpose of, the carrying out of an otherwise lawful activity."[94] To obtain an incidental take permit, the owner needed to prepare a habitat conservation plan (HCP) for her property explaining the impacts of land-use activities on listed species, steps she would take to limit or mitigate those impacts, reasons why she did not have reasonable alternatives, and explanations of how she would meet any other requirements the secretary imposed. In order to approve the plan and issue a permit, the Interior secretary needed to determine that it would not appreciably decrease the listed species' chance of survival. These 1982 exemptions were similar to those provided for federal agencies in the ESA's 1978 amendments, which authorized incidental take permits for agency actions so long as the FWS or the NMFS determined that the action was not likely to jeopardize a listed species' survival.[95]

The FWS had only issued fourteen incidental take permits during the Reagan and Bush administrations, which Babbitt saw as evidence that they never engaged in a good-faith effort to make the act work. Secretary Babbitt made this exemption process the cornerstone of significant administration reform almost immediately after taking office, and he had approved roughly 200 HCPs by the end of President Clinton's first term.[96]

Secretary Babbitt used the exemption options almost immediately to address ESA crises. In March 30, 1993, he approved listing the California gnatcatcher as a threatened species. Looking back, Babbitt recollects that "what I had not paid sufficient attention to, was the land use implications. . . . By placing the gnatcatcher on the endangered species list, we had, by operation of law, dropped a blanket moratorium on [250,000 acres] of the remaining developable land in the fast growing real estate market in California."[97] To soften the blow, Secretary Babbitt also proposed a special rule, known as a 4(d) rule, that would allow incidental taking of the gnatcatcher to landowners who participated in California's Natural Community Conservation Planning (NCCP) program. This rule recognized that participation in the NCCP program resulted in something comparable to an HCP. This, he explained, rewarded property owners who were doing their due diligence to protect ecosystems on which the gnatcatcher depended, achieving far better results for the gnatcatcher than traditional regulatory enforcement.

The listing and proposed rule took place just before President Clinton, Vice President Gore, Secretary Babbitt, and other cabinet members headed to Oregon for a summit on the northern spotted owl crisis, where the president faced intense opposition to the ESA. There, President Clinton pledged to find a scientifically credible way to break the gridlock over logging in the spotted owl's habitat. What emerged was the most robust ecosystem management plan to date for nearly 25 million acres of federal land, designed to give the owl a high probability of survival and to reduce the burden that private landowners would bear for the owl's protection. At the end of 1993, the Interior Department proposed a new special rule for timber harvesting on private land in northern spotted owl habitat. Under the rule, the Weyerhauser Corporation agreed to prepare a HCP for 100,000 acres of its timber land in order to receive incidental take permits to continue logging. Assistant Secretary George Frampton explained, "We will enforce the laws and regulations that exist. But we are also committed to exercising our flexibility under the Endangered Species Act to achieve reasonable and balanced land use, based on strong science."[98]

Over the next three years, Secretary Babbitt worked through the FWS and the NMFS to develop a more comprehensive reform package to make the ESA more flexible and more user-friendly and to win over some of the act's less vociferous detractors. The overarching concepts of this reform were outlined in a 1995 FWS report, "Protecting America's Living Heritage: A Fair, Cooperative, and Scientifically Sound Approach to Improving the Endangered Species Act."[99] This report placed special emphasis on "flexibility to minimize socio-economic effects and assure fair treatment for landowners [and] reducing delay and uncertainty for States, local governments, private industry, and individuals";[100] and it outlined ten principles for ESA reform and implementation, including "minimize social and economic impacts" and "provide quick, responsive answers and certainty to landowners."[101]

The specific policy reforms that emerged in 1994–1996 made it clear that these principles were more than rhetoric. In the summer of 1994, the FWS and the NMFS announced a series of policy changes intended to make the ESA more "user-friendly." On July 1, the departments published the official proposals in the *Federal Register.* Regulatory agencies would now be required to solicit expert peer review of the science used in listing decisions and recovery planning to give landowners greater confidence in the underlying science used in ESA implementation.[102] The agencies would also be required to "identify, to the extent known at the time a species is listed, specific activities that will not be considered likely to result in violation of [S]ection 9."[103] In other words, upon listing a species, the agencies would give landowners permission for certain activities and reduce the cloud of uncertainty regarding the potential impacts of a listing or recovery plan decision. And more generally, the departments pledged to involve stakeholders directly in decision making, "to minimize the social and economic impacts of implementing recovery plan development," and to redouble their efforts to meet deadlines imposed by the ESA.[104]

In August, the administration announced a more significant policy change to make the HCP provision a central mechanism for ESA implementation on private land. The problem in the past, Babbitt argued, was that landowners were hesitant to put the necessary time and money into HCP preparation because they didn't have confidence in the durability of the plans: "The key issue for non-federal landowners is certainty. . . . They want to know that if they make a good faith effort to plan ahead for species conservation, and do so in cooperation with the relevant agencies, then their plan won't be ripped out from under them many years down the road."[105] Under the "no surprises" policy, the FWS or the NMFS would approve single or multiple-species HCPs based on the best available science. If the status of species covered in an approved HCP were to change, the agencies pledged not to impose any additional property use restrictions or mitigation costs for the life of the plan. Obviously the agencies could not authorize express violations of the ESA in the future, so if changes were legally required in the future, the agencies would bear responsibility for them.[106] As Secretary Babbitt explained, "We're telling landowners that a deal is a deal . . . This 'No Surprises' policy says that if . . . you invest money and land into saving species, we won't come back ten years from now and say you have to pay more or give more."[107] The ESA reforms thus attempted to reconcile the inevitably open-ended nature of scientific recommendations with the more binding regulatory power of legislation, offering some certainty to landowners while maintaining flexibility for changing ecosystems and changing ecological science.

The FWS/NMFS joint handbook on HCP preparation emphasized this system of assurance. It also detailed an ecosystem perspective on HCP planning. While this wasn't binding, it encouraged landowners as much as possible to "consider as large and comprehensive a plan area as is feasible and consistent with their land or natural resource use authorities" and "to include as many proposed and candidate species as can be adequately addressed and covered by the permit."[108] Thus, through recommendations in lieu of regulations, the agencies were working to expand broad, ecosystem-oriented planning to private and state lands.

In November, the FWS issued draft guidelines for another flexible tool in the ESA called Candidate Conservation Agreements (CCAs). Since vulnerable species and candidate species receive no explicit protection under the ESA, private property owners have every incentive to eliminate those species before the ESA's jaw closes on them. CCAs are voluntary agreements that private landowners make with the FWS to manage their lands to protect candidate species. If the agency viewed the measures of these agreements as sufficient to address threats to the species, it would opt not to list the candidate species. In other words, using the threat of listing as the incentive, the FWS could work collaboratively with landowners to avoid species listing.[109] This later expanded into CCA "Assurance" agreements, in which the agency would commit in the agreement to issue an incidental take permit in the future if the landowner maintained her agreement and the species was still listed.

In 1995, the FWS also announced a program called "Safe Harbors," which encourages landowners to manage their lands in ways that support threatened and endangered species. In a Safe Harbor Agreement (SHA), the FWS establishes baseline conditions on the owner's property. If the owner enhances her property in ways that attract additional listed species or allow existing populations of listed species to expand, the property owner will only be responsible for protecting the baseline number of each listed species. They will essentially be granted an incidental take permit for all species above the baseline.[110]

In July 1995, the FWS announced a new proposed rule that would provide exemptions for small landowners (owning less than five acres or for activities impacting less than five acres) and "low impact activities that are presumed to individually or cumulatively have little or no lasting effect on the likelihood of survival and recovery of threatened species."[111] The ESA allows this type of flexibility for threatened species only, so the proposed rule would not apply to cases involving endangered species. The FWS also indicated that it would support legislation providing the same flexibility for endangered species in the future.[112]

Taken together, the ESA reforms of Secretary Babbitt achieved modest, yet significant, improvements in the protection of biodiversity and the mollification of private property interests. They spawned a wave of conservation efforts, including some landscape-scale, multispecies HCPs that amounted to ecosystem management of private lands. One project, the San Diego Multi-Species Conservation Plan, "addressed eighty-five species on several hundred thousand acres and involved negotiations with FWS, two state agencies, nine city and county jurisdictions and representation of several thousand landowners and local environmental and community groups."[113] The plan provided protection for vulnerable species while allowing lucrative development to move forward. More importantly, it was genuine land-use planning of private lands. It is true that these types of plans slowed toward the end of the Clinton administration, in part because the planning process was lengthy, expensive, and complex; yet compared to what had occurred in the 1980s, the Clinton administration could claim significant progress.

The reforms turned down the volume of ESA criticism from private property interests by providing a new system of risk sharing between private landowners and the federal government. This shifted the incentive structure for private property owners, making it more attractive for them to protect or in some cases even improve the habitat of listed species. Owners who agreed to an approved HCP or CCA would enjoy significantly greater certainty regarding their land-use restrictions in the future.[114] Perhaps the most significant testimony to their success with private property interests is the fact that they remained largely intact during the George W. Bush administration.[115]

The ESA reforms also illustrated the Clinton administration's approach, advanced most clearly by Secretary Babbitt, to integrating science, democracy, and federal coercion. The ESA and other regulatory statutes established ecosystem management goals, and they also provided a legal stick that the administration could use to force recalcitrant parties to act. The science of ecology not only defined the goals more clearly, as required by law, it vetted a variety of tools and options for legal compliance. Genuine democratic deliberation could then take place within clear constraints, focusing not on the goals as much as on the means of achieving them in the least economically and socially disruptive ways. This is what one critic called Babbitt's strategy of "coercive harmony."[116]

Conclusion

Despite the fact that the stage was set during the Bush administration, the Clinton administration could genuinely take credit for adopting ecosystem management. Under its leadership, ecosystem management became official policy for all federal land and resource agencies, and the administration invested both time and funding in ecosystem management research, experimentation, and promotion. It chose to keep the meaning of ecosystem management somewhat ambiguous and fluid, making it a more difficult political target for Republicans to hit.

The meaning of ecosystem management, then, had to be discerned not so much from lofty policy statements as from concrete administrative decisions—the NBS, Rangeland Reform '94, and ESA reforms—and ecosystem management projects: the Chesapeake Bay ecosystem, the Florida Everglades Ecosystem, Rocky Mountain ecosystems, the Sierra Nevada Ecosystem, the northwest forest ecosystem on the west side of the Cascades, and the interior Columbia Basin east of the Cascades. What these revealed, not surprisingly, was that the Clinton administration did not have a developed ecosystem management paradigm that it forced into service. Instead, what the administration called ecosystem management was an emergent set of scientific and normative concepts that administration officials worked pragmatically to adapt and implement in various legal and political contexts.

If the most common principles of ecosystem management were managing along ecological rather than political boundaries, managing for dynamic ecological processes and functions rather than steady resource outputs, and developing adaptive and collaborative forms of management, the administration chose to emphasize different principles in different contexts. In some cases, the administration emphasized the protection of ecological processes and functions over other concerns, pursuing what here is being called a substantive model of ecosystem management. In other contexts, the administration emphasized collaborative management, pursuing what here is being called procedural ecosystem management.

The Clinton administration and Congress launched a variety of ecosystem management projects in the 1990s that can be placed at various points on this substantive/procedural spectrum, ranging from a scientific assessment in the Sierra Nevada Ecosystem Project to the Comprehensive Everglades Restoration Plan. The effects of these projects varied considerably based on different landownership patterns, specific resource conflicts, and the legal/political forces.[117] Surveying all of these projects, as well as many nonfederal ecosystem management projects, would reveal points of commonality and divergence, demonstrating the diversity of meanings that ecosystem management had.[118] However, the final chapters in this book take a different approach in order to (1) focus more narrowly on an ecosystem approach specific to federal lands and resources and (2) highlight the two very different models or interpretations of ecosystem management that the administration pursued.

Analyzing two of the earliest and largest federal ecosystem management projects in the 1990s—the Northwest Forest Plan (NWFP) and the Interior Columbia Basin Ecosystem Management Project (ICBEMP)—serves this purpose. The projects both dealt almost exclusively with federal land and resources; they were responsive to the same federal statutes; they were initiated around the same time; and they were contiguous projects in the northwestern United States. At the same time, they were initiated for different reasons, and they had radically different outcomes. NWFP and ICBEMP, then, illustrate two different models or interpretations of federal ecosystem management. NWFP was a substantive ecosystem management project driven by the relatively clear, enforceable goals of the ESA and designed primarily by experts in the biological sciences rather than by public deliberation. Here, ecological protection was the predetermined priority for federal ecosystem management, and other goals were accommodated secondarily. ICBEMP, by contrast, was a procedural model driven by NEPA and its requirements for rational decision making informed by the best available science and extensive public participation. Here, ecological protection was one important goal among many for federal ecosystem management, and prioritizing these goals was just one of the core elements of the process.

6 | The Northwest Forest Plan: Substantive Ecosystem Management

Introduction

The Northwest Forest Plan (NWFP) (see Fig. 6.1) was the Clinton administration's first and highest-profile ecosystem management project, in large part because it addressed an ecological and political crisis that had defined the presidential campaign in the Pacific Northwest. As James Pipkin, the Interior Department's director of policy analysis, explained, "this was a virtually ideal situation in which to attempt [ecosystem management]: lawsuits had created a stalemate; many of the relevant parties were ready for a new solution; the issues were clearly defined; [and] 24 million acres of relatively homogeneous federal forest land defined the area to be addressed."[1] Furthermore, Judge Dwyer had essentially ordered the US Forest Service and the BLM to abandon their narrow focus on resource outputs and to develop an ecosystem protection plan instead: "The duty to maintain viable populations of existing vertebrate species," he had concluded, "requires planning for the entire biological community—not for one species alone."[2]

The plan that emerged in 1994 was an ecosystem management landmark in many respects, not least of which was its scope, scale, and rapid adoption. Never before had scientists attempted to protect biodiversity over such a large region, including over 24 million acres of federally owned land. Never before had ecological protection been made such a high priority for large areas of multiple-use lands otherwise open for timber production, resulting in drastic reductions in the probable sale quantity of timber. And never before had so many agencies—the Bureau of Indian Affairs, the Bureau of Land Management (BLM), the Environmental Protection Agency, the US Fish and Wildlife Service (FWS), the Forest Service, the National Park Service, the Soil Conservation Service, and others—collaborated in the preparation and implementation of an ecosystem management plan. And what makes the NWFP even more

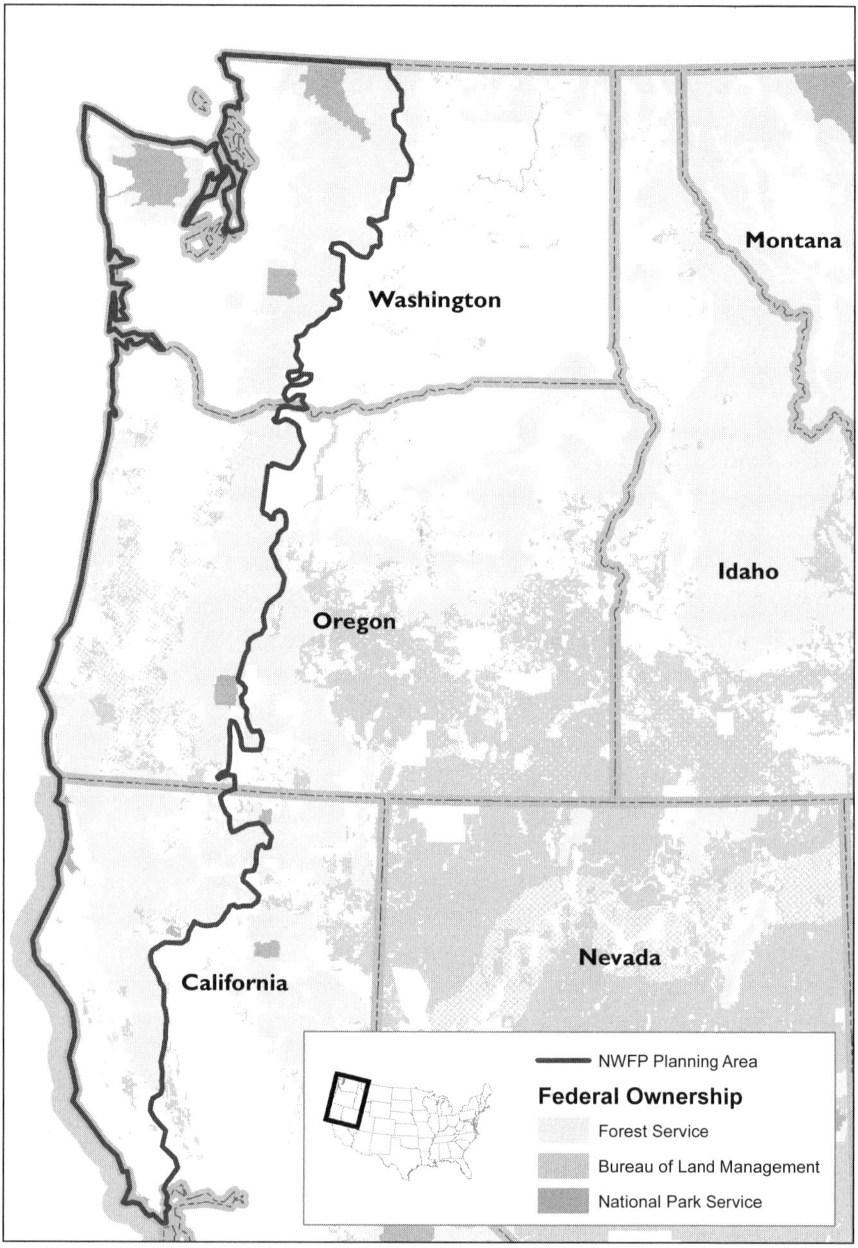

Figure 6.1: The Northwest Forest Plan Area. Map prepared by Cam Patterson.

remarkable is that it withstood legal and political challenges for almost a decade and continues to guide federal management in the region.

The NWFP was driven by a variety of legal forces but principally by the substantive requirements of the Endangered Species Act (ESA), which requires the protection of listed threatened and endangered species, and of the National Forest Management Act (NFMA), which requires the Forest Service to protect viable populations of all vertebrate species. In this context, the NWFP was an unprecedented opportunity for scientists to test relatively new scientific concepts coming out of ecology and conservation biology on a grand scale. In particular, it was the first large-scale test of adaptive management in the political and legal context of federal lands, and it was one of the most significant tests of the promise that ecosystem management created win-win opportunities for environmental protection and economic development. A decade after its implementation, the NWFP had still not delivered on all of its promises, particularly the promises of timber outputs and adaptive management, but it had certainly provided an unprecedented level of ecological protection for federal lands in the Pacific Northwest.[3] This was the model of ecosystem management that environmental organizations were pressing the administration to adopt; this was the model of ecosystem management that the timber industry and other natural resource developers feared; and this was the model of ecosystem management that the George W. Bush administration would later reject.

The Legal and Political Context of the NWFP

As explained in previous chapters, the northwest forest crisis developed during the George H. W. Bush administration and came to a head during the presidential campaign of 1992. On May 28, Judge Dwyer had enjoined Forest Service timber sales in the range of the northern spotted owl and ordered the agency to prepare a new supplemental environmental impact statement (SEIS) that complied with the National Environmental Policy Act (NEPA) by August 10, 1993.[4] On June 22, Judge Frye had likewise enjoined BLM timber sales in the range of the northern spotted owl until it completed an SEIS that complied fully with NEPA.[5] Both agencies appealed to the 9th Circuit and lost. Judge Schroeder upheld the lower courts' rulings "because the Forest Service's EIS rests on stale scientific evidence, incomplete discussion of environmental effects vis-à-vis other old-growth-dependent species and false assumptions regarding the cooperation of other agencies and application of relevant law"[6] and "there are 'no defenses to excuse the BLM from complying with NEPA.'"[7]

The forest crisis featured prominently in the 1992 presidential campaign. President Bush's campaign advisors instructed him to support the timber industry and timber-dependent communities over and against environmentalists in order to win back frustrated Republicans and garner support from rural Democrats. President Bush and Vice President Quayle once again campaigned on a platform of "balanced" environmental policy, but they argued that the Endangered Species Act mandated unbalanced policy that prioritized owls over people. They promised to restore balance in the region by seeking amendments to the ESA, reducing northern spotted owl protection, and increasing timber production: "It's time to fight for jobs, families and communities."[8] President Bush vigorously distanced himself from "the spotted owl crowds or the extremes in the environmental movement" that he said Governor Clinton courted.[9]

Governor Clinton also campaigned on a platform of "balanced" environmental policy, countering that President Bush was presenting voters with a false choice between owls and jobs. The real problem, Clinton charged, was the fact that President Bush had largely ignored the crisis. "Presidential leadership on the environment," Governor Clinton charged, "has become an endangered species."[10] Clinton did not offer any details about how he would resolve the crisis, other than reducing the number of logs that were exported and diverted from US mills. What he did promise, over and over again, was decisive presidential leadership. If elected, he pledged, he would return to the region and personally mediate among the relevant federal agencies, the timber industry, environmental organizations, and other interested parties. Through presidential leadership and democratic deliberation, he assured voters, they could find an equitable resolution to the crisis that could save owls *and* jobs.[11]

After winning the November election, President-Elect Clinton immediately established a Timber Summit Transition Office to organize the meeting he had promised. The office sent letters to over 1,500 organizations, state and local officials, tribal leaders, and other interested parties, to gather information for the meeting and to ensure that all interests were represented.[12]

While the administration planned for its timber summit, the Forest Service's Scientific Analysis Team (SAT) completed the "Viability Assessments and Management Considerations for Species Associated with Late-Successional and Old-Growth Forests of the Pacific Northwest." The SAT had evaluated a total of 667 species closely associated with old-growth forests and recommended a number of mitigation measures beyond those specific to the northern spotted owl: "large, interim reserves for marbled murrelets, riparian reserves to protect aquatic systems, and buffers established for certain terrestrial species that were thought not to be adequately protected by other measures."[13] The broadness of these recommendations confirmed the need for major changes in federal land and resource management if the Clinton administration hoped to meet ESA and NFMA requirements.

On April 2, 1993, just four and a half months before the court-imposed deadline for a new Forest Service EIS, the president hosted a "forest conference" in Portland, OR. The conference was a show of executive force, including the vice president, the secretaries of agriculture, interior, labor, and commerce, the EPA administrator, the environmental policy director, the deputy director of management and budget, and the science advisor. Together they listened for eight hours to fifty-two witnesses.[14]

Those who participated in the conference offered a wide range of perspectives. Representatives of the timber industry called for amendments to the ESA and insisted that the region's economy depended on continued logging of old-growth forests: "We cannot stop cutting old growth without creating a huge vacuum that private timber supplies cannot fill. We cannot fill the nation's building material needs. We will have massive unemployment. There is no way to make a transition to second growth." Environmentalists, by contrast, asked the president to enforce environmental laws and to halt all logging in old-growth forests: "We must disturb no more of the last remaining centers of biodiversity [old-growth forests]. These are the refuges and the seed sources for tomorrow's forest, tomorrow's wildlife, and tomorrow's economy."[15]

Although diametrically opposed on management outcomes, the two sides shared one common rhetorical commitment: ecosystem management was the best way forward in the Pacific Northwest. It was, many argued, comprehensive enough to achieve balanced management rather than forcing a zero-sum choice between jobs and owls. As one representative of the timber industry explained,

> we believe there is only one way to break that gridlock, and that is to embrace the exciting and innovative concepts that you heard described here today called ecosystem management. . . . If we can do that, we can avoid the economic catastrophe that is otherwise going to happen. We can maintain healthy forest ecosystems in a manner that we've never been able to before, and most importantly, we can have healthy watersheds. We can have fish and wildlife habitat. We can have recreational opportunities, maintain diverse ecosystems, and still produce the wood product needs that this country demands.[16]

The challenge, of course, was that the word "balance" and the phrase "ecosystem management" meant very different things to different participants in the meeting. Environmentalists assumed that ecosystem management required a complete prohibition on logging in old-growth forests, while the timber industry and the Forest Service viewed it as the possibility of expanding timber harvesting through new management techniques.

Throughout the day, President Clinton and Vice President Gore reiterated their own conviction that one did not need to choose between these two positions: "A healthy economy and a healthy environment," the president asserted confidently, "are not at odds with each other. They are essential to each other. Here in the Northwest, as in my own home state, people understand that healthy forests are important for a healthy forest-based economy." And the vice president echoed this conviction, pronouncing, "the days when this debate was defined by either/or choices are over. This isn't about saving jobs or saving the environment. It's about saving jobs and saving the environment. We can't do one without doing the other; certainly not in the long term."[17]

At the end of the conference, President Clinton promised to develop "a balanced, a comprehensive, a long-term policy" that adhered to five broad principles:

> First, we must never forget the human and the economic dimensions of these problems. . . . Second, as we craft a plan, we need to protect the long-term health of our forests, our wildlife and our waterways. . . . Third, our efforts must be, insofar as we are wise enough to know it, scientifically sound, ecologically credible, and legally responsible. Fourth, the plan should produce a predictable and sustainable level of timber sales and non-timber resources that will not degrade or destroy our forest environment. And, fifth,

to achieve these goals, we will do our best, as I said, to make the federal government work together and work for you.[18]

These principles were the essential scaffolding of the administration's broad and open definition of ecosystem management that retained space for the divergent interpretations outlined by the timber industry and environmentalists.

Rhetorically, President Clinton's pledge fulfilled the call that Lynton Caldwell had made in the late 1960s for an ecosystems approach to federal land management: "An ecosystems approach is essentially a total systems approach. It therefore includes in its purview many things omitted in less comprehensive systems."[19] The spatial and temporal breadth of ecosystem management, the president implied, avoided the need for many either/or choices, creating space for win-win solutions: "Where sound management policies can preserve the health of forest lands, [timber] sales should go forward." The president also acknowledged the challenges that Caldwell had pointed out decades earlier: an ecosystems approach "would impose constraints upon single-purpose approaches to the environment and would arouse hostility among individuals whose single-purpose pursuits would thereby be constrained."[20] In cases where timber harvesting conflicted unavoidably with forest health, the president promised that his administration would "do our best to offer new economic opportunities for year-round, high-wage, high-skill jobs."[21] Put another way, an ecosystems approach required remarkable levels of administrative discretion to balance competing values at a variety of scales.

The work that President Clinton directed, then, was a grand experiment in the promises and pitfalls of an ecosystems approach to land and resource management. Going beyond the broad principles of comprehensive, long-term, and balanced management that everyone supported, the Clinton administration was wading into ill-defined territory. As the lead scientist and later Forest Service Chief Jack Ward Thomas warned, "Ecosystem management is in vogue. It's the new means of natural resources management. . . . However, that approach is not going to be simple. It's not going to be cheap. . . . There may be not more than a hundred or so people in the entire world that are geared up to really think about what ecosystem management means."[22] What lacked definition were precisely the specific goals or principles that would direct agency discretion under ecosystem management.

Foundations of the NWFP

At the end of the Forest Conference, President Clinton promised to provide a new plan within sixty days that would break the impasse in the Pacific Northwest and strike a balance between environmental protection and economic opportunity. The task was divided among three main planning teams, which had been assembled in the weeks leading up to the conference. The Forest Ecosystem Management Assessment Team (FEMAT) outlined management options for forests in the range of the northern spotted owl that were scientifically credible, legally sufficient, and sensitive to social and economic dynamics of the region. The Labor and Community Assistance Team (LCAT) assessed and proposed mitigation measures for negative economic impacts that the plan would invariably have on the region. The Agency Coordination Team designed the administrative structures necessary to coordinate the actions of numerous federal agencies across wide spatial and temporal ranges. Simultaneously, the administration set up a NEPA team to prepare the SEIS ordered by Judges Dwyer and Frye. The Clinton administration had staked considerable political capital in the project, so perhaps no other single land-use plan had ever received so much direct oversight and support from the White House.

FEMAT: The Natural Science

FEMAT's work was unquestionably the most important, since the NEPA team was incorporating it into an SEIS that had to pass judicial muster before the Forest Service and the BLM could resume timber harvesting. The scope of FEMAT's mandate was incredibly broad, following the trajectory of past scientific committees and Judge Dwyer's exhortation to protect old-growth forest ecosystems. President Clinton directed FEMAT to consider:

> maintenance and/or restoration of habitat conditions to support viable populations, well-distributed across their current ranges, of species known (or reasonably expected) to be associated with old-growth forest conditions; maintenance and/or restoration of spawning and rearing habitat on Forest Service, Bureau of Land Management, and National Park Service lands to support recovery and maintenance of viable populations of anadromous fish species and stocks and other fish species and stocks considered

> "sensitive" or "at risk" by land management agencies, or listed under the Endangered Species Act; and, maintenance and/or creation of a connected or interactive old-growth forest ecosystem on the federal lands within the region under consideration.[23]

In addition, FEMAT was directed to seek "maintenance of sustainable levels of renewable natural resources, including timber, other forest products, and other facets of forest values; and maintenance of rural economies and communities." And recognizing the massive data gaps that the team faced, it was instructed to "include suggestions for adaptive management that would identify high priority inventory, research and monitoring needed to assess success over time, and essential or allowable modifications in approach as new information becomes available."[24] And FEMAT was to do all of this in a two-month period.

FEMAT began work immediately after the Forest Conference, and the White House made sure that its work was the top priority for all of the agencies involved. As one participant recalled, "With White House authority used liberally to 'grease the skids,' the governmental 'red tape' was reduced greatly. . . . As with a critical fire situation, money was no problem—whatever was needed was purchased or rented."[25] Still, the sixty-day timeframe made FEMAT's work a daunting task. By all accounts, the teams worked more than twelve hours a day, seven days a week in order to meet the strict deadline, and they also tended to work independently because there simply wasn't sufficient time for consistent integration of their tasks.[26]

FEMAT was divided into six assessment teams addressing terrestrial ecology, resource analysis, aquatic/watershed analysis, spatial analysis, economic assessment, and social assessment. The economic and social teams conducted new research and consultations with the affected communities, but the natural science teams simply drew on the massive amount of data and management proposals already compiled by the Interagency Science Committee in 1990, the Gang of Four in 1991, and the Scientific Analysis Team in 1993. And since FEMAT included many of the scientists from these earlier committees, the team had considerable experience not only in research science but in the application of science to management questions. In all, FEMAT included over 100 natural and social scientists and hundreds of support staff.

The team leaders first reviewed over fifty existing management proposals for forests in the Pacific Northwest, judging them against five biological criteria: "(1) viability of northern spotted owls, (2) viability of marbled murrelets, (3) viability of at-risk fish species and stocks, (4) viability of other species associated with old-growth forests, and (5) maintaining an interacting late-successional/old-growth forest ecosystem."[27] In doing so, FEMAT considered the viability of over 1,100 species and species groups that were associated with late-successional/old-growth forests, and the team used over 500 different measurements of sustainability to filter existing and new management proposals. Its 1,000-page report was essentially a comprehensive risk assessment for biodiversity in the range of the northern spotted owl over a 100-year timeframe.[28]

The viability criterion was particularly challenging for FEMAT's natural scientists, since they lacked reliable population and distribution data on a significant number of species. From a scientific perspective, this was an opportunity for innovation. As Jack Ward Thomas and biologist Bruce Marcot later reflected: "the FEMAT work forced more clearly the issue of using new, uncertain, and disparate sources of information in evaluating large-area effects on ecosystems and species."[29] Data limitations forced FEMAT "to lump geographic areas and species into large groups and utilize information sources that had not been the subject of independent peer review and referee evaluation," as well as to rely in many cases on expert opinion rather than quantitative analysis in viability assessments.[30] In other words, FEMAT worked at the cutting edge of ecology and conservation biology, where irreducible uncertainty was an accepted feature of analysis.

FEMAT initially narrowed the management choices down to eight options using a coarse-filter approach that Jack Ward Thomas and other scientists had been advocating since the late 1970s. This approach rests essentially on the inverse of assumptions used in studying indicator species. Ecologists cannot identify and monitor every ecological process or function that they believe is required for maintaining ecosystem integrity, so they often monitor a species such as the grizzly bear or the northern spotted owl, which they believe depends on high levels of ecosystem integrity. If that species is doing well, ecologists infer that the ecosystem as a whole is functioning well. Inversely, a coarse-filter approach to biodiversity conservation assumes that ecologists do not have the time, funding, or methods needed to monitor every species of interest. Instead, then, ecologists identify the collective habitat needs of the species involved and work to protect that habitat within its historic range of variability. They assume that if the habitat is intact, the variety of species associated with it will remain viable.[31]

So, rather than trying to develop management strategies for all 1,100 species associated with old-growth forests, particularly since they lacked the necessary data and time, FEMAT took the position that protecting significant, well-distributed areas of old-growth forest and key watersheds would in itself provide sufficient habitat for the 1,100 associated species. In reviewing various management options, then, FEMAT was really searching for options that would protect the "viability" of old-growth forests and watersheds in the Pacific Northwest while adding additional criteria for specific species such as the northern spotted owl.[32]

The core conservation strategy in FEMAT's various options was a system of forest and riparian reserves in which logging would be prohibited or severely restricted. Late-successional reserves would include old-growth/late-successional forest stands as well as some younger forest stands that were anticipated to develop old-growth forest characteristics. These reserves, coupled with congressional withdrawn areas such as national parks and wilderness areas, would serve as refugia for old-growth-dependent species. Of the initial eight management options that FEMAT considered, Option 1 provided the largest system of late-successional reserves—8.5 million acres—while Option 7 provided the smallest system of reserves—4.5 million acres.[33]

To protect aquatic and riparian organisms, FEMAT developed an Aquatic Conservation Strategy: a "region-wide strategy seeking to retain, restore, and protect those processes and landforms that contribute habitat elements to streams and promote good habitat conditions for fish and other aquatic and riparian-dependent organisms."[34] The strategy prioritized watersheds based on the presence of vulnerable aquatic species and water quality, and it included a system of riparian reserve areas, which excluded logging to protect riparian species and to protect aquatic habitat from sedimentation, thermal alterations, and other factors that might threaten aquatic species. FEMAT developed three reserve strategies—Riparian Reserve 1, 2, and 3—which differed primarily regarding the width of the reserve area on the edges of water bodies. Riparian Reserve 1 resulted in a system of riparian reserves totaling 1.9 million acres, while Riparian Reserve 3, by contrast, resulted in a system that totaled 1.5 million acres.[35] Seven of the eight management options in the FEMAT report had an associated riparian reserve strategy. Option 1 applied the more extensive Riparian Reserve 1, while

Option 8 applied the least extensive Riparian Reserve 3, and Option 7 did not apply any of the three.

To provide a sustainable supply of timber for the regional economy, all eight options included a system of "matrix lands," which were to be managed for multiple uses, particularly timber production. This came the closest to the BLM's and the Forest Service's existing understanding of multiple-use management oriented primarily toward resource outputs, which had led to the northern spotted owl crisis in the first place. The total area of matrix lands ranged from 2.8 million acres in Option 1 to 8.5 million acres in Option 7.[36]

Although President Clinton's goal was a plan that would protect *both* ecological *and* economic values, the initial eight options provided rather stark trade-offs. Option 1, for example, offered a high probability of meeting FEMAT's five biological criteria—viability of the northern spotted owl, viability of the marbled murrelet, viability of fish species, viability of other old-growth-associated species, and maintenance of late-successional forests—yet, with timber production coming primarily from just 2.8 million acres of matrix lands, it promised to yield less than 200 million board feet of timber. This was less than 5 percent of the harvest levels in the 1980s. Option 7, with three times as many acres in matrix designation and with some timber harvesting in late-successional and riparian reserves, promised 1.8 billion board feet of timber, roughly 40 percent of the harvest levels in the 1980s, but it would have almost certainly violated the conservation standards of the ESA and NFMA.[37]

Late in the process, FEMAT added two additional options. Option 9 was a compromise solution that provided far more timber harvesting than Option 1 but provided higher viability ratings than Option 7.[38] Option 9 designated roughly equal acreage as late successional/old-growth reserves and as matrix lands, placing almost a quarter of the remaining late-successional forest in the matrix category, to serve as a primary source of timber during the NWFP's initial years of implementation. Option 9 also included a unique category of land called "adaptive management areas" (AMAs) totaling 1.5 million acres. These areas were designated near communities hardest hit economically by the plan, and they were intended to be laboratories in which the Forest Service and the BLM could experiment with management techniques that they believed promised *both* conservation and economic benefits. As FEMAT explained in its report,

> The overarching objective for Adaptive Management Areas is to learn how to do ecosystem management in terms of both technical and social challenges, and in a manner consistent with applicable laws. It is hoped that localized, idiosyncratic approaches that may achieve the conservation objectives of this plan can be pursued. These approaches rely on the experience and ingenuity of resource managers and communities rather than traditionally derived and tightly prescriptive approaches that are generally applied in management of forests.[39]

Management in these areas was required to include monitoring and assessment so that the agencies could actively learn from successes and failures.

The AMAs were important in and of themselves, because they were areas where the agencies and the public were to have freedom and flexibility to balance ecological protection and economic development on a case-by-case basis rather than through the more general terms of the reserves and the matrix areas. The AMAs, it should be emphasized, were laboratories not just for management techniques but for experiments in collaborative management that integrated science and open, public deliberation in creative ways. The AMAs were also intended to illustrate the potential of the NWFP's broader commitment to adaptive management. "Ecosystems," Jack Ward Thomas had said at the forest conference, quoting ecologist Frank Egler, "are not only more complex than we think, they're more complex than we can think."[40] And the broader social, economic, political, and legal systems involved were similarly complex: "Our understanding of the underpinnings (supporting science, ecological constructs, legal interpretation, and societal acceptance) of natural resource management is in rapid flux and deals with imprecise concepts such as 'ecosystem management' itself and sustainable development as a means of achieving ecosystem management."[41] In the face of this complexity and uncertainty, the AMAs were designed to test "a continuing process of planning, monitoring, researching, evaluating, and adjusting management approaches."[42]

FEMAT also emphasized the need for a tiered approach to planning, which reflected an understanding of ecosystems as hierarchical, nested systems that interact across various scales. FEMAT recommended (1) a "regionwide conservation strategy" that provides general direction and does "not set quantitative goals for goods and services"; (2) a "physiographic province (or river basin) conservation strategy" that provides intermediate guidance; (3) a "watershed level analysis" that provides site-specific information for project decisions. FEMAT's work on a regional strategy was assumed to be only the first step in a much larger decision-making structure: "Although the regionwide plan provides a method for standardizing processes across provinces, the physiographic province is intended to become the focal point for ecosystem planning and is expected, ultimately, to replace the current National Forest and Bureau of Land Management District plans."[43] This was, indeed, a significant new experiment in federal land and resource management.

FEMAT worked at a remarkable pace on its final report, working under enormous pressure and behind closed doors. This proved to be both a benefit and a liability for the team. On the one hand, it ensured that federal forest management in the region would not suffer from "process paralysis," as would undoubtedly have happened if the natural scientists had been asked to fill in all of the gaps in their scientific knowledge, dramatically reduce the uncertainties in their viability assessments, and provide opportunities for public participation. On the other hand, it limited FEMAT's analysis in important ways, and it seemed to undermine the president's promise that a solution in the Pacific Northwest would be collaborative. And because FEMAT included several university scientists rather than just federal scientists, its secrecy was later found to be a violation of the Federal Advisory Committee Act.[44]

LCAT: The Social Science

Although FEMAT scientists reviewed the socioeconomic impacts of each management option, allowing the president to choose a particular balance between environmental protection and socioeconomic interests, it was clear from the outset that any option complying with existing federal law would reduce timber harvesting significantly from its high levels in the 1980s. A second team, based in Washington, was responsible for mitigating the economic impacts of whatever level of timber production the president selected. Research indicated that the regional economy would be stable despite the lost timber revenue, but rural, timber-dependent communities and counties would suffer significantly because of the reduced timber program. The economic team was not constrained, like FEMAT, by detailed legal mandates, and its members worked closely with the governors of Washington, Oregon, and California as well as with the congressional delegations from those states.

The team developed an economic transition program designed to help the states, counties, and communities in the plan area make the transition to less timber-dependent economies. The Northwest Economic Adjustment Initiative promised $1.2 billion in assistance over five years. Some of the money was allocated for unemployed timber workers to retrain them for different employment; some was allocated for the counties, which by long-standing agreement had received 25 to 50 percent of all federal timber receipts to use for education, roads, and other key services; and some was allocated for ecosystem investment. The Forest Service and the BLM would not have as much commercial timber to offer for sale, and the additional money would allow the agencies to hire timber workers to do forest improvement projects, such as forest thinning. The team recommended that Congress end a federal subsidy for whole log exports from the region, which hurt local mills, and reallocate that money to the Northwest Economic Adjustment Initiative. Congress supported the initiative and committed to helping the region through its economic transition, although it later refused to appropriate all of the promised funding.[45]

The Labor and Community Assistance Team (LCAT) also faced considerable data gaps and uncertainties in their work. Just as FEMAT took a coarse-filter approach that assumed a causal connection between maintaining well-distributed blocks of old-growth forest habitat and maintaining species viability, the LCAT relied on the working assumption that maintaining predictable timber harvests was key to maintaining stable rural communities. And given the limited time and data available, the LCAT used aggregated county data instead of analyzing every rural community.[46]

Interagency Coordination

As the first two teams worked on the substance of a forest plan, a third team focused on the interagency coordination and cooperation that would be necessary for such a complex project. Although the BLM and the Forest Service were the primary land management agencies, a total of seven departments and sixteen agencies had responsibilities for lands and resources within the range of the northern spotted owl.[47] These agencies had difficulty working effectively together because of their diverse legal mandates and judicial rulings, past administrative direction, cultures, constituencies, and congressional oversight and appropriations. Indeed, during the Bush administration, the Forest Service and the BLM took very different approaches to the northern spotted owl because the Forest Service had to follow NFMA's more demanding viability requirement. As President Clinton lamented at the outset of the Forest Conference, "I was mortified when I began to review the legal documents surrounding this controversy to see how often the departments were at odds with each other, so that there was no voice of the United States. [We] have had our own agencies suing one another in courts, often over issues which are hard to characterize as monumental."[48]

The coordination team reviewed these obstacles, focusing on organizational changes necessary for adaptive ecosystem management. In the spirit of the vice president's national performance review, the coordination team recommended ways to make agencies more flexible, more responsive, and more accountable. The coordination team recommended new committees and structures to integrate agency work, from centralized oversight in the White House to analysis at the watershed level. They also recommended standardizing agency Geographic Information System capabilities so that the agencies could share information more easily.

The coordination team also waded into areas of complexity and uncertainty that it could not resolve immediately. Implementing ecosystem management in the Pacific Northwest required new types of coordination across agencies and new relationships among planning efforts at the regional, watershed, and site levels, but these innovations had to be layered on top of the existing agency planning systems. For example, to achieve *ecosystem* management, the team recommended planning for physiographic provinces throughout the range of the spotted owl. In other words, it recommended planning along ecological boundaries rather than political boundaries. But the agencies still needed to fulfill their legal requirements to plan along and manage within their political boundaries.

NEPA

The NEPA team worked concurrently with the other three teams, preparing the official decision document—an SEIS—to implement the Clinton administration's new forest plan. The new SEIS, Judges Dwyer and Frye had made clear, needed to take an ecosystem approach and needed to incorporate all of the recent science, which FEMAT was compiling at a feverish pace. This meant that the NEPA team had to continually adapt to FEMAT's evolving report.

The NWFP Takes Shape

By late June, the White House had relatively complete drafts of the FEMAT report and the Draft SEIS (DSEIS), which mirrored one another. FEMAT's ten management options were duplicated with some revision in the DSEIS—there labeled alternatives. Thus, like FEMAT's report, the DSEIS provided alternatives ranging from Alternative 1, "the green dream," to Alternative 7, "the brown bomb."[49] The White House selected Option/Alternative 9, and on July 1, 1993, President Clinton announced "The Forest Plan: For a Sustainable Economy and a Sustainable Environment."[50] Flanked by Vice President Gore and the cabinet members who had joined him for the forest conference, the president offered the new plan as an alternative to "the failed policies of the past," under which "no timber cutting at all could occur because of litigation, and still environmentalists believed that the long-term concerns of the environment were not being addressed."[51] With Option/Alternative 9's trade-offs between ecosystem protection and economic outputs, he insisted, "the plan is fair and balanced. I believe it will protect jobs and offer new job opportunities where they must be found. It will preserve the woodlands, the rivers, the streams that make the Northwest an attractive place to live and to visit."[52]

The president's plan promised to redraw the planning and decision-making boundaries to ensure an ecological focus in federal land management. Watersheds were "the fundamental building block" of this new perspective, and the plan would create new "provincial-level teams that would develop analyses for physiographic provinces and particularly watersheds. . . . the objective would be to involve all affected parties in discussions on the biological, timber, community, and other needs."[53] Thus, the plan would not only affect federal agencies, it would reshape the politics of federal land and resource decision making.

The plan was, the president admitted, complex, and it offered far less commercial timber production than anticipated (about 1.2 billion board feet per year). This represented a 75 percent reduction from harvest levels during the Reagan administration, translating into considerable economic costs for the region. However, the president promised that the economic initiative would fill in the financial gap during the economic transition and that he was signing a bill into law that would stop the export of raw logs from federal lands and redirect them to local mills.

The DSEIS, identifying Alternative 9 as the preferred alternative, and the final FEMAT report were published in late July, followed by the ninety-day public comment period required by NEPA. During the comment period, the Clinton administration reorganized to implement the plan. First, in late October, the administration signaled a major shift in the Forest Service by forcing Chief Dale Robertson and Associate Chief George Leonard to resign and appointing Jack Ward Thomas in the chief's place. The selection of Thomas broke the long-standing tradition of selecting a new chief from among those who had moved up through the agency's management ranks. Soon after the announcement of Thomas's appointment, Senator Hatfield complained, "I find this practice troublesome because these positions, which have traditionally been filled by lifetime career personnel, provide the agency with a sense of continuity, institutional knowledge and insulation from the shifting tides of Washington D.C.'s political culture."[54] Selecting Thomas, a PhD research wildlife biologist who was not in the senior executive service, was a clear statement that the Clinton administration did not want continuity in the Forest Service and that the administration was willing to push scientific authority rather than bowing to conventional political processes.[55]

Second, the Clinton administration created a broader infrastructure to coordinate the complicated new approach to forest management in the Pacific Northwest and ensure White House control. On October 8, 1993, Katie McGinty, Secretary Babbitt, Secretary Espy, and several other key members of the Clinton administration signed the "Memorandum of Understanding for Forest Ecosystem Management" "to establish a framework for cooperative planning, improved decision making, and coordinated implementation of the forest ecosystem management component of the President's Forest Plan."[56] The MOU established an Interagency Steering Committee, chaired by McGinty and based in Washington, to provide overall policy guidance for the plan. It also established the Regional Interagency Executive Committee and the Regional Ecosystem Office in the Pacific Northwest to provide direct oversight of the plan, a Research and Monitoring Committee to develop some of the plan's protocols, and twelve provincial teams to oversee public involvement and specific projects throughout the plan area. Other MOUs outlined collaborative plans for economic assistance and the relationship of federal agencies to state and local governments. And in December 1993, reflecting the political stakes of the project, the White House established an additional layer of coordination in the Pacific Northwest: the Office of Forestry and Economic Development. This office was created to represent the White House in the region and ensure direct White House oversight of the plan's implementation.[57]

By the end of the DSEIS comment period on October 28, the Forest Service and the BLM had received over 100,000 public comments, which included a wide range of criticisms and calls for either more stringent environmental protection or more timber production. The sheer volume of feedback overwhelmed the NEPA team, and Judge Dwyer granted a three-month extension on the Final SEIS (FSEIS). In addition to the public comments, the agencies received feedback from the Fish and Wildlife Service and other relevant agencies, and the White House received significant pressure from interest groups and members of Congress.

The feedback put the White House in a difficult position. It had chosen Alternative 9 because of its balance between ecosystem protection and timber, but the DSEIS feedback raised significant concerns that it did not provide enough ecosystem protection to pass judicial review. To address these concerns, the White House directed the NEPA team to revise Alternative 9.[58] Some of the changes were basic corrections, data additions, and a shift in the Aquatic Conservation Strategy that doubled the width of riparian corridors along ephemeral streams. But the FSEIS also included a new protocol—Survey and Manage (S&M)—that redesigned Alternative 9 in a fundamental way. Under Jack Ward Thomas's direction, FEMAT had taken a coarse-filter approach to ecosystem protection, which assumed that well-distributed and representative reserves would ensure the protection of old-growth-dependent species, and the NEPA team had replicated this in the DSEIS. The coarse-filter approach was necessary in part because FEMAT scientists lacked adequate population and distribution data for a significant number of amphibians, bryophytes, mollusks, vascular plants, fungi, lichens, and arthropods and thus simply could not provide detailed viability assessments for all 1,100 species they considered in their report. In other words, FEMAT had been unable to "guarantee" that Option/Alternative 9 would ensure the viability of all species in the planning area. FEMAT emphasized adaptive management as a way to deal with this uncertainty, but officials in the administration decided that this was too great a legal liability.

The White House, through the NEPA team, therefore added a fine filter process to Option 9 that would increase protection for these species and increase the likelihood of court approval: Survey and Manage.[59] The S&M guidelines identified 404 species that did not have high viability ratings under the coarse-filter approach, and for each species, the guidelines added one or more out of four additional requirements: (1) protect known occurrences of many species, regardless of whether or not they were located within a reserve; (2) conduct site-specific surveys for many species prior to any ground-disturbing activities; (3) conduct extensive surveys for individual species in order to identify high-priority areas for protection; and (4) conduct general regional surveys for certain other species.[60] This was a significant shift in Alternative 9's risk management. The original design assumed that the coarse-filter approach provided adequate viability protection and assumed that agencies would add mitigation strategies for species as they moved forward with management projects; the S&M guidelines, by contrast, required the agencies to prove that their actions would not jeopardize any species before moving forward with management projects.

The updated Aquatic Conservation Strategy and the S&M guidelines certainly promised to reduce the risk to species viability, but they did two other things as well. First, they essentially added an entirely new layer of environmental assessment to federal land management on top of existing layers required by ESA, the Federal Land Policy and Management Act (FLPMA), NFMA, and NEPA to ensure that the agencies avoided further threats to species viability. Second, and more controversially, these changes reduced the area open to timber harvesting drastically, at least until the agencies could complete their analysis work, and therefore reduced the volume of available timber well below the DSEIS projected sale quantity of 1.2 billion board feet. Presumably because the added restrictions would be lifted after the agencies completed their analysis work, the FSEIS still indicated a projected sale quantity of 1.1 billion board feet. At an Interagency Steering Committee meeting chaired by Katie McGinty, Jack Ward Thomas later recalled arguing that the effects of these changes would reduce the total sale quantity to around 200 to 250 million board feet of timber a year rather than 1.1 billion board feet, and in his estimation, the White House had "manipulated Option

Nine to, essentially, become Option One."[61] After the briefing, he wrote in his journal that what frightened him most was "that the politicians who are cutting the deals will have promised more, much more, than we can deliver."[62]

Despite these concerns, the agencies published the FSEIS on February 28, 1994, and on April 13, the secretaries of agriculture and interior signed a record of decision that formally adopted Alternative 9 as the president's Northwest Forest Plan and formally amended land-use plans for nineteen national forests and seven BLM districts.[63] The administration forwarded the FSEIS to Judge Dwyer and Judge Frye for review and turned to the new wave of litigation that followed.

The lawsuits were consolidated before Judge Dwyer, who heard a variety of substantive and procedural challenges to the NWFP. He upheld the plan against all challenges, and two rulings were particularly important for adaptive ecosystem management. First, responding to the Northwest Forest Resource Council's challenge that the agencies lacked authority to prepare a regional ecosystem plan, Judge Dwyer argued, "Given the current condition of the forests, there is no way the agencies could comply with the environmental laws *without* planning on an ecosystem basis." Second, responding to the Seattle Audubon Society's challenge that the agencies had violated NEPA by not considering the scientific judgment that any additional loss of old-growth habitat constituted an undue risk to the spotted owl, he argued that the agencies had provided solid reasons for believing that the spotted owl and the marbled murrelet had better than an 80 percent chance of survival and that they would not be held to a "degree of certainty that is ultimately illusory." But he qualified this to a certain extent: "The FSEIS adequately discloses the risks and confronts the criticisms as required by NEPA," but given the uncertainties outlined in the FSEIS, "careful monitoring will be needed to assure that the plan, as implemented, maintains owl viability. New information may require that timber sales be ended or curtailed."[64] Thus, in reviewing the Northwest Forest Plan, Judge Dwyer reiterated his ruling that environmental laws essentially required ecosystem management and that to achieve their goals the agencies needed to take an adaptive approach grounded in solid scientific monitoring and research.

While legal challenges were resolved for the time being, political challenges to the NWFP began in earnest after Republicans took control of Congress in the 1994 elections. Responding to a heavy forest fire year in 1994, Congress added an "Emergency Salvage Timber Sale Program" to an appropriations bill in June 1995. It established a category of emergency timber sales to "achieve, to the maximum extent feasible, a salvage timber sale volume level above the programmed level to reduce the backlogged volume of salvage timber." The program would accomplish this essentially by exempting salvage timber sales from various environmental laws through 1997 so that they could (1) be sold quickly and (2) avoid administrative appeal and judicial review.[65] This was very similar to the FY 1989 appropriation rider, Section 318, explained in chapter 4. What is more, the 1995 rider directed the Forest Service and the BLM to release old Section 318 timber sales that had been offered but then withdrawn to protect the marbled murrelet within the NWFP area. In all, the 1995 rider directed the agencies to offer 650 million board feet of green timber in addition to actual salvage sales.[66]

President Clinton initially vetoed the spending bill based on a number of objections. Of the salvage rider, he said, "Nobody has worked any harder than I have to start logging again in our country's forests in an appropriate way. Suspending all the environmental laws of the country for three years is not the appropriate way."[67] But in late July, he signed a revised spending bill into law that included the salvage rider, expressing concern that it would undo aspects of the NWFP. Writing to relevant cabinet members, he reluctantly directed them to "move forward expeditiously to implement these timber-related provisions in an environmentally sound manner, in accordance with my Pacific Northwest Forest Plan, and existing environmental laws, except those procedural actions expressly prohibited" by the rider.[68]

The rider put an enormous strain on the BLM and the Forest Service, which were just starting their work with numerous other agencies to implement the Northwest Forest Plan. They had to develop new protocol for environmental assessment and consultation specifically for salvage sales and advance that program while simultaneously implementing a plan that placed severe restrictions on timber harvesting. The rider highlighted differences of opinion within the agencies, supporting those who resisted the steep cuts in timber production, and it warned that Congress was impatient with low timber production in the Pacific Northwest.

Implementing the Northwest Forest Plan: Adaptive Ecosystem Management

Amidst disruptions caused by the salvage rider, the agencies began the long and complicated task of implementing the president's plan. As noted above, the plan provided a regional conservation strategy, so the agencies needed to step down from the written plan to the level of twelve provinces, hundreds of watersheds, and thousands of individual projects. With oversight from the Interagency Steering Committee and the Office of Forestry and Economic Development, decisions were coordinated by the Regional Interagency Executive Committee (RIEC) and supported by the Regional Ecosystem Office (REO). The RIEC included representatives from the Army Corps of Engineers, the BLM, the Bureau of Indian Affairs, the EPA, the FWS, the Forest Service, the US Geological Survey, the Park Service, and the Natural Resources Conservation Service.

The RIEC began work in the late summer of 1993. Since the NWFP was unprecedented in many respects, the committee was working in uncharted territory. "A new planning framework embracing an interagency, multi-scale, hierarchal process designed to deal with multiple values, scales and disciplines is required," one of the committee's working groups wrote, "if we are to efficiently and effectively serve the ecosystem based management principles envisioned in the President's Plan."[69] Such a framework would take considerable time to prepare, so the RIEC recognized that the first few years of the plan would be a transitional period in which the agencies moved "from the present planning strategy of the 'boundary/resource/output' concept to an 'ecosystem/condition/outcome' perspective as outlined in the President's Plan."[70] Of particular concern, existing laws continued to push the agencies toward the boundary/resource/output perspective, so they would "be one of the most significant issues to be addressed along with interagency collaboration. The transition task may take as long as 4 or 5 years."[71]

Some aspects of the plan had been worked out in previous scientific reports—by the Interagency Science Committee, the Scientific Analysis Team, FEMAT, etc.—and were fairly straightforward; and some of the basic land allocations—late-successional reserves, adaptive management areas, administratively withdrawn areas, and matrix—had already been clearly mapped. Some complications arose from the fact that riparian reserves and key watersheds were layered over the other allocations, and Congress intervened in plan implementation in the 1995 rider that exempted a number of timber sales from environmental restrictions. But the real challenges, as the RIEC recognized in its early meetings, stemmed from the NWFP's requirement that the agencies develop a new, truly adaptive ecosystem management framework that required a host of new analysis, monitoring, and collaboration protocol.

As explained in chapter 1, scientists advocated iterative, adaptive management because they believed that the traditional standards of rational decision making—assemble *all* of the relevant data, compare *all* of the management alternatives, and make optimal choices—were simply impossible to meet. The best that agencies could do, they argued and the NWFP required, was to plan based on existing information, monitor the planned activities, evaluate the outcomes, and revise the plan according to new goals, knowledge, technology, and inventories.

Thus, adaptive ecosystem management involves a very different, iterative process of rational decision making that looks far more like an extended program of scientific research than like traditional policymaking. The initial plan serves as a hypothesis, management activities are experiments, and the results lead to revised hypotheses.[72] The model works well for individual researchers or research teams committed in advance to the process and working with relative autonomy. It does not work as well in the legal and political context of federal land and resource management. In the case of the NWFP, employees in the BLM and the Forest Service were hardly analogous to dedicated researchers, since they worked in different resource programs and had different things to gain and lose by following adaptive ecosystem management. More importantly, agency planners and managers had limited autonomy and discretion to carry out the adaptive cycle for at least two reasons. First, many federal laws were designed expressly to constrain flexibility and adaptation in order to protect against real and perceived abuses of administrative power. Agencies face a variety of checks and balances designed, though not all effectively, to hold them accountable for fixed courses of action. Second, congressional appropriations, court injunctions, political confrontations, and other exogenous dynamics, regularly stalled, vetoed, or redirected agency actions, limiting their ability to pursue adaptive ecosystem management even if they wanted to. As a result, researchers involved in the NWFP explained, "Scientists generally see adaptability as essential to managing complex, poorly predictable ecosystems. Thus, most ecosystem assessments call for adaptive management. But when the scientists look back to assess the results of applying adaptive management, they are often disappointed in what managers have been able to implement."[73] The scientific ideal of adaptability ran into the political realities of management implementation.

In addition, the agencies faced significant ambiguity about who should be involved in the adaptive process and what form their involvement should take. In particular, the agencies struggled early on determining how to engage the public in this new project. Certainly NEPA, NFMA, and FLPMA all had strict requirements for public participation in major planning decisions, but the adaptive management cycle was meant to be a fairly continuous process and it was difficult to engage the public at such regular intervals. The Federal Advisory Committee Act (FACA) also required the agencies to work with formal advisory groups whereas the adaptive management cycle worked better with more flexible models of consultation. After the March 1994 court ruling that FEMAT had violated FACA by working with nonfederal employees behind closed doors, the agencies were particularly cautious about the less formal participation envisioned in adaptive management. The Office of Forestry and Economic Assistance lamented that "many federal officials in the region were beginning to make progress working with state and local governments, tribal officials, and with various partnership and community groups [but the FEMAT case] impaired these relations."[74] This was a particular challenge for the Provincial Interagency Executive Committees. The NWFP envisioned shifting from traditional agency planning areas to the twelve physiographic provinces as the focal point for interagency planning, yet the infrastructure for doing so was a work in progress.

The issue of public participation was pressing for federal employees, who understood quite clearly that dual purpose of public participation, particularly in the hostile context of the NWFP. The first purpose of public participation, according to social scientists writing in descriptive terms, is determining the "social acceptability" of management decisions. During the 1980s, for example, Forest Service researchers worked on the "Limits of Acceptable Change System" to understand how the agency could better understand "socially acceptable" goals for management.[75] And throughout the 1990s, agency scientists studied social acceptance of ecosystem management goals.[76] In a 1994 report on adaptive management, Forest Service researchers argued for improved public participation this way: "The democratic ideal of majority rule with protection of minorities rights based on the best available information can be approached most effectively through participation by a diversity of people. The more people participate in shaping decisions, the more they will accept the outcomes and take responsibility for them."[77]And researchers in the RIEC's adaptive management working group emphasized that "Goals must be set by society because successfully implemented management—adaptive, sustainable, or otherwise—depends on societal approval."[78]

Clearly, though, the courts and FEMAT had set the major goals of the NWFP without significant public participation, and it was clear from responses to the draft and final SEIS that the plan was not socially acceptable to many people in the region. This leads to the second purpose of public participation, namely selling management decisions to the agencies' constituencies. In this case, the agencies were fighting a rearguard action to *make* management plans like the NWFP socially acceptable. It isn't surprising, then, that minutes from early RIEC discussions of adaptive management include things like "Goals set by society have limitations both legally and biologically. . . . How and where will societal values be incorporated? How will national, regional and local values be used and balanced? Societies' values tend to be short-term and species needs tend to be long-term. How do we rectify?"[79]

Essentially, the RIEC was wrestling with the tension inherent in the administration's descriptions of ecosystem management between a substantive model with predetermined goals, in which scientists are the primary decision makers, and a procedural model in which the goals are established and pursued through open, democratic deliberation. The RIEC also wrestled with the adaptive management components of the NWFP, on which scientists said the plan's success depended.

Aquatic Conservation Strategy

Option/Alternative 9 established an Aquatic Conservation Strategy (ACS) with several key components: key watersheds, riparian reserves, watershed analysis, and watershed restoration. It also required an effectiveness monitoring program. The area of the NWFP was divided into key watersheds and non-key watersheds. Key watersheds provided either habitat for at-risk fish species or high-quality water, and totaled over 9 million acres of land.[80] The standards and guidelines prohibited new road construction in roadless areas of key watersheds and mandated an overall reduction of road mileage within key watersheds.

Riparian reserves consisted primarily of aquatic ecosystems and land "at the margins of standing and flowing water, intermittent stream channels and ephemeral ponds, and wetlands," and the reserves were initially estimated to total 2.6 million acres. The standards and guidelines indicated that the exact boundaries of riparian reserves—essentially the distance from a water body or intermittent stream channel—would eventually be determined based on detailed, site-specific analysis, but provided fixed interim boundaries that FEMAT believed would provide adequate temporary protection: for fish-bearing streams, lakes, and ponds, the reserve boundaries extended a minimum of 300 feet on either side of the stream channel, though they could extend further based on the outer edges of the floodplain, the outer edges of riparian vegetation, etc.; for permanently flowing non–fish-bearing streams, the boundaries were a minimum of 150 feet on either side of the stream channel; and for intermittent streams, the boundaries were a minimum of 100 feet.[81]

Since riparian reserve land ran parallel to rivers and streams, a map of the reserves looked like a great network of veins stretching out over the NWFP area. This vascular system was designed first and foremost to protect aquatic and riparian species, including anadromous fish, but it had wider-ranging purposes as well:

> maintain and restore riparian structures and functions of intermittent streams, confer benefits to riparian-dependent and associated species other than fish, enhance habitat conservation for organisms that are dependent on the transition between upslope and riparian areas, improve travel and dispersal corridors for many terrestrial animals and plants, and provide for greater connectivity of the watershed. The Riparian Reserves will also serve as connectivity corridors among the Late-Successional Reserves.[82]

To achieve this, the standards and guidelines placed significant management restrictions on the riparian reserves, including a prohibition against most timber harvesting.

The enormous challenge of implementing key watersheds and riparian reserves was not the designation of the categories themselves but the core analysis required as the third component of the ACS: watershed analysis. The SEIS acknowledged that diverse conditions throughout the plan area made uniform management prescriptions inappropriate in achieving the NWFP goals, so one of the adaptive measures it provided was permission to modify restrictions on timber harvesting and other "ground disturbing activities" in key watersheds and the boundaries of riparian reserves if site-specific watershed analyses showed that those adjustments would not undermine the ACS goals. The standards and guidelines provided only general direction for watershed analyses—"systematic procedure for characterizing watershed and ecological processes to meet specific management and social objectives"—and it did not describe or require specific watershed conditions for the plan area. In theory, this reflected the recognition that the agencies would have to establish both the process and the goals of watershed analysis based on specific ecological conditions, public demands, and management objectives.[83]

The RIEC immediately began developing a protocol for watershed analysis, granting some additional flexibility on the ACS's fourth component, watershed restoration, as they worked. It approved draft guidelines and established a pilot watershed analysis program with fifteen pilot projects in the spring of 1994. With preliminary information from these projects, the RIEC approved a final set of guidelines in August 1995. It was clear that watershed analysis was not simply aquatic and riparian analysis; it was "essentially *ecosystem analysis at the watershed scale*."[84] The guidelines provided six steps to watershed analysis, from "characterization of watershed" to "recommendations," and each watershed analysis was required to address erosion processes, hydrology, vegetation, stream channel, water quality, species and habitats, and human uses. The analysis was to be done by interdisciplinary, interagency teams that were competent to address all of the core topics. The analysis was to be "issue-driven," and the issues could either "be known or suspected before undertaking the analysis or . . . discovered during the analysis." And the analysis was "not a decision making process. Rather it is a *stage-setting* process" that would provide information for later decision making.[85] Three participants in an early watershed analysis workshop later reflected, "Watershed analysis was intended to provide the generalized understanding of the processes and interactions that shape the physical, biological, and social landscapes in an area, and it was intended to provide an overtly interdisciplinary view of the ecoscape."[86]

The benefit of this broad definition of watershed analysis and the resulting broad guidelines provided by the RIEC was consistency with the standards and guidelines in the SEIS, as well as flexibility and adaptability throughout the NWFP area. The cost, of course, was a lack of precise direction for teams conducting the analysis: "Watershed analysis is new," wrote two Forest Service researchers and a manager. "Federal agencies have no history of such analyses, but they are now required to produce hundreds of analyses over the next several years. A humbling array of analytical, cultural, and procedural barriers will need to be overcome before watershed analysis becomes recognized as a routine phase of management planning."[87] Another challenge was a lack of clear direction for agency policy and decision making. The RIEC guidelines required immediate cooperation among multiple federal agencies at multiple scales, even though those agencies had different mandates and priorities; the iterative nature of watershed analysis delayed immediate answers to concrete questions. "The fundamental questions facing land managers and policy makers at this juncture," wrote two other Forest Service researchers in 1994, "are how do we use [watershed analysis] to make better decisions on the ground, and is the improvement in our decisions worth the cost."[88]

As expected, watershed analysis started slowly in the first two years—1994–1995—but increased significantly over the next three years. By 1998, the agencies had done watershed analyses for most of the 610 watersheds in the NWFP area.[89] At the same time, a number of complaints surfaced about the analyses.[90] The guidelines allowed enormous flexibility in terms of the management issues driving the analyses, so they varied in detail, scope, and impact. Several Forest Service researchers complained that many of the watershed analyses constituted mere "watershed characterizations" and fell far short of meaningful synthesis and management recommendations.[91] Just completing the analyses was difficult: the different agencies struggled to prioritize watersheds; there was an enormous learning curve for agency specialists; staffing and funding fell far short; analysis teams were often uncertain about the level of detail required; analysis teams struggled to include public participation in their work, which was recommended but not required; and perhaps most importantly, the watershed analyses did not lead to significant changes in riparian reserve boundaries and slowed timber harvesting in key watersheds.[92] Indeed, ten years after the NWFP was implemented, riparian reserve boundaries had been altered in only an estimated 14 percent of watersheds in the plan area.[93]

This leads to a second key challenge that the agencies faced in implementing the ACS: monitoring. The NWFP required the agencies to establish three types of monitoring programs to support adaptive management: implementation monitoring, effectiveness monitoring, and validation monitoring. Implementation monitoring helped determine whether or not the standards and guidelines of the NWFP were followed, and validation monitoring allowed the agencies to determine whether or not the assumptions underlying the standards and guidelines were valid. In between, effectiveness monitoring provided information necessary to determine if agency actions were succeeding in meeting the goals outlined in the plan.

The RIEC's Monitoring Design Group (MDG) began work in 1994 on an overarching effectiveness monitoring strategy for the NWFP and, within it, strategies for five priority elements: the northern spotted owl, the marbled murrelet, the S&M program, late-successional/old-growth forests, and riparian/aquatic ecosystems. (The RIEC developed a separate strategy for socioeconomic monitoring.) As with the RIEC's approach to the NWFP, the MDG argued that it would be necessary to "transition from an intensive, individual species-resource focus to a more extensive, ecosystems approach" in effectiveness monitoring."[94] The general design emphasized monitoring "measurable, biotic and abiotic structure and composition" of ecosystems, which the group argued would both reveal underlying ecological processes and assessments of biodiversity and ecosystem integrity. In this context, monitoring of the northern spotted owl and the marbled murrelet were relatively easy to develop, since methods were well established. Monitoring of old-growth forest conditions, though more complex, had received considerable research over the last decade, so the group was also able to develop an effectiveness monitoring program fairly quickly.

But riparian/aquatic ecosystems and the S&M program posed an enormous challenge for three reasons. First, the objectives were in some cases vague, making it difficult to monitor how effectively the agencies were achieving them. Second, as illustrated by the challenges of watershed analysis, the group found "inconsistent, changing, or nonexistent methods" to measure aspects of riparian/aquatic ecosystems. Third, the group found that it lacked the background information needed as a baseline for monitoring. Compounding these challenges, the agencies involved disagreed on objectives and methods because of their divergent mandates and cultures.[95] As a result, the MDG did not complete an effectiveness monitoring program for riparian/aquatic ecosystems or the S&M program until 1999.

It remains difficult to gather the information needed to assess the quality and impact of watershed analyses and to make a robust assessment of the Aquatic Conservation Strategy's effectiveness. The impact that emerged in public debate, however, was clear: the ambiguity of watershed analyses and the difficulty of staffing and funding robust analyses caused the agencies to take a conservative approach to the ACS. Watershed analyses rarely led to expansion or reduction of riparian reserve boundaries, for example, because the agencies were unclear about the burden of proof they needed to meet in order to justify these changes. The result was that after ten years, the ACS appeared to have improved the ecological conditions of aquatic and riparian systems only to a modest extent while it clearly led to lower timber harvest levels than promised in the NWFP. The trade-off between ecological protection and timber harvesting had been greater than promised.[96]

Survey and Manage (S&M)

The ACS caused challenges enough, but without question the most difficult and hotly contested element of the NWFP was the S&M program, which the Clinton administration added between the draft and final versions of the SEIS. It, more than any other part of the NWFP, reduced the volume of federal timber harvest in the Pacific Northwest. As explained above, FEMAT had reviewed over 1,100 species thought to be closely associated with old-growth forest habitat. The science team had sufficient data to conduct population viability assessments for ESA-listed species and for hundreds of other prominent plants and animals. Their comprehensive review, though, also addressed more than 400 species for which there was no detailed population or distribution information. In some cases, these species were extremely difficult to detect in the field without taxonomic expertise or even genetic testing.

S&M was also perhaps the most remarkable element of the NWFP. The ESA imposes strict obligations only for individual species listed as threatened or endangered, and NFMA requires the Forest Service to maintain viable populations of vertebrate species. No enforceable law mandated that the agencies work to protect *all* species associated with old-growth forests. S&M, then, dramatically exceeded the letter of conservation law and advanced instead the spirit of Judge Dwyer's ruling that the agencies needed to take an ecosystem approach to land-use planning and management. In some ways, all species in the plan area became surrogates for measuring the trend of ecosystem health in the Pacific Northwest.

The S&M guidelines outlined in the NWFP established an adaptive approach to protecting 404 species and 4 arthropod groups, placing each in one or more of four mitigation and survey categories: (1) protect known occurrences of many species, regardless of whether or not they were located within a reserve; (2) conduct site-specific surveys for many species prior to any ground-disturbing activities; (3) conduct extensive surveys for individual species in order to identify high-priority areas for protection; and (4) conduct general regional surveys for certain other species.[97] Over 250 species were placed in the first category, and the agencies were required to identify and begin mitigation for all known occurrence sites by 1995. Since mitigation requirements for known occurrences were the same regardless of the land allocation, they forced modest reductions in timber available for harvest within the matrix areas. For 76 species in the third category, the agencies were required to conduct extensive surveys, and for 284 species in the fourth category, the agencies had to conduct broad regional surveys. Though this work was significant in scope, it did not delay other management activities, and the agencies had until 2006 to complete it.[98] In retrospect, S&M Categories 1, 3, and 4 produced significant new information, provided meaningful benefits to hundreds of rare species, and generally did not interfere dramatically with other federal land uses.

The real controversy with S&M came from the requirements for 77 species, particularly mollusks, in category 2. The agencies were required to conduct surveys and mitigate sites occupied by 5 species of salamander and the red tree vole before any ground-disturbing activities beginning on October 1, 1996. They were required to begin developing survey protocol for the other 71 species immediately and to implement surveys and mitigation prior to ground-disturbing activities beginning on October 1, 1998.[99]

The task was daunting for a number of reasons. First, when the RIEC requested information on known sites of category 1 species, they learned just how little information was available for some of the species in category 2. The initial estimate was that they had known site occurrences for just 5 percent of the 43 mollusk species that were in both category 1 and category 2.[100] Second, with limited scientific data and expertise on a number of species, the RIEC's S&M work group could not develop all of the necessary survey protocols on schedule. Third, even where the agencies had developed survey protocol, they fell far behind schedule in the survey work because of funding and staffing limitations. One report requested by the administration painted a sobering picture: For several species, the author wrote,

> the agencies have been unable to develop survey protocols, conduct surveys, and develop recommendations for managing habitat. The difficulties have arisen in part due to the fact that only a few experts can identify some of the species using laboratory techniques, the lack of in-depth knowledge about the methods to identify and detect the presence of the species, and the need in the case of some species to conduct surveys over a number of years. Moreover, some species appear to be improperly classified.[101]

It is not surprising, then, that the agencies did not meet the category 2 S&M deadlines.

This was particularly troubling politically, since the survey work that was being done had already slowed or halted timber sales throughout the NWFP area. The surveys revealed that some species were indeed extremely rare—over 40 percent of the species had been found in ten sites or less—while others were far more common than anticipated. Forest Service biologists Bruce Marcot and Randy Molina reported that in the first decade of the S&M program, "5 percent of the species account[ed] for most (two/thirds) of the sites and likely are not rare."[102] They also reported that since field managers tended to postpone or cancel activities around these sites to protect the surveyed species, the S&M program reduced timber production in the matrix areas dramatically. Full implementation of category 2 S&M standards became a less and less likely political option for the Clinton administration.

Officials in the Agriculture Department, the Interior Department, and the White House began to reassess the S&M program, looking for ways to uphold its general goals while reducing its direct and indirect costs. To buy time, the Forest Service and the BLM issued a joint memorandum that in various ways exempted planned timber sales from S&M requirements. When the Oregon Natural Resources Council sued in 1999, Judge Dwyer once again enjoined a number of federal timber sales, ruling that the "actions by the federal defendants are arbitrary and contrary to the plain language of the" NWFP. "Far from being minor or technical violations, widespread exemptions from the survey requirements would undermine the management strategy on which the [NWFP] depends."[103] Judge Dwyer's decision launched a decade of conflict over the S&M standards. The agencies prepared new SEISs in 2001, 2004, and 2007, first trying to modify the standards and then trying to drop them altogether. Each SEIS was followed by litigation that ended in settlement agreements or injunctions. After a decade of implementing the NWFP, the S&M standards for category 2 species had become the most contentious provisions of the plan.[104]

Adaptive Management Areas

Whereas the ACS and S&M both advanced adaptive approaches to species and habitat protection, the adaptive management areas (AMAs) totaling 1.5 million acres were designed for true experimentation within the broad aims of the NWFP. Said another way, whereas ACS and S&M both focused to a considerable extent on adaptive restrictions on management activities, the ten AMAs distributed throughout the NWFP area were intended to provide fewer restrictions to creative and innovative management experiments. They were, according to the NWFP, created "to develop and test new management approaches to integrate and achieve ecological and economic health, and other social objectives," and they were designed "to facilitate collaborative efforts, partnerships, mutual learning and innovation."[105] What the NWFP also made clear was that the ten AMAs had been located near communities hardest hit by the decline in federal timber sales precisely so that managers and their partners could find innovative ways to produce timber. Thus, AMA plans were required to have short-term and long-term assessments of timber outputs.

In essence, the AMAs were designed to demonstrate the capacity of adaptive ecosystem management to create win-win solutions that advanced ecological protection and economic development with minimal trade-offs. It is worth quoting the NWFP's standards and guidelines here at length:

> The primary social objective of Adaptive Management Areas is the provision of flexible experimentation with policies and management. These areas should provide opportunities for land managing and regulatory agencies, other government entities, nongovernmental organizations, local groups, land owners, communities, and citizens to work together to develop innovative management approaches. *Broadly, Adaptive Management Areas are intended to be prototypes of how forest communities might be sustained.*"[106]

In a legal and political context that stifled creativity, the AMAs were to be an exception, a creative outlet, and hence the premier test of "new forestry," "new perspective," and "ecosystem management."

The AMAs, however, also proved far more challenging than anticipated. The NWFP had declared by fiat that the AMAs would allow for flexibility, collaboration, and experimentation, but this language did little to change existing law, staffing and funding shortages, political conflict among interest groups, and agency cultures. Indeed, when Forest Service social scientists George Stankey and Bruce Shindler reviewed the AMAs two years into the NWFP, they found "a disquieting feeling among some observers, both within and outside the agencies, that the ideals expressed in FEMAT about AMAs have been compromised or lost altogether." And even among those who still believed in the AMAs, Stankey and Shindler found enormous concern about ambiguous guidelines and a lack of institutional support.[107]

Initial enthusiasm about management flexibility and experimentation, despite meaningful work in some of the AMAs, quickly waned as AMA managers found that the promise of freedom and discretion was somewhat hollow. In a litigious context shaped in part by the ESA and NFMA, it was difficult to shake the agency's risk aversion. Stankey argued that the agencies faced a Catch-22: "Experimentation is not permitted until sufficient evidence is available to predict confidently that the treatment will not have an adverse impact, but until such experiments can be undertaken, it is not possible to generate such knowledge."[108] In the end, Jack Ward Thomas reflected, the AMAs were "managed no differently than other areas covered by the NWFP. This was probably due to the reluctance of regulatory agencies to approve habitat modifications or departures from the overall plan as related to short-term impacts on threatened species."[109] While this was perhaps an overstatement, it certainly captured how many inside the Forest Service felt that the AMAs fell grossly short of their promise.

Evaluating the NWFP after Ten Years

When the agencies stepped back and reviewed the NWFP's effectiveness in its first ten years, they found a variety of mixed results. The plan was so ambitious in scale and scope and so experimental in design that this was hardly surprising. Monitoring found that the reserve system had indeed stemmed the loss of old-growth habitat from trends in the 1980s, watershed conditions showed slight improvements, collaboration among the federal agencies had improved considerably, and the monitoring program had achieved a number of its stated goals. What is more, the agencies had made enormous strides in information management through the REO.

Some of the flagship goals of the plan, however, had not been met. In particular, northern spotted owl populations continued to decline, the plan had fallen well short of its projected timber targets, and the envisioned adaptive approach to ecosystem management had not taken root. The plan had made a substantive promise to stem the northern spotted owl's decline and recover it to the point where it could be delisted. The initial recovery plan for the northern spotted owl had assumed a continued decline for some period of time, but it was assumed that the owls' decline was related directly to the rate of habitat decline. Monitoring from 1990 to 2003, however, showed that owl populations had declined more steeply than the owls' habitat. The most plausible explanation is that the northern spotted owl faced greater than anticipated competition from the barred owl, which was expanding into its range.[110]

Of far greater political consequence, at least for the congressional delegations from Washington, Oregon, and California, the plan had provided a substantive promise of 1.1 billion board feet of federal timber each year and had delivered only about half of that amount in the plan's first ten years. After Judges Dwyer and Frye lifted their injunctions, timber harvesting resumed at a fairly low rate for the first two years. It then increased rapidly, due in large part to the 1995 appropriations rider, peaking in 1997 at 880 million board feet. After the rider expired, requirements of the ACS, S&M, and the AMAs all reduced timber production dramatically to 308 million board feet in 2001.[111] There was, of course, a direct relationship between the NWFP's conservation successes and its timber production failures, affirming Thomas's earlier conviction that the final NWFP plan was hiding very real land-use trade-offs.

When George W. Bush took office in 2001, then, one of his administration's primary goals for the NWFP was to find ways to boost timber production. The Bush administration worked to eliminate S&M requirements, to reduce riparian reserve areas, to expand timber harvested through significant fire reduction treatments, and to revise resource management plans, particularly for the BLM. Many of these efforts were blocked by litigation, blunting changes to the NWFP. Still, after twenty years of implementation, the conflict between ecological protection and economic development remains a central piece of the NWFP.

The plan's substantive promises to achieve both ecological protection (owls) and economic development (jobs) had rested on a broader procedural experiment in *adaptive* ecosystem management. Rather than calculating all trade-offs up front, the plan had promised that an adaptive approach would create new knowledge and new opportunities to balance ecological protection and economic development on various temporal and spatial scales. The agencies also fell short of this promise, though this was not particularly surprising.

The concepts of adaptive ecosystem management, as explained in chapter 1, grew out of contemporary ecology and public administration. In both cases, scholars argued that the earlier focus on controlling ecological and administrative systems through fixed prescriptions had failed because they were based on fundamental misunderstandings of how those systems worked. Adaptive ecosystem management was offered as an alternative to address the dynamism and uncertainties of land and resource conflicts. Ecologists, using probabilistic population viability analysis and complex modeling, emphasized the need to plan, act, monitor, and adjust. New public administration scholars and the Clinton administration's reinvention program emphasized the need to jettison somewhat rigid hierarchical controls in favor of flexible and responsive service to the public. Yet these emphases had not changed the legal infrastructure, the political demands, or the agencies' cultures, so adaptive management never put down roots in the Pacific Northwest, and advocates of the approach were extremely disappointed in this aspect of the NWFP.[112] In other words, the envisioned process simply didn't fit the legal and political realities of federal land and resource management.

Those reviewing adaptive ecosystem management in the NWFP have had a variety of explanations for its success and failure. Two Forest Service researchers in the natural sciences wrote in 2004 that adaptive ecosystem management would require "a closer alignment of adaptive management with organizational goals, a demonstrated organizational commitment and will to act in an adaptive fashion, increased capacity (skills, resources), a clear, shared set of terminology, agreement on expectations within and outside the management organizations, a reasonable likelihood of continuity to ensure the process had a fair chance to succeed, clear performance benchmarks, and formal and explicit documentation."[113] Yet this sounds almost tautological; if all of the forces were aligned in support of adaptive ecosystem management, it would work. The conclusion is not necessarily to abandon adaptive management; rather it is to think further about what adaptive management might look like within the existing constraints.[114]

Conclusion

Despite the challenges outlined above, the NWFP remains the most durable of all the large-scale federal ecosystem management projects initiated by the Clinton administration, and it has achieved some significant ecological goals. It also illustrates one understanding of ecosystem management—the one most often advanced by environmentalists: ecosystem management is a substantive management framework that makes biodiversity protection and ecosystem integrity planning and management prerequisites. Certainly the NWFP did not uniformly privilege biodiversity protection and ecosystem integrity over other interests, and certainly this was a contentious understanding that provoked backlash, but this was written into the basic architecture of the NWFP in threatened and endangered species protection, the ACS, and S&M guidelines.

This understanding of ecosystem management conflicts to a considerable extent with commitments to open public participation and deliberation. It would be difficult to overstate the fact that what drove this model of ecosystem management was not democratic participation, deliberation, and collaboration; what drove the NWFP were the fairly rigid regulatory requirements of the ESA and NFMA and by extension the strict procedural requirements of ACS and S&M. The sustained commitment to the ecological goals of the NWFP was not maintained by a public process or innovative partnerships; it was maintained by the somewhat rigid restraints of laws and policies that Congress and the agencies created years earlier without having been able to anticipate their full implications.

This single example—the NWFP—would not itself support any broad conclusions about the role of democratic participation and collaboration in ecosystem management, but it does match the findings of more systematic studies of this broader relationship. Political scientist Judith

Layzer argues in one of the more thorough empirical studies of ecosystem management that "only in cases where public officials circumscribed the planning process by articulating a strong, pro-environmental goal and employing regulatory leverage are the resulting policies and practices likely to conserve biodiversity or restore damaged ecosystems."[115] Another study comparing the Park Service and the BLM found an inverse relationship between biodiversity protection and collaboration. Neither of these studies denies the possibility that open, collaborative planning can lead to substantive achievements in ecological protection, but they do undermine assertions that the two are necessarily linked.

The NWFP also illustrates the challenges of substantive ecosystem management grounded in contemporary ecology. Political ecologist Ron Herring argues that "the power claim inherent in environmental 'management' is typically a claim that some public good will result from following a set of rules placed in motion by public authority, rules that are an embodiment of scientific knowledge."[116] Traditional scientific management in the United States was premised on exactly this type of claim, namely that federal agencies had scientific expertise and authority to manage ecological systems for clear public goods: timber, minerals, range, recreation, etc. They used this claim of scientific authority to justify the broad delegations of political authority they received from Congress. The NWFP rested on the same claim, namely that agencies relying on solid science could deliver both ecological and economic public goods. Yet the very sciences supporting the plan caution against these kinds of claims, and the plan's successes and failures only reaffirm the need for predictive humility. The challenge for federal agencies is that this type of predictive humility does not build political support for subsequent projects.

7 | ICBEMP: Procedural Ecosystem Management

Introduction

The Northwest Forest Plan (NWFP), adopted in 1994, continues to guide federal land management within the range of the northern spotted owl, and it remains, despite various failures, a flagship example of substantive ecosystem management driven by the ESA. What has received less attention is its sister project, the Interior Columbia Basin Ecosystem Management Project (ICBEMP).[1] Like the NWFP, ICBEMP was an ambitious landscape-scale planning effort that crossed many jurisdictional boundaries. Unlike the NWFP, though, ICBEMP never led to direct revisions of land-use plans for national forests and public lands managed by the BLM. Indeed, whereas the NWFP moved quickly from scientific assessment to an environmental impact statement (EIS) and a record of decision (ROD) in approximately one year, ICBEMP continued for a decade and ended in a memorandum of understanding (MOU) stating primarily that agencies would use the information gathered by the project in their land-use planning.

ICBEMP was not, for this reason, a failure in any final sense, since the information gathered has supported federal land-use planning and the agencies involved have improved their information sharing and collaboration. ICBEMP can best be understood as a comprehensive experiment in procedural ecosystem management. Following the National Environmental Policy Act (NEPA) model more than the ESA model, federal agencies committed themselves first and foremost to a planning process rather than to specific outcomes from the start. What is more, ICBEMP was a truly collaborative project, involving tribal, state, and county governments; federal regulatory and management agencies; and interest groups across the spectrum. In this sense, it was closer to what Chief Robertson had envisioned for the US Forest Service when he announced an ecosystem management approach to multiple use in 1992 than was the NWFP.

The plan's failure—ending without a record of decision and revised management plans—is complicated. This chapter does not suggest that its procedural (as opposed to substantive) orientation made this inevitable, nor that the process design was the problem, as many social scientists and policymakers might be tempted to conclude.[2] What it does assume is that a procedural ecosystem management project that takes democratic participation and collaboration seriously must, by definition, be open to a broad range of outcomes that cannot be predicted at the outset. In the end, then, ICBEMP leaders failed to reach a ROD not because the project deviated from some idealized model of ecosystem management; rather, its fate was contingent on a variety of factors. Three stand out as particularly significant.

First and foremost, the project was trapped between conflicting demands for certainty. Regulatory agencies—the National Marine Fisheries Service (NMFS) and the US Fish and Wildlife Service (FWS)—driven by the ESA and backed by the courts, demanded high levels of scientific confidence, even certainty, that the final plan would protect and restore listed and candidate species in the basin. The Pacific Northwest congressional delegation, governors in the affected states, and Republicans broadly demanded assurance that the final plan would reduce fire risk and produce a continuous flow of natural resources. Failing to satisfy the ESA would have resulted in court injunctions of actual management decisions under the plan; failing to satisfy the Pacific Northwest congressional delegation would have resulted, and almost did result, in the complete loss of project funding. Thus, the agencies were trying to maximize two variables—ecosystem protection and economic development—simultaneously without a clear strategy for reconciling conflicts between the two.

Second, the project was driven at the outset by a wide range of issues, explained below. As the project dragged on past its initial timeline, policymakers and interest groups simply shifted their focus elsewhere to resolve many of them. Wildland fire risks, roadless area protection, threatened and endangered species, etc., each became a node of political or judicial action in and of itself and was disaggregated from the comprehensive perspective of ICBEMP.

Third, and closely related, the Clinton administration's commitment to the project declined significantly beginning in 1998, and project leaders lost the political capital they needed to push the project to a definitive conclusion. This might not have killed the project if Vice

President Gore had won the presidential election in 2000 and retained some of the key political appointees in Washington; but the transition to a Republican administration effectively ended any serious chance of project completion.

In the end, ICBEMP is instructive for another important reason, namely that it was the procedural model of ecosystem management that survived past the Clinton administration and into the Bush administration, though often under other titles. The Bush administration rejected any approach to ecosystem management that presupposed the priority of species or ecosystem protection or that framed land and resource management primarily in ecological terms. Rather, the Bush administration once again framed management issues primarily in political terms that emphasized collaboration and cooperation with land and resource users, and this shifted the emphasis from ecosystem protection back to land and resource development.

Background

Unlike the NWFP, where judicial rulings under ESA, NEPA, and the National Forest Management Act (NFMA) clearly defined the management issues and provided substantive direction at the outset, ICBEMP was initiated to address a wide range of concerns about forest, range, and aquatic health proactively: declining timber harvests, the growing risk of catastrophic wildfire, catastrophic tree mortality stemming from insects and disease, the spread of rangeland invasive species, declines in wildlife species, and the potential listing of various species as threatened or endangered.[3] The concerns that drove ICBEMP can be separated into broad concerns about forest health and more focused concerns about particular species.

Forest Health

Forest health concerns in eastern Oregon and Washington (hereafter "Eastside") began in earnest during the 1980s when the Forest Service completed its first and second rounds of Land and Resource Management Plans under NFMA. The first round of planning, which ended for many Eastside forests in 1982, took place during a timber recession, when local communities were struggling with falling timber demand and therefore falling timber-related income.

The second round of planning wrapped up for many Eastside forests in 1990, and by this time broader concerns about forest health dominated public debate for a number of reasons. First, given the growing influence of "ologists" hired by the agency to meet its NEPA and NFMA obligations and given the growing public interest in the amenity values of Eastside forests, the forest plans gave considerably more attention to noncommodity values than the first round of plans. In particular, battles over old-growth forests on the west side of the Cascades had elevated the agency's attention to biodiversity concerns. Second, and more importantly, mounting scientific and anecdotal evidence indicated that the productivity of Eastside forests for both timber *and* biodiversity was in rapid decline. Indeed, after almost a decade of drought and severe outbreaks of insects and diseases, the Forest Service estimated that 63 percent of all trees in the region were damaged, dying, or dead.[4] When taken together, the Forest Service was promising to satisfy additional interests in its 1990 forest plans at the same time that the capacity of Eastside forests to meet those demands was declining. As a result, the Forest Service faced a wave of litigation over its 1990 forest plans, some of which led to timber injunctions.[5]

The Blue Mountains in northeastern Oregon and southeastern Washington were representative of these problems. Past fire suppression, overgrazing, and timber harvesting practices had, over the last fifty years, transformed the Blue Mountains forests. Many open ponderosa pine stands had been replaced by or filled with dense, disease-prone fir, exposing the old-growth ponderosa that remained to stand-replacing fire. Riparian areas were heavily damaged in some places, threatening a variety of aquatic species, including anadromous fish. Wildfires were becoming more costly and destructive. Here, as in the Eastside region generally, traditional forestry's collateral damage to forest productivity and biodiversity was becoming clearer, to the alarm of the timber industry and environmental groups alike. As a result of forest health decline and the subsequent lawsuits and injunctions, the timber production in the Blue Mountains national forests—Wallowa-Whitman, Malheur, and Umatilla—plummeted from over 700 million board feet in the late 1980s to just 100 million board feet in 1993.[6]

Driven in part by the New Perspectives program, discussed in chapter 4, the regional forester for the Pacific Northwest directed an overall assessment of forest health in the Blue Mountains. Completed in April 1991, the "Blue Mountains Forest Health Report" painted a relatively grim picture of forest conditions: "Scientific evidence suggests that we have encouraged conditions that may result in catastrophic levels of tree mortality. . . . Much of the forest today is characterized by over-stocked, uneven aged stands; large populations of stagnated true fir occupy the sites where ponderosa pine once stood."[7] Authors of the report also highlighted the role that past forestry practices had played in reducing forest health over the years. The problem wasn't just the amount of fire suppression or the particular timber harvest methods. As environmental historian Nancy Langston explains, "These policies developed in large part because the central mission of the Forest Service was to make the forest more productive, eliminating waste by cutting old growth and simplifying forest structure. Early foresters, like many current foresters, had believed in the ideology of efficiency."[8]

The authors argued that the current trends would undermine many of the Forest Service's explicit goals outlined in each forest management plan and that a new approach was necessary. The new approach followed the emerging principles of what was variously called new forestry, new perspectives, and ecosystem management. It included an emphasis on forest resilience, which the authors argued could be achieved best by managing for biodiversity. In other words, rather than working toward a simplified forest structure and a particular timber flow, the agency should manage for the "diversity of species, diversity of successional stages in both space and time, and diversity in the size, shape, and distributions of openings across the landscape."[9] Given the agency's past management practices, the report thus advanced a fairly radical shift that required "an attitude and philosophy that seeks ways to mimic both the natural heterogeneity of the physiographic setting and the natural pattern of disturbances inherent in the landscape."[10] Such an approach, the authors acknowledged, would be difficult to sell to an already skeptical public, particularly when it entailed reductions in timber harvests and increased use of prescribed fires, and it would require significantly more funding than Congress was likely to provide.

In 1992, just as the Forest Service was preparing to adopt ecosystem management as official policy, members of Congress initiated their own studies of forest ecosystem health, this time of Eastside forests as a whole. For communities in eastern Oregon and Washington, the northern spotted owl crisis was demonstrating the worst possible scenario, and members of Congress from the region were eager to avoid a replay. There was no spotted owl on the Eastside, but growing concern over anadromous fish and some old-growth-dependent terrestrial species had the potential to cause the same effect if these species were listed as threatened or endangered.

In May 1992, members of the House of Representatives convened the Eastside Forests Scientific Society Panel to conduct a full review of Eastside forests in Oregon and Washington. The primary purpose was to gather information necessary to include Eastside forests in any old-growth or ancient forest legislation that Congress might pass. In September 1993, the panel issued a report outlining protection measures for Eastside late-successional forests, fisheries, and watersheds. Their assessment of forest health was also grim: "Many ecologists believe that the combined effects of logging old growth and fire prevention have significantly increased the vulnerability of Eastside landscapes to catastrophic disturbances, further threatening what are already severely reduced habitats."[11] The panel recommended interim protection measures that would have, among other things, halted logging in old-growth forests and in sensitive riparian areas, prevented livestock from grazing in riparian areas without extra protection measures, and established a second panel to prepare a restoration plan for the region, emphasizing "protecting the health and integrity of regional biotic elements as well as the processes on which they depend."[12]

Contemporaneous with formation of the Eastside Forests Scientific Society Panel, Representative Tom Foley (D-WA) and Senator Mark Hatfield (R-OR) requested that the Forest Service prepare a separate, comprehensive assessment of the Eastside forests. The agency assembled three teams of scientists in September 1992, which completed a five-volume *Eastside Forest Ecosystem Health Assessment* in May 1993. The three teams included one assessment team as well as teams responsible for crafting a broad, strategic framework and an implementation framework. Collectively, their report provided an overall assessment of ecosystem sustainability under current Forest Service management, outlined various principles and practices of ecosystem management, and offered five general management options ranging from mitigation of catastrophic risks alone to full restoration of ecosystem sustainability.[13]

As the lead editors explained in the first volume, interest in restoring Eastside forests stemmed from two related developments: the publics served by the Forest Service had developed broader expectations than traditional forestry could meet and Eastside forests faced growing risks, including insect outbreaks, tree diseases, and catastrophic wildfire. At the outset, the editors and teams of scientists accepted that traditional forestry would not resolve all of the problems of the Eastside forests and worked explicitly within the Forest Service's new ecosystem management strategy, which they described as emphasizing "conservation of biodiversity, long-term productivity, and the capacity for sustained flows of renewable resources."[14]

The teams' work echoed earlier reports in a number of ways. They found that some, but not all, "eastside ecosystems are stressed and unstable" and that "some management practices of this century have reduced diversity . . . and long-term productivity . . . and have thereby diminished the capacity to ensure a sustainable flow of renewable resources."[15] The problems had been caused by fire suppression, insect and disease suppression, selective logging, road building, and other factors. In short, the teams "found solid evidence that most management practices were applied to objectives that were inappropriate to conserving biodiversity and long-term productivity." On the ecological side, they argued, restoring ecosystem sustainability required sustaining or restoring "the historical range in variability of ecosystem patterns and processes at multiple hierarchical scales."[16] But the teams also recognized the importance of human interests and impacts within Eastside ecosystems, so "the greatest challenge to ecosystem management is to ensure that public expectations are compatible with ecological potential."[17]

To this end, the teams proposed a framework that "laces together societal values and ecological capacity."[18] The teams thus confronted the basic challenge of trying to understand, and indeed manage, two complex, interrelated systems: human and nonhuman. On the ecological side, scientists acknowledged the complexity and uncertainty inherent in ecological systems that made it difficult to predict fully which ecosystem processes were the most important and sensitive. This led to twin emphases on more aggressive management to reduce uncertainty in some cases and more passive management that relied on natural processes. On the human side, social scientists also acknowledged the complexity of social, economic, and political dynamics in the region that made it difficult to predict how ecological changes would impact human communities. This also led to twin emphases that are particularly interesting. On the one hand, the report emphasized the need for active education and outreach to convince the public to support ecosystem restoration; it emphasized the need to manage public opinion. On the other hand, the report emphasized the need for more passive public participation and collaboration intended to allow public opinion to direct forest management; it emphasized the need to learn from public opinion.

Given all of the challenges involved, the report repeatedly emphasized the need for an adaptive, experimental, and iterative process between the Forest Service and the ecosystems and people they managed: "Ecosystem management must be implemented as an experiment because of our evolving theoretical base, uncertainties concerning society's expectations, and an incomplete knowledge of ecosystem structures and processes."[19] Or as outlined in the final volume of the report: "We conclude that management must incorporate more science and more societal processes in the system. . . . The lacing model integrates scientific, societal, and managerial information through an iterative, rather than linear, process of defining objectives. An iterative exchange of information is central to developing creative solutions to complex natural resource issues."[20]

"*What remains to be done,*" the editors of the report concluded, "*is to clearly assess the rate of investment in ecosystem management that is congruent with social expectations, values, and economic interests.*"[21] The lowest level of investment, which presumably would enjoy broad public support, would focus on avoiding catastrophes. With this level of investment, the Forest Service would carry out its most basic responsibilities to prevent species extinction, catastrophic fires, and catastrophic tree mortality. Monitoring would be limited to the most hazardous sites. The highest level of investment, which presumably would enjoy limited public support, would focus on restoring Eastside ecosystems to full health. This would include restoring populations of threatened and endangered species to the point of delisting and restoring terrestrial and aquatic ecosystems to their full range of variability. This would also require monitoring of all ecosystems in the region.[22]

Endangered Species

In addition to the broad challenges of forest health on the eastern side of the Cascades, the Forest Service and the BLM faced a more focused, looming crisis: declining stocks of salmon, steelhead, and bull trout.

The National Marine Fisheries Service (NMFS) had considered listing salmon species under the ESA as early as 1978, but had suspended its review, first pending passage of the Northwest Power Act and then pending completion of the newly formed Northwest Power Planning Council's program to mitigate the impacts of hydropower on fish and wildlife. The program, however, had little effect on the status of anadromous fish, and in the spring of 1990, the Shoshone-Bannock tribe and then a coalition of environmental groups petitioned the NMFS to list the Snake River sockeye and the Snake River chinook salmon under the ESA. NMFS listed the Snake River sockeye in late 1991 and Snake River chinook in 1992.[23]

The agency identified four main threats to the salmon: dams/hydropower, which blocked spawning fish passage upstream and caused increased mortality among juvenile salmon moving downstream; fish hatcheries, which threatened the genetic diversity of wild stocks; commercial, recreational, and subsistence harvesting; and habitat loss. Most of the initial attention in 1992 went to the hydropower system, hatcheries, and habitat on the west side of the Cascades, shaping the NWFP's aquatic conservation strategy. The Forest Service and NMFS agreed that they did not need to consult over the land and resource plans that the Eastside forests had completed in 1990, arguing that the nature of the plans "does not allow for an analysis that would provide a substantive discussion relative to potential effects to a given population of salmon."[24] Instead, the regional forester for Region 6 issued interim guidelines—Eastside Screens—in August 1993 for old-growth forest areas as well as riparian areas and fish habitat. The guidelines required an evaluation of all timber sales to determine their potential impacts on salmon and other species, and they led to an immediate reduction in timber available for harvest. The Forest Service also agreed to review range allotments and suspend any that were likely to harm the listed salmon.[25]

The Pacific Rivers Council sued, arguing that the Forest Service was in violation of the ESA for not consulting with NMFS on the impacts of its land and resource plans on the listed salmon, and in October 1993 the federal district court for Oregon enjoined all further timber harvests in two Blue Mountains national forests until consultation with NMFS was completed. One detail in Judge Malcom Marsh's opinion is particularly important. In deciding the case, he relied in part on *Lane County Audubon Society v. Jamison*,[26] which struck down the Jamison Strategy for the spotted owl: "I can find little to distinguish the [land and resource management plans] in this case from the Jamison Strategy analyzed in *Lane County Audubon.*" The judge essentially warned the Forest Service that it was heading down the same path that had led them to a complete injunction of Westside timber harvesting.

Launching ICBEMP

In July 1993, having launched what became the NWFP on the west side of the mountains, President Clinton directed the Forest Service to begin a similar project for Eastside forests. Like the NWFP, it was to be a "scientifically sound and ecosystem-based" strategy for the Eastside forests as a whole. Specifically, he directed the Forest Service to incorporate the Eastside Forest Ecosystem Health Assessment. BLM state directors, regional foresters, and research scientists recognized that the BLM would eventually face the same ESA litigation as the Forest Service, and they agreed on the need for a broader, interagency planning effort that would cover all Forest Service and BLM lands in the Interior Columbia Basin. This was particularly valuable for the BLM, which had almost no funding or staff in research science to address basin-wide management issues. In January 1994 the two agencies signed the Interior Columbia Basin Ecosystem Management Project charter and committed to developing a joint (1) framework for ecosystem management, (2) full assessment of Interior Columbia River Basin (ICRB) ecosystems, and (3) full EIS, later separated into two.[27] The project charter did not identify the specific policy questions that ICBEMP needed to resolve, but instead assumed that these questions would emerge through the framework, assessment, and scoping for the EIS.[28]

The ICBEMP charter established an executive steering committee (ESC), which included the relevant regional foresters, research station directors, and BLM state directors. Curiously, the team did not initially include representatives from NMFS or FWS, even though ESA issues loomed. The charter also set up a management team. The ESC decided to split the project into two separate EISs. One team in Walla Walla, WA, worked on what was called the Eastside EIS, and a second team in Boise, ID, worked on what was called the Upper Columbia River Basin (UCRB) EIS. The charter also set up a Science Integration Team (SIT), comprised entirely of government scientists (see Fig. 7.1).

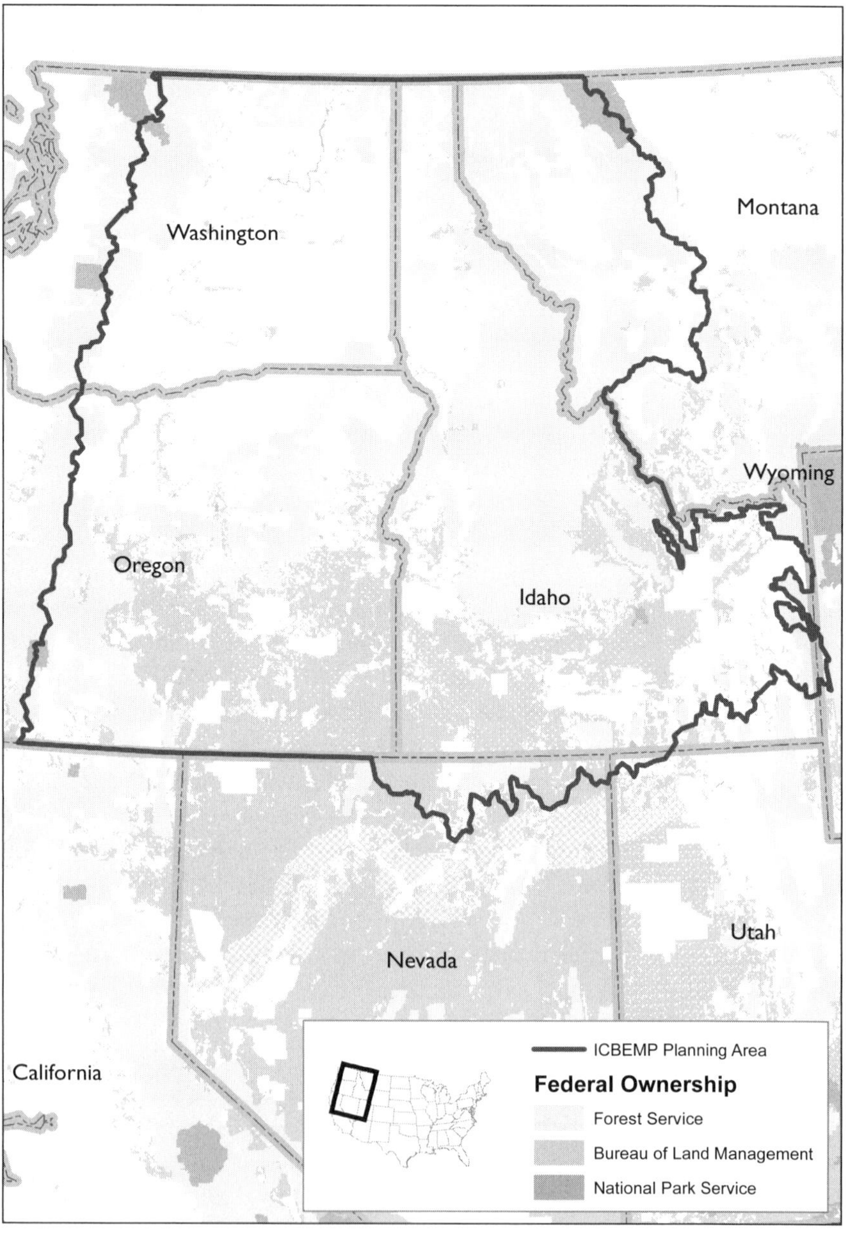

Figure 7.1: The Interior Columbia Basin Ecosystem Management Project Planning Area. Map prepared by Cam Patterson.

The project's geographic scope was remarkable. The total planning area included 145 million acres in Idaho, Montana, Nevada, Oregon, Utah, Washington, and Wyoming. Through the project, the agencies planned to make management recommendations for almost 75 million acres of Forest Service and BLM land—almost 25 percent of the national forest system and almost 10 percent of the public lands. The EIS(s) would cover forest and rangeland presently governed by approximately 70 different land and resource management plans within 42 administrative units. Amending these plans and changing management practices would impact people in 100 counties and 22 Indian tribes.[29] If the scope of the NWFP had looked daunting, ICBEMP dwarfed it.

The scope of issues in ICBEMP was likewise vast. Certainly, *Pacific Rivers Council v. Robertson* had important influences on ICBEMP, and the EISs would need to address a staggering number of species, subspecies, and fish stocks with ESA status. But, as outlined above, ICBEMP really focused on broader questions of forest health and rangeland health that affected economic and environmental interests, and without the kind of judicial mandate that drove the NWFP, the project began without clear prioritization of these issues. Wildfire, tree mortality, endangered species, clean drinking water, tribal treaty rights, etc., were all important.

This lack of clear prioritization also meant that all of the institutions and constituencies involved in the project were important, since none had control over the process. From the outset, then, ICBEMP leaders pursued a very different strategy than the one followed for the NWFP; they made the project open, participatory, and collaborative, consulting frequently and extensively with everyone from tribal leaders and state governors to environmental organizations and interested citizens. Indeed, in addition to the formal consultations with particular groups, ICBEMP leaders held more than 80 public meetings in the project's first year.[30]

The wide range of parties involved represented not only conflicting interests, such as jobs vs. fish, but also divergent types of power and influence. What most of the parties had in common, however, was a demand for certainty or at least very high levels of predictability in the project's outcome. The regulatory agencies demanded that the project provide assurances that federal land management would not jeopardize ESA listed and candidate species, and their demands were backed by possible court injunctions. The county governments, the Pacific Northwest congressional delegation, and resource users demanded that the project ensure a steady flow of timber, continued grazing levels, and other economically significant resources, and these demands were backed by threats of appropriations cuts and riders.

This was a critical test of ecosystem management's promise to avoid a zero-sum contest between environmental protection and economic development. This promise rested on the assumption that a broader perspective could more effectively find room for both sets of interests within the vast ICRB: "If the goal is systemwide—such as restoring the overall health of the watershed—then the proper solution becomes obvious."[31] The real hope for federal agencies, however, wasn't that either a systemwide perspective or EIS would itself determine management choices on the ground. Rather, the hope was that if the courts, Congress, and the interest groups could all agree on broad management goals for the ICRB, land managers would regain professional discretion and flexibility to balance competing interests at the local level.[32]

But before the agencies could move forward toward a plan emphasizing flexibility on the ground, they had to deal with the inflexible requirements imposed by the ESA. In the spring of 1994, the agencies drafted "Interim Strategies for Managing Anadromous Fish-Producing Watersheds" within the plan area. The interim strategies, known as PACFISH, placed new restrictions on a wide range of activities, such as logging, road building/maintenance, grazing, mining, firefighting, etc., all designed to protect riparian areas and water quality. The strategies effectively amended all forest plans and BLM land-use plans in the ICRB. As an interim strategy, PACFISH was established for an eighteen-month period, giving the agencies time to develop permanent guidelines through the ICBEMP process.[33] NMFS issued an initial biological opinion approving PACFISH, but in March 1995 issued another biological opinion finding that eight forest plans, as amended by PACFISH, could jeopardize listed salmon and would require additional amendment. In other words, PACFISH as an overarching strategy was acceptable, but the agencies would still need to evaluate their actions at the project level to comply with the ESA.

In June 1995, in response to the FWS's pending decision to list the bull trout, the Forest Service added interim guidelines known as the Inland Native Fish Strategy, or INFISH. INFISH expanded the range of many PACFISH guidelines, focusing on bull trout habitat. As with PACFISH, the FWS issued a biological opinion determining that INFISH would not jeopardize bull trout but that it was not a long-term solution. In 1997, the federal district court for Oregon affirmed that INFISH met the Forest Service's viability obligations under NFMA, but only as an interim rather than a permanent plan. "At some point," the judge wrote, "if the Forest Service does not adopt the ICBEMP on its current schedule, INFISH will be inadequate as a long-term fulfillment of the Forest Service's viability responsibilities to the bull trout."[34]

The Eastside Screens, explained earlier, along with PACFISH and INFISH, reflected the complicated relationship between the broader planning project and regulatory compliance. On the one hand, ICBEMP was an open, participatory project, in which stakeholders had genuine influence in shaping its goals and strategies. On the other hand, that flexibility was trumped legally by other federal statutes that imposed strict, substantive requirements on one or both of the agencies. This was true of NFMA, with its requirement to protect viable populations of vertebrate species. But in the end, it was the ESA that posed the greatest challenge, because it was a moving target throughout the project. As NMFS and FWS listed additional species in the ICRB, the EIS teams and the SIT had to respond to changing regulatory demands.

With the regulatory agencies temporarily satisfied by interim measures like PACFISH, INFISH, and the Eastside Screens, ICBEMP began to take shape.[35] In ICBEMP science and management had an iterative relationship. The SIT and the EIS teams began working at roughly the same time, so the EIS teams were already beginning their scoping work for the EISs as the SIT developed the ecosystem management framework to be employed and the scientific data needed to evaluate different management options. This had certain advantages, since it created a genuine dialogue between science and management, which proponents of ecosystem management endorsed, but it caused significant delays and frustrations as well. The main problem was that the pace of work for the two EIS teams and the SIT did not always line up, so the teams were often waiting for one another to finish various tasks before continuing their work.[36]

The Scientific Framework . . . or . . . What Is Ecosystem Management?

The first major product from ICBEMP, "A Framework for Ecosystem Management in the Interior Columbia Basin and Portions of the Klamath and Great Basins," was completed in June 1996 and served as the blueprint for ICBEMP planners.[37] What makes this document so interesting is that while it was intended to serve as a binding guide for a political process, its authors from the SIT repeatedly eschewed "value choices" and "normative judgments" as things beyond their scientific purview, working to articulate a neutral planning process for managers to follow. Thus, the framework merely highlighted the challenge of using science to determine legitimate goals for ICBEMP.

The greatest tension in this framework stemmed from the fact that the authors addressed humans as both components of ecosystems that required scientific study and as the normative decision makers beyond the reach of scientific analysis. There was, then, an inherent tension between the authors' emphasis on studying "socially acceptable goals" for ecosystem management as part of the scientific data that informs management decisions and their insistence that "normative judgments about desired outcomes and goals are determined through the established democratic and institutional processes." The authors acknowledged, then, that the fundamental challenge of ecosystem management was integrating "biological science with social science and with philosophical views" as three separate reference points.[38]

The framework emphasized four key scientific assumptions about ecosystems: "ecosystems are dynamic, evolutionary, and resilient; ecosystems can be viewed spatially and temporally within organizational levels; ecosystems have biophysical, economic, and social limits; and ecosystem patterns and processes are not completely predictable." As a result, the authors wrote, "No one landscape condition will be 'best.' Selecting the desired condition is a social decision that can be made by understanding ecological limits."[39]

To give the ICBEMP process some organization and direction while ostensibly avoiding the establishment of any particular goals at the outset, the authors started with a goal so broad that they could assume it enjoyed public consensus: ecosystem integrity. This goal, they explained, included both "maintaining ecological integrity and maintaining the resiliency of social and economic systems."[40] Based on this working assumption, the team outlined a series of more specific goals that the agencies could use as guides and as measures of success:

- maintain evolutionary and ecological processes;
- manage ecosystems using multiple ecological domains and evolutionary timeframes;

- maintain viable populations of native and desired nonnative species;
- manage ecosystems to encourage social resiliency;
- manage ecosystems for the human sense of "place"; and
- manage ecosystems to maintain the mix of ecosystem goods, functions, and conditions that society wants.[41]

The framework outlined an overarching goal, and a subset of goals, that had no trade-offs.

The framework then outlined a four-step process for ecosystem management—monitoring, assessment, decision making, and implementation—in which the authors attempted to separate goal-neutral science from goal-driven decision making. The first two steps were to be purely scientific and descriptive. Monitoring involved basic data collection on a range of biophysical, social, and economic outcomes. While the choice of what to monitor was goal driven, monitoring was assumed to be a purely descriptive enterprise. The assessment step involved synthesizing and evaluating scientific knowledge to inform management decisions. The third and fourth steps, the framework authors argued, went beyond their purview as scientists and into the realm of policy decisions. Decision making was described as establishing goals, comparing alternatives, and selecting alternatives that were most likely to achieve the desired goal. The final step was implementing the decisions.[42]

In many ways, the framework reproduced the traditional process of rational decision making endorsed by proponents of scientific management in the first half of the twentieth century, though it updated this model in two important ways. First, to address the high levels of complexity and uncertainty emphasized by the sciences involved in ecosystem management, it outlined a continuous cycle of decision making intended to allow ongoing adaptation to changing ecological, social, economic, and political conditions. In this sense, the framework was more relevant to ongoing decision making at the forest and district levels than the more fixed and linear EIS process of ICBEMP. Second, to address ecosystem management's emphasis on democratic deliberation, the framework attempted to frame all value or goal choices within the context of public participation: "Defining a desired future is fundamentally a value choice, a choice best explored interactively with stakeholders before managers make decisions on federally administered land. . . . Traditional forms of rational, comprehensive planning have operated poorly . . . because they have often failed to accomplish meaningful stakeholder involvement."[43] The authors explained that stakeholder participation was essential as a tool of information gathering, mutual learning, and consensus building. It said almost nothing about how the agencies should use public participation to address existing, legally enforced goals of ESA, nor did it explain what should happen when stakeholder participation produced conflicting information, political conflict, and legal and political threats. This was a situation that neither scientific authority nor the presumed authority of the participatory process could resolve.

The framework document was not overly sanguine, however. It highlighted the technical challenges involved in trying to integrate monitoring, assessment, decision making, and implementation at a variety of scales simultaneously. This would require enormous flexibility in management rather than formulae, and it assumed that new knowledge would reduce uncertainty in the future and make management more effective. But more importantly, it pointed to the problems that would eventually sink ICBEMP, namely the challenge of setting concrete goals for management:

> This strategy depends on clearly defined goals. Ecosystem management is not an end in itself, it is a means to achieving goals. Defining the goals for ecosystem management also defines the purposes for which federally administered lands are managed. The discussion presented here regarding potential goals and current debates about property rights and Federal land management, highlight how difficult it may be to select goals that are acceptable to the Congress and the public [not to mention the courts]. Yet, the selection of goals is fundamental to moving forward with ecosystem management on federally administered lands. Additional development of process related to implementing ecosystem management cannot supplant the need to confront the issue of goal selection and priority setting.[44]

This highlighted both the enormous challenge of ICBEMP and one of the things that differentiated it so clearly from the NWFP. The agencies had no clearly defined and prioritized goals. Worse, they had manifold goals and demands from the White House, Congress, the courts, and various constituencies, which all retained certain types of veto power in the process (injunctions, funding cuts, employee reassignments, lawsuits, etc.). In this context, how could the agencies move forward with a project governing almost 75 million acres of federal land?

The Scientific Assessment

The framework document was followed in the fall of 1996 by a scientific assessment of the ICRB. The sheer scale of this assessment was remarkable, for it included all 145 million acres of federal, state, tribal, and private land. The assessment reiterated the four key ecological principles outlined in the framework document: ecosystems are dynamic rather than static; ecosystems and relationships among ecosystem components exist at a variety of temporal and spatial scales that are often difficult to integrate; ecosystems have limits, both biophysically and socially; and ecosystem prediction is limited.

The scientific assessment included wide-ranging reports on ecology, economics, sociology, etc., totaling more than 2,000 pages. The SIT was committed to a definition and model of ecosystems that included human social, political, and economic activity. Indeed, they defined ecosystems as "places where all plants, animals, soils, waters, climate, people, and processes of life interact as a whole" or as "places where biophysical and social components interact as a whole."[45] They highlighted the shifting human values of the region, particularly in the late twentieth century, as more members of the public had come to value the public lands for their amenity values rather than their resource development values and as the interests of Native Americans had gained political significance. But the heart of the scientific assessment was of the nonhuman ecosystem components in the ICRB, including thirty-six endangered, threatened, and candidate species listed under the ESA. Human activity had dramatically altered species habitat, such that only 24 percent of the land *could* still be managed for its historic range of variability, thought to support many of these species.

Furthermore, the assessment highlighted how limited existing scientific data really was. Scientists had documented 17,186 distinct species in the basin but estimated that the real total was closer to 43,000. And for many of the known species, scientists lacked basic inventories and other data: "About 86 percent of arthropods, 67 percent of fungi, and 51 percent of mollusk species estimated to occur in the assessment area have not been studied, surveyed, or, in some cases, even identified."[46] This is, perhaps, not surprising, given the scope of the project, but that didn't make it any less of a problem for planning and management.[47]

Following the framework document, the assessment reflected the assumption that maintaining or restoring "ecosystem integrity" was a socially acceptable goal of ecosystem management, and the authors outlined different management approaches to that end. Ecosystem integrity as defined in the assessment was "the degree to which all ecosystem components and their interactions are represented, functioning, and able to renew themselves," and this included both the nonhuman and human components.[48] Integrity assessments for nonhuman components of the basin ecosystems varied but, as with earlier assessments of forest health, were grim. The researchers classified only 16 percent of the basin as having high integrity and 24 percent as having moderate integrity, classifying a full 60 percent of the basin as having low ecological integrity. The estimated socioeconomic integrity, or resiliency, was mixed.

While the assessment addressed both human and nonhuman concerns, it did not consider the agencies' range of legal and political responsibilities for low integrity/resiliency areas. Certainly stabilizing communities was a stated political goal of federal planning in the region, and Congress had traditionally reinforced that goal through the appropriations and oversight processes. The BLM also had legal obligations under the Federal Land Policy and Management Act (FLPMA) to coordinate its planning efforts with local governments. But the agencies had far more specific legal obligations to restore nonhuman ecological integrity under NFMA and ESA. After all, the courts had interpreted the mandates of these statutes as binding, regardless of available funding or socioeconomic cost, and they had recently interpreted them as requiring broad ecosystem protection. In other words, agency responsibility for various ecosystem components varied considerably both by source and depth of commitment.

The scientific assessment was not itself a decision document, but the authors did outline three management options that could be considered in the pending EISs. Option one was essentially the no-action option. It would maintain the existing forest and public lands plans that emphasized timber and forage production coordinated with other multiple uses. Option two would be an about-face, emphasizing aggressive restoration of ecosystem integrity through active management. Importantly, though, restoration would focus on nonhuman ecosystem integrity, rather than socioeconomic resiliency. Option three would address ecological integrity by a zoning system that included extensive reserves, not unlike the NWFP. Within these reserves, the agencies would practice more passive management, depending more on natural processes to restore integrity.[49]

The SIT essentially ruled out option one, since it would likely fail to meet basic ESA requirements and eventually fail to meet socioeconomic goals as well. They explained that option three would do a better job of protecting listed and vulnerable species and their habitat, but it would have greater socioeconomic risks. In the end, they suggested that option two would be the most favorable in meeting nonhuman ecological goals and in meeting public demands. Most importantly, it would maintain managerial flexibility and allow for maximum management activity across the region.[50]

Drawing Up a Plan: The EIS Process

As the SIT worked on the framework and assessment documents, the EIS teams began work on two separate EISs: the Eastside EIS for lands in Oregon and Washington and the Upper Columbia River Basin (UCRB) EIS for the remaining ICRB area in Idaho, Montana, Wyoming, Utah, and Nevada. The NEPA process provided the primary structure for public participation and consultation with tribes, local governments, and state governments. During the early summer of 1994, the EIS leaders held fifteen public scoping meetings in Oregon and Washington to identify key issues for the Eastside EIS, and in January 1995, they held a teleconference that included thirty locations in Idaho, Montana, Wyoming, Utah, and Nevada to identify key issues for the Upper Columbia River EIS. The EIS leaders also consulted with the affected state governors' offices, with tribes, and with organizations such as the Eastside Ecosystem Coalition of Counties (EECC, representing approximately 100 counties in the basin).[51]

The initial plan was to complete both draft EIS documents by the fall of 1995, leading to final EIS documents (FEIS) and a record of decision (ROD) by the spring of 1996, but the EIS process was far slower than anticipated. One of the factors was the pace of the SIT's work on the framework and scientific assessment. Existing ecological and socioeconomic data had been more robust on the west side of the Cascades when the Forest Ecosystem Management Assessment Team (FEMAT) began its work, and the Eastside science teams scrambled to fill in key gaps in the overall data set.[52] Another factor was the evolution of EIS alternatives, which the SIT was required to evaluate for consistency with the available science. The EIS teams completed early draft statements in 1995, but the SIT found that alternatives were too vague to evaluate rigorously and that they contained numerous internal inconsistencies.[53]

The EIS teams completed a revised set of drafts in February 1996, which went to the SIT as well as the regulatory agencies. EIS leaders also consulted with the counties, the states, and the tribes, seeking feedback on the seven alternatives outlined in the Draft SEIS (DEIS). At this point, the EECC, governors of the four most significant states (WA, OR, ID, MT), and a number of interest groups still supported the project. Representatives of the regulatory agencies, however, expressed concerns about the alternatives with respect to listed fish species. The overarching concern was that the standards and objectives for management in the DEIS alternatives did not offer clear management priorities and therefore did not tell managers what to do when the dual goals of ecosystem integrity and people's needs conflicted. The regulatory agencies also expressed concern about whether or not Congress would fund the restoration activities outlined in the DEIS that were essential for meeting conservation goals and concern about whether or not the alternatives would meet water quality standards under the Clean Water Act of 1972.

ESC members initially believed that they could address these concerns and release the DEIS by September 1, but by the end of August, when concern from the regulatory agencies had grown rather than diminished, they chose to revise the standards and objectives and delay public release of the DEIS.[54] Forest Service Chief Jack Ward Thomas wrote in his journal that September, "With the decision to pull back the EIS from the printers . . . the political situation is now about to come unhinged."[55]

The problem wasn't simply a matter of ESA compliance; the problem was ICBEMP's position between the ESA on one side and a Republican-controlled Congress determined to amend or repeal the ESA on the other side, particularly after what had happened with the northern spotted owl. The regulatory agencies overseeing ESA implementation were backed by the courts, which could block implementation of any EIS alternative that did not receive their approval. Yet the more immediate threat was the Republican-controlled Congress, which came close several times to cutting funding for the project. As Chief Thomas saw it, though, the Clinton administration pushed too hard at the early stage on the side of environmental protection, making it more difficult to defend the project to irate western members of Congress.[56]

And the threat from Congress was real. In the summer of 1995, the new Republican majority in Congress singled out ICBEMP for attack. Republicans from the Interior Columbia River Basin states saw the project, particularly the EIS process, as a tool for expanded White House control not only over federal lands in the basin but over private lands as well. Representatives George Nethercutt (R-WA) and Helen Chenoweth (R-ID) had both entered Congress in the 1994 Republican landslide, and both represented districts with deep suspicion of the federal government. They were joined on the Senate side by Slade Gorton (R-WA), who saw the whole project as cover for reducing logging and other resource development activities in the ICRB. In the FY 1996 budget negotiations, which eventually led to a government shutdown, the Republicans attempted to functionally kill ICBEMP. The house budget, Chief Thomas recalls, granted only $60,000 for peer review and publication of the scientific assessment, eliminated the EISs altogether, and prohibited expenditures for PACFISH and INFISH implementation.[57] With a veto threat from the president, the ICBEMP cuts and other provisions were dropped from the final appropriations bill, but the threat remained palpable throughout the project.

Thus, struggling with the pace of science as well as the relationship between the science and the EIS process, the complexity of massive public participation and consultations, criticism from NMFS and FWS, and a highly critical Congress, the two DEISs promised in 1995 were not released until May 1997.[58] It was clear in the selection of alternatives that, unlike in the NWFP, the ESA was only a shaping rather than a controlling force in the ICBEMP project. For the NWFP, biodiversity protection trumped other concerns. In ICBEMP, by contrast, biodiversity protection competed with other forest health concerns throughout the analysis. Most obviously, the tension was over the goals of ecosystem management: protecting endangered species; reducing forest fire, disease, and insect risks; producing timber, forage, and other resources; etc. But closely related to the question of goals was a question about the means of management. Most often, advocates of biodiversity protection called for interconnected systems of reserves that limited management manipulation, while advocates of improving forest health called for even more intensive and expensive silvicultural practices. As the Wilderness Society put it in an assessment of the EISs, "Throughout these documents, future management of the Columbia Basin is presented as a choice between active and passive management."[59]

The EIS alternatives reflected tensions over both goals and methods. The first two alternatives, including the no-action alternative, made no changes or relatively minor changes to federal land management in the region. The other four alternatives, while all embracing various principles of ecosystem management, ranged both in the goals and the level of active management they emphasized. They ranged from Alternative 5, which emphasized active management to produce goods and services within the region's ecological limits to Alternative 7, which emphasized the use of reserves and far more limited management intensity.[60]

The agencies selected Alternative 4 because, according to the scientific analysis, it promised the greatest improvements in forest and rangeland health and very high levels of riparian/aquatic health, species protection, fulfillment of tribal treaty obligations, and production of timber and forage. It also called for aggressive, *and expensive*, management activity. Under the alternative, the agencies would burn and thin forests, increase protection for salmon, and maintain existing levels of livestock grazing while working aggressively to limit invasive species in rangeland. This fit the historic model of Forest Service management, namely using intensive silviculture and other management techniques to maximize multiple uses.

Despite the promises of Alternative 4, Chief Thomas had been correct. The delay of and revisions to the DEIS imposed a serious political cost. In January 1997, BLM state directors for Oregon/Washington, Idaho, and Montana announced new rangeland health standards and guidelines. Because of the delayed DEIS, they decided to move forward with Draft Environmental Assessments and public comment to implement these standards by August of that year. This shifted political focus away from ICBEMP with respect to range management.[61] But the real political fallout was with the EECC, and by extension, Congress. The EECC viewed the changes with suspicion. Specifically, they viewed the delay and revisions as a sign that the regulatory agencies and their preservation interests were hijacking the project. The question, as the ESC heard after meetings with EECC leaders, was about whether public participation would be honored: "Will the Regional Executives work with the counties to address and resolve their issues in a substantive way, as they did with the regulatory agencies?"[62] The EECC leaders and their congressional representatives argued that in a genuinely open, collaborative project, the counties should be equal partners with the regulatory agencies.

Other concerns emerged in the final months of DEIS preparation. The ESC was concerned about the government to government relationship they had with Indian tribes and how to weigh their input versus public input. Again, setting aside the political rhetoric, this relationship had distinct legal foundations that couldn't be ignored. But many of their concerns focused on balance in two key respects. First, the DEIS preparers knew they would face criticism over the balance between the production of goods and services and the protection/restoration of ecological integrity. Second, they were concerned with the balance between certainty provided by detailed standards and guidelines and the flexibility for adaptive management.[63] Based on some of these concerns and internal disagreements, the Upper Columbia Basin EIS team leader left the project for a position with the Idaho Department of Fish and Game, and the Eastside EIS team leader returned to his position as forest supervisor of the Umatilla National Forest. They were replaced by a single EIS team.

A public comment period on the DEIS began in June 1997 and, under direction from Congress, lasted almost a full year. During that time, project leaders and employees hosted dozens of public meetings throughout the basin to discuss the DEISs and gather feedback. Project leaders, for example, hosted forty open house meetings throughout the basin in July and August 1997, and they organized an interactive teleconference broadcast to 60 communities in the basin. They consulted with states, tribes, and counties, and they had a steady march of other public meetings in all seven states. They also began to work toward formal ESA consultation on the preferred alternative with NMFS and FWS.

If Alternative 4 had something for everyone, it also included things that angered almost everyone. Indeed, the only people that seemed pleased with the DEIS were ranchers who were pleased that it did not call for grazing reductions. Environmentalists generally argued that the agencies had ignored the scientific assessment and its implications for biodiversity protection. They also complained that the alternative lacked binding objectives and standards. The timber industry, the counties, and western Republicans all complained that the agencies had collected insufficient and inaccurate socioeconomic information and that ICBEMP was shifting federal land management away from multiple use and toward preservation. As Slade Gorton (R-WA) complained to project leaders, "what your scientists and what your managers and your administration wants with our publicly owned lands is a state of nature."[64]

Geoff Middaugh, the deputy project director, later recalled, "Environmentalists did not like the lack of a terrestrial component, hard standards, and spatial identity, and resource developers felt that the weak standards were still too biased towards protection without any consideration of regional and local economic conditions. Both were probably right."[65]

The EECC and members of Congress such as Nethercutt pressed their complaint about socioeconomic data during the comment period. Specifically, they complained that their interests were underrepresented because the socioeconomic analysis had been conducted at an inappropriately broad scale at which the economic growth and diversity of larger cities in the region obscured the resource dependency and vulnerability of smaller, rural communities. In response, the SIT prepared a new, more detailed socioeconomic study that met the EECC's expectations during the winter of 1997–1998. The ESC then extended the public comment period from February 6 until April 6 to allow time for additional comment on the new socioeconomic information.[66] The study regained support from the EECC, but other groups were not satisfied. In February 1998, the Idaho Association of Counties passed an internal resolution recommending that ICBEMP end without a ROD and that planning be turned over immediately to local managers.

In the final months of the comment period, the Forest Service moved forward with a separate decision that reignited county and congressional frustration. In January 1998, the Forest Service announced that it would place a temporary moratorium on new road construction within the national forest system. The Forest Service had been directed by President Clinton to develop a new road policy for the national forest system that was based on the best available science, and some of the science cited in the Forest Service moratorium decision came from the ICBEMP scientific assessment, which found that roads have direct and indirect negative effects on fish and wildlife. Thus, just as the comment period for the DEISs was wrapping up, it appeared to critics that the Forest Service was doing an end run around the ICBEMP planning process and making major land-use decisions affecting the basin. As Rick Hill (R-MT) asked Forest Service Chief Mike Dombeck rhetorically in a hearing, "are you aware of the potential impacts that this [road moratorium] will have on the collaborative effort that has gone into the Interior Columbia Basin Study?"[67] And when the agencies reviewed public comments on the DEIS, they found that road management and roadless area management received overwhelming public input on both sides of the issue.

In the end, the DEISs garnered 83,000 comments, virtually all of which were negative.[68] Congress was quick to follow up with oversight and appropriations hearings that were essentially indictments of the project. The House Subcommittee on Forest and Forest Health held hearings in Washington and near Boise, ID. Helen Chenoweth (R-ID) opened the first hearing with the inauspicious introduction: "We have now invested 5 years and some $40 million in a project that is not authorized by law and is simply too big to work. . . . the Preferred alternative described in the Draft EIS imposes hundreds of new, vague and conflicting management standards on land managers, creating an atmosphere of uncertainty and confusion for managers and the public alike, leading to excessive and costly delays in decisionmaking."[69]

In the more extensive Senate hearings, ICBEMP received an even more diverse and thorough beating. Senator Gorton expressed his ongoing concern about how ICBEMP would impact private property rights. His central fear seemed to be that the project's scientific assessment and EISs would provide ammunition for the regulatory agencies to restrict private property use in order to protect listed species under the act. But a diverse range of witnesses representing governors in the basin states, Indian tribes, county governments, and interest groups all criticized the plan as either too detailed and restrictive or too vague and unrestrictive. Even Steve Mealey, who had directed the UCRB EIS process until 1997, testified against Alternative 4 in the DEIS. He had left the project because of disagreements with the NMFS and the FWS over how prescriptive the standards and guidelines in the preferred alternative needed to be. He complained that their interpretations of key regulatory statutes, particularly the ESA, made federal management "nearly impossible" and "extremely difficult."[70] The witnesses were divided as to whether or not the agencies should move forward with the DEIS.

And internally, the agencies were having doubts about the project's future. Dombeck had testified during the hearing that "the Project is the best management tool to create a common vision for the long-term management of the Interior Columbia Basin. I believe the Project is a wise investment in the future of the Basin, and that we will complete this effort."[71] Yet Dombeck later recalled that it had become increasingly clear that the project was not going to survive: "It was too big a bite over too diverse an area in a part of the country that was as strongly anti-federal government as you could get," he explained. At some point in 1998, Dombeck remembers, he and Secretary Babbitt "agreed we ought to let this thing die."[72] Project leaders pressed on in their jobs, but they were also losing confidence as public support evaporated over the DEIS. Middaugh wrote in his annual performance report that "technically, the ICBEMP remains the premier ecosystem management strategy for the nation. . . . Externally, however, the plan has not been received well by the general public, by our regulatory partners, by the science team who helped write it, and by internal review of the agencies."[73] Project leaders also began to feel acutely the loss of support from Washington. Agriculture Undersecretary Jim Lyons had told ESC pointedly that the administration had saved ICBEMP from the appropriations axe in 1998, but that ICBEMP was on its last legs.[74] And as the ESC prepared for briefings in Washington that summer, they were effectively asking the administration, "Are you disowning us? If so, why and what do we need to do about it?"[75]

In the summer of 1998, the FY 1999 appropriations bill markups included language to end ICBEMP altogether, and the agencies scrambled to redefine the project in ways that would mollify the Pacific Northwest congressional delegation. In October, Secretaries Babbitt and Glickman sent a letter to Congress explaining that the agencies would prepare a supplemental EIS (SEIS) with new management alternatives. The new SEIS would have a more limited scope of issues, focusing on those issues that required broad-scale attention—"long-term viability for wide-ranging fish and wildlife species, water quality, the rapid spread of noxious weeds, dangerous and unnatural large fire events, and the social and/or economic needs of people"—and deferring the remaining issues to local Forest Service and BLM units.[76] Later, the EIS also dropped the 12 million acres of land in Utah, Wyoming, and Nevada, which for the BLM also meant eliminating their consultation requirement with three governors.[77]

Even with reductions in scope, the SEIS process was slow and difficult. The new preferred alternative would have to address demands from the regulatory agencies for greater specificity in the objectives and standards as well as more detailed explanations of how the agencies would "step down" from the basin scale to individual forests and BLM districts. Representatives from Montana saw this as a critical weakness:

"It has been our experience and observation that the real driving force behind management decisions today are the actions taken by the regulatory agencies. . . . the EIS does not reveal how those regulatory agencies will adapt their individual and collective regulatory regimes" to the project.[78] And Frank Walker, Chair of EECC, wrote to Commerce Secretary William Daley expressing concern about the need for regulatory compliance: "As the Project enters its final phase, we are concerned that [NMFS] has not made a commitment to endorse the Project's final work product. In effect, it seems to us that NMFS has veto power over the whole process, yet has made no clear commitment to it."[79]

While the regulatory agencies complained that the SEIS needed greater specificity, others complained that it was too detailed in its prescriptions. As representatives from Idaho complained, "Many of the proposed standards in the DEIS appear to be fine-scale, quite detailed and specific, and inconsistent with the landscape scale data on which they should be based . . . any effort on the part of the Executive Steering Committee to proceed with a [FEIS and ROD] that contains detailed, fine-scale perspective standards and objectives will be strongly opposed by the state of Idaho."[80] They recommended that the SEIS focus on very broad-scale issues and allow flexibility at the local level. Project managers felt the tension between these two sets of demands acutely. Susan Giannettino, the new project leader, explained to the ESC that "the long-term strategy must deal with the potentially contradictory requirements of: (a) *specific direction* to address the critical broadscale issues and provide for more consistent and predictable outcomes throughout the Basin, and (b) *appropriate flexibility* to allow for local variability."[81] However, it was difficult to achieve both flexibility and sufficient detail. As the EECC argued, "There are 166 Standards in the DEIS. Many constrict reasonable and monitored flexibility, work directly against the theme, and clearly indicate a lack of trust in the abilities of professional field staff."[82] In preparing the SEIS, project managers refocused on broad-scale issues, supplementing this with a carefully designed protocol for "stepping down" to finer-scale management decisions and with some specific areas of fine-scale direction as required by the ESA, NFMA, and the CWA. In addition to these balance challenges, the SEIS also had to address concerns that Alternative 4 had been too dependent on increased funding from Congress and that funding uncertainty could undermine the plan's goals. "Budget implications loom intimidatingly large," the EECC wrote.

Initially scheduled for completion in September 1999, the supplemental draft EIS (SDEIS) was finally released in April 2000. At almost 650 pages, it was virtually a stand-alone EIS, rather than simply a series of updates to the two DEISs released in 1997. It contained three alternatives, including the agencies' new preferred alternative, Alternative 2, and it included significant new spatial components prioritizing areas for restoration work. The preferred alternative "focuses on restoring and maintaining ecosystems across the project area and providing for the social and economic needs of people, while reducing short- and long-term risks to natural resources from human and natural disturbances."[83] The proposal had four main elements: integrated management direction, a step-down process, adaptive management principles, and monitoring and evaluation. Essentially, it provided broad goals for "landscape dynamics, terrestrial source habitats, aquatic species and riparian and hydrological processes, and social-economic and tribal governments," and it outlined a "step down process" for translating these broad goals into specific management activities. In the process, the agencies would conduct additional ecosystems reviews at the subbasin (about 1 million acre drainage areas) level and then at the watershed (about 50,000 to 100,000 acres) level to determine how best to scale down management decisions from the ROD to management activities. It openly endorsed a flexible, adaptive approach that could assimilate new information, and it outlined a monitoring strategy to measure success or failure. It also outlined different levels of restoration activity depending on available funding.

The new preferred alternative was indeed a compromise between regulatory specificity and local flexibility; it was an attempt to integrate conflicting demands that would be enforced by the courts (ESA, NFMA, and CWA) and Congress (production of goods and services). It is important to understand how the preferred alternative attempted this delicate balance. The project managers were not trying to evade regulatory obligations; rather they believed that many of the regulatory decisions could be deferred to the local level in full compliance with the law. As Giannettino wrote in a briefing to the ESC, "I presume we are not trying to ensure that ALL 'critical ESA and CWA issues are adequately incorporated into [this] document.' As you know, there are two more, finer scale, points at which land managers comply with NEPA and provide opportunities for consultation."[84] Essentially, the preferred alternative focused on broad outcomes rather than specific land-use activities within the planning area, leaving it up to local managers to determine how they would meet those outcomes in compliance with the ESA and other laws: "Instead of saying, for example, 'Prohibit timber harvest in riparian conservation areas [RCA]' . . . we are saying 'No management activity shall be approved . . . that is projected to prevent attainment of or degradation of RCA objectives over the long term.'"[85]

This highlighted another point of tension in the SDEIS, which was the tension between short-term and long-term risk, whether to species or economic development. The ecosystem management alternative was, by design, focused both spatially and temporally on broad-scale goals. Thus managers were willing to accept short-term risks to species if it meant longer-term recovery of species and their habitats. Likewise, managers were willing to accept short-term reductions in timber outputs if it meant longer-term increases in timber productivity. Yet most parties in the negotiations were unwilling to accept these short-term risks because they didn't trust the agencies to bring about longer-term benefits. As Jack Ward Thomas later reflected, "the most significant language or cultural barrier between actions of multiple use agencies and regulatory agencies is that when dealing with sensitive, threatened, endangered species the multiple use agencies will opt for a higher level of risk over a shorter period than the regulatory agencies, who will opt for a lower or no risk over a longer period."[86]

ICBEMP leaders provided the standard 90-day comment period for the SDEIS. ICBEMP staff continued an aggressive public meeting and consultation schedule throughout the comment period, trying to build support for the preferred alternative, but it was an uphill battle made even more difficult by the looming presidential election. In the end, with all of the controversial issues going on outside of the ICBEMP process, the SDEIS just didn't receive much attention. The Forest Service roadless decision, the BLM's rangeland health standards, national wildfire debates, etc., had all become political focal points, and the draft EIS was no longer seen as the most important decision point. It elicited only 543 comments, compared to 83,000 for the DEISs.[87]

This did not mean, however, that ICBEMP had a smooth path to a FEIS and ROD. The feedback that the agencies received was deeply divided, and many respondents said that the best decision for ICBEMP would be no decision at all.[88] The two agencies were also struggling to reconcile their own differences. For example, the Forest Service argued that the BLM's rangeland health standards, which were to be applied throughout the basin, did not meet its more rigorous viability standards under NFMA. Their concern had been heightened by a FWS finding in August 2000 that the Western Sage-Grouse, which lives through the ICRB rangeland, warranted listing under the ESA.[89] The two agencies also had differences stemming from the Forest Service's moratorium on road building.

Yet the most significant obstacle remained regulatory compliance. A summary of SDEIS comments from the regulatory agencies—NMFS, FWS, and EPA—included concerns that the "adequacy and effectiveness of the aquatic conservation strategy is uncertain," "the FEIS and ROD should include additional measures to prevent the decline of sagebrush shrub steppe habitats and the species that depend on these areas," "rewrite base level objectives and standards to be more clearly risk aversive for listed fish habitat," etc. Perhaps most pointedly, agency comments argued that "to meet ESA requirements outcome based management direction must . . . prioritize ESA and water quality goals and objectives above other uses where necessary to ensure predictable implementation."[90]

Once again, this went to the heart of the debate between ICBEMP leaders and the regulatory agencies, since they had very different ideas of what the FEIS and ROD would need to accomplish. Two of the greatest challenges were dealing with the issues of scale and certainty. ICBEMP leaders believed that the Biological Assessment they prepared and the consultation with NMFS and FWS they initiated in June 2000 should be focused on the scope and intent of the project at that point. They viewed their work as providing the broad framework for amending national forest and public lands planning documents, leaving many critical decisions to the forest and BLM district level. It was at this level, they believed, that the regulatory agencies should make detailed determinations of jeopardy. The NMFS and the FWS, by contrast, were evaluating the preferred alternative by asking whether or not it was sufficient to protect listed species. These fundamentally different perspectives, driven by law, funding, and administrative direction, were irreconcilable.

At the end of the comment period, Mike Crouse, Assistant Regional Administrator for the NMFS, submitted an initial comment on the preferred alternative. "The outcome-based approach, presented in the SDEIS," he wrote, "offers great potential for ecosystem management. This innovative approach, however, is untested and presents specific challenges."[91] The problem from the NMFS's perspective was that, whereas the original DEISs in 1998 had "activity-specific standards" aligned with PACFISH/INFISH, the SDEIS's broad approach did not have "sufficient clarity and specificity to consistently lead to actions that avoid jeopardy/adverse modification." For example, the NMFS was concerned that the preferred alternative did not provide sufficient safeguards within the riparian conservation areas (RCAs) and argued that "Management in RCAs should be risk aversive. Timber harvest and roading in RCAs should be preceded and supported by interagency-agreed-upon [ecosystem analysis at the watershed scale] and site specific information that indicates de minimis risk from these activities."[92]

The regulatory agencies were backed by multiple judicial decisions that had upheld PACFISH and INFISH only as interim strategies for protecting ESA listed fish species. The judges had said explicitly that their affirmation was dependent on the promise that these strategies would be replaced by the ICBEMP ROD.[93] The regulatory agencies made it clear that the ICBEMP FEIS and ROD needed to replace every essential aspect of these interim strategies.

From the perspective of ICBEMP leaders, the regulatory agencies were asking too much of the FSEIS. As they saw it, the FEIS covered such a vast and diverse area that it served primarily as a framework or guide to subsequent planning decisions. The FEIS's broad scale meant that it could not provide final assurance of species protection that the regulatory agencies demanded under the ESA, but ICBEMP leaders assumed that "*all decisions subsequent to ICBEMP ROD will become 'ripe' for ESA consultation at the appropriate scale.*"[94] As one ICBEMP staff member wrote, the EIS could establish

> an EXPLICIT terrestrial vertebrate habitat network across the entire Interior Columbia Basin, complete with EXPLICIT rigorous management direction. . . . Please tell me how that satisfies an integrated (and I mean integrated with social-economic issues of the Basin) ecosystem-based strategy, where trade-offs are disclosed and should be expected, rather than just a terrestrial habitat-based strategy that is focused on 'optimizing' terrestrial habitat only. . . . the SDEIS is not supposed to be the Interior Columbia Basin Terrestrial Species Viability Project, is it?[95]

In other words, the ICBEMP FEIS was not meant to be another NWFP developed almost exclusively to meet species protection requirements; it was meant to establish terms on which more detailed species protection decisions would be made.

The ICBEMP staff quickly finished the FEIS in November 2000, not yet knowing the outcome of the presidential election, and they scrambled to get to a ROD before the presidential transition in late January 2001. Once again, public and even congressional opposition were not the primary concern. At this point, ESA consultation with the NMFS and the FWS remained the most significant hurdle. The FEIS did include some adjustments to meet ESA concerns, but heading into the final weeks of consultation, a number of key challenges remained. For example, the regulatory agencies were concerned that necessary conservation and recovery of listed species was dependent on additional appropriations of $67 million, which the BLM and the Forest Service could not guarantee.[96]

The regulatory agencies' budgetary concerns were far from a minor point in this case. The science advisory group (formed from the SIT) that reviewed the preferred alternative noted that "outcomes are least responsive to budget levels where management direction restricts human action." In other words, future funding levels were least likely to affect regulatory policy. But ICBEMP, as it was revised in the SDEIS, was *not* primarily a regulatory plan. It certainly identified reserve areas of terrestrial and aquatic habitat in which human uses would be limited, but even there the plan was designed to give managers some flexibility within the confines of the ESA. The core of the strategy, designed to maximize *both* conservation and economic values, was a strategy of active restoration. It included forest thinning and burning as well as stream restoration work. And these activities were highly sensitive to future appropriations.[97]

The land management and regulatory agencies continued to work through the consultation process as the public comment period for the FEIS wrapped up. With a party shift in the White House, ICBEMP leaders were particularly concerned with finishing a ROD before the inauguration on January 20, but they simply ran out of time. The Forest Service determined that it would need another six to eight weeks to work through the protests that had been filed against the FEIS and ROD. ICBEMP staff were also completing a biological assessment of the preferred alternative, and they could not complete a ROD before the regulatory agencies reviewed it and issued a formal biological opinion on the plan. They were in almost daily conversation with representatives from the NMFS and the FWS in December 2000 and January 2001.

On January 16, 2001, however, ICBEMP hit a wall. Project leaders learned that instead of waiting to issue a biological opinion on the FEIS and ROD as the long-term replacement for PACFISH and INFISH, the regulatory agencies were planning to issue a supplemental biological opinion (SBO) on PACFISH and INFISH, essentially reauthorizing them. ICBEMP leaders reviewed drafts of the SBO over the next two days, but their input had little impact on the document. On January 19, 2001, the day before President Bush's inauguration, the regulatory agencies issued the final SBO.

On its face, the document would not seem to challenge completion of ICBEMP, since it was merely reaffirming PACFISH and INFISH as temporary measures until completion of ICBEMP. However, the SBO was effectively a vote of no confidence in the preferred alternative outlined in the FEIS; the regulatory agencies were functionally calling jeopardy on the FEIS before it was finished. Or, as one BLM staff member put it, "the document appears to be a veiled attempt to continue to raise the issue that there are essential pieces lacking in the ICBEMP strategy, and those pieces are found in PACFISH/INFISH."[98] Immediately after the inauguration, ICBEMP leaders fought the SBO, arguing that the regulatory agencies had no legal grounds for issuing it and arguing that the hastily drafted document contained inaccurate and insufficient information. At least one group representing the timber industry filed suit to block the SBO, also arguing that it violated the ESA.[99]

It is difficult to reconstruct the decision process to determine what drove the hasty SBO release. One reasonable interpretation is that it was a last minute effort to shore up ESA protection for listed fish before the Republican administration of George W. Bush took office. It is also difficult to know whether the decision was related specifically to ICBEMP or if it was part of the larger effort on the part of the regulatory agencies to develop legally sufficient salmon recovery plans. Whatever the reasons, the impact was clear. Deputy Project Manager Middaugh later wrote, "So a project that once had the vision of providing ecologically adaptable and determined standards for managing lands never could get over the perceived regulatory necessity for stream buffers. All of the science reports (costing about $31 million) that we used could not overcome the dogged deference to the administrative record of the regulatory agencies."[100]

Letting Go of the EIS

With the full transition of cabinet members and agency leaders under way, the SBO had essentially killed ICBEMP. The ESC continued to map out the steps necessary to complete a ROD: resolving FEIS protests and completing consultation with affected governors, the biological assessment, consultation with Indian tribes, and the administrative record. The stated goal was to have a signed ROD by March 2001, yet the project was clearly hamstrung and the new Bush administration certainly did not provide new political energy or capital to complete the plan. Even after the NMFS and the FWS withdrew the SBO in March 2001, the project as it was originally conceived could not regain its footing.

ICBEMP leaders began to consider their options. With multiple court rulings making clear that PACFISH and INFISH needed to be replaced, the agencies couldn't simply walk away from almost eight years of work. The ESC considered a number of different options, but they were all some variation or combination of two basic approaches. The first approach was to press ahead with the ROD regardless of the opposition. The ROD itself could contain additional decisions that refined aquatic and terrestrial habitat protection measures and made them legally sufficient. This had the advantage of bringing better closure to the project and at least realizing the vision of a comprehensive plan for the basin, but it was clear that the project lacked sufficient financial and political support for that approach. The other basic approach was to stop short of a ROD and find ways to use the scientific information and even aspects of the preferred FEIS alternative at the local level. This had the advantages of greater flexibility, lower project cost, and political feasibility. It had the obvious drawbacks of giving up on consistent conservation and restoration objectives for the basin and potentially returning to the previous rounds of litigation that had helped to start ICBEMP in the process.[101] What is more, this approach would require the very thing that the NMFS and the FWS were trying to avoid, which was a much higher number of ESA consultations at a smaller scale. Nonetheless, this was the direction that the project took.

After another two years of negotiations, largely at the Washington, DC, and state levels, the Forest Service and the BLM signed a Memorandum of understanding with the NMFS, FWS, and EPA to implement what they called "The Interior Columbia Basin Strategy." The press release read simply, "instead of a formal, basin-wide decision from the

Project . . . the regional decision makers adopted a strategy of incorporating the science into ongoing land use planning efforts."[102] The strategy was essentially a framework that directed forest supervisors and BLM district managers to utilize the ICBEMP scientific assessment in the eventual revision of land-use plans. It included the step-down process of connecting local conditions to regional trends and patterns. It was, in other words, a formal way of guiding what the forest and public lands planners needed to do anyway under NEPA, FLMPA, and NFMA, which was to utilize the best available science in their planning process.[103]

Conclusion

The Interior Columbia Basin Ecosystem Management Project remains the largest and most ambitious ecosystem management project for federal lands to date, and it illustrates a procedural model of ecosystem management driven by NEPA. Rather than beginning with a predetermined set of goals, as did the NWFP, federal agencies initiated ICBEMP intending to discover and prioritize goals through an open, participatory process. The fact that it ended in 2003 without formally revising Forest Service and BLM management plans was anticlimactic, but the information assembled and analyzed for ICBEMP was used for planning and management in subsequent years.

The point of this analysis is not to compare ICBEMP with some ideal model of ecosystem management but rather to understand how some of the ideals of ecosystem management worked in a particular legal, political context. After all, ICBEMP clearly did *not* fail to produce a plan for lack of science nor lack of participation/collaboration called for in the framework document and by advocates of ecosystem management. Certainly, some of the scientific data could have been more robust and some of the participants complained that they were not always kept informed, but ICBEMP rested on some of the most comprehensive scientific assessment and public participation/collaboration ever produced for a single land-use plan. Nor could it be said that ICBEMP was hijacked by any particular interest group.

In the case of ICBEMP, stated explicitly in the framework document, ecosystem management was a process rather than a set of goals. Or perhaps more correctly, it was a process involving a host of goals that the agencies had not, to date, been able to balance. The Forest Service's justifications for it in congressional hearings emphasized the fact that the agency was buried in planning processes at the local level and in litigation to the point where it could not actively manage its land and resources. It supported ecosystem management precisely because it was an alternative *process* through which it hoped to balance trade-offs more effectively and avoid or win litigation more often. Those who control the process often control the outcome, and the Forest Service and BLM hoped that the large-scale ecosystem management approach would allow them to control the process and move on more quickly to land management.

The first important fact about ICBEMP is that it was a Forest Service/BLM project. It was, in other words, a vastly expanded and ecologically informed version of multiple-use management. Like traditional Forest Service and BLM planning, the agencies began with a land area, surveyed the host of natural resources and public values, and worked to achieve the "greatest good of the greatest number in the long run." Those more sympathetic to the Forest Service might even say that ICBEMP was an attempt to achieve Gifford Pinchot's utilitarian vision more holistically and ecologically than ever before. As a Forest Service/BLM project, it was governed legally by FLPMA and NFMA, so the project was driven partly by a legal, multiple-use mandate.

The Forest Service and the BLM had traditionally managed multiple-use lands in the basin for natural resource outputs, such as timber, forage, and minerals. The proposal in ICBEMP was not necessarily to reduce that management focus as much as it was to bolster other management foci that had evolved over the last twenty years: recreation, fish and wildlife beyond game species, ecological integrity, etc. Thus, instead of optimizing for one set of values, the agencies set out to optimize many sets of values simultaneously. On its face, this would be mathematically impossible, but like one interpretation of multiple-use management, the hope was that this could be achieved by creating a broad framework within which different values could be optimized at various spatial and temporal scales.

The second important fact about ICBEMP is that it was an ecosystem management project torn between flexible, adaptive management and both the legal and political demands for certain guaranteed outcomes. This is where the very environmental sciences that formed the foundation for ICBEMP also created significant challenges. The scientific assessments talked about restoring the "natural range of variability" and about the inherent uncertainties involved in planning for ecological and socioeconomic systems. The goal of the science was to explain how the agencies could cope and adapt over time to hit the uncertain and moving target called ecological integrity. Thus, from a scientific perspective, ICBEMP was a process intended to establish another process or posture for management.

This conflicted fundamentally with the legal and political authorities involved. One scientist who worked on the project reflected, "Adaptive management is certainly a defensible approach in a world of ecological uncertainty. But it can work, and be defensible, only if simultaneously adopted in the sociopolitical world."[104] But clearly this was not and is not the case. As outlined above, the regulatory agencies had authority under the ESA to demand guarantees that species would not be jeopardized in the short term or long term; and Congress had constitutional authority to insist, as it did, that ICBEMP guarantee economic stability for local communities in the project area. And both sides had sanctions to wield if their demands weren't met. The regulatory agencies were backed by the possibility of court injunctions of any plan that didn't comply with the ESA, and congressional demands were backed by a Republican-controlled appropriations process that could shut ICBEMP down entirely.

The third important fact about ICBEMP is the simple reality of contingency. The project was launched at a time when a number of factors made ICBEMP the primary vehicle for decision making. ICBEMP, in other words, had a gravitational force drawing a wide range of issues into its framework. With the specter of the northern spotted owl crisis fresh in the political memory, the ESA initially aided ICBEMP. Fish listed under the ESA and lawsuits that identified ICBEMP as the longer term solution to ESA issues in the Basin lay at the heart of ICBEMP's gravitational pull. The other issues orbiting the project—tree mortality, wildland fire risk, vulnerable terrestrial species, etc.—had no primary policy or management home, and ICBEMP gave them a common home.

Over time, however, as different parties lost confidence in ICBEMP, these issues broke free from ICBEMP and shifted to other decision-making arenas. Salmon recovery, for example, ultimately required NMFS to deal not just with habitat issues but with hydropower, hatcheries, and harvest issues as well. These issues were centered primarily on the west side of the Cascades, and they involved a variety of agencies and interest groups that were not involved in ICBEMP. After a number of destructive fire seasons in the West, not least of which was the deadly fire season in 1994, the politics of fire management shifted to a national level rather than a regional level. Likewise, as rangeland health and road policy broke free from ICBEMP and became national controversies in and of themselves, the project no longer served as the decision framework in which those issues would be resolved. Said another way, ICBEMP suffered from ecological and political problems of scale.

The NWFP was markedly different in this regard. First, the NWFP was driven primarily rather than partly by the ESA. So while both projects involved court rulings that tied ESA compliance with the evolving management plan, the ESA was not in the driver's seat of ICBEMP in the same way. But more importantly, the court injunctions on the west side of the Cascades *shut down* the entire federal timber program, causing enormous economic hardship, whereas the timber injunctions on the east side only affected two forests, and these injunctions were lifted through PACFISH. The net result was that injunctions on the West side ensured that the NWFP was the only decision space available for interested parties, and it made federal scientists the primary decision makers in forest planning.

These three facts or factors should be sufficient to demonstrate that the failure of ICBEMP was not procedural; the ICBEMP leaders did not practice "bad ecosystem management." Rather ICBEMP started with an open, procedural definition of ecosystem management as the integration of ecological science with multiple-use decision making, and it ended that way. Furthermore, the strategy points to the most important legacy of the project, which is the unprecedented scope of its scientific assessment that not only aids land-use planners today but helped reframe some of the public debate at a broader ecological scale.

Conclusion

The George W. Bush Administration

When Vice President Gore narrowly lost the 2000 election and President George W. Bush took office in 2001, environmentalists braced for a fight over the public lands. Both the president and Vice President Cheney had complained on the campaign trail that the Clinton administration had locked up the federal lands and resources from essential development activities, running "willy-nilly all over the West."[1] Immediately after taking office President Bush established the National Energy Development Group, chaired by the vice president, which recommended dramatic increases in energy development on federal lands. The administration then told federal land agencies that their top priority was the concrete goal of energy development rather than the more nebulous goals of ecosystem management.[2]

The Bush administration also quickly attacked a range of presidential proclamations and agency rules that the Clinton administration had completed during its final few months, threatening to undo national monument proclamations on BLM lands and other land-use restrictions. The Bush administration's top priority was undoing the US Forest Service Roadless Area Conservation Rule, published on January 12, 2001. The new rule, which had been influenced in part by scientific research conducted for the Northwest Forest Plan (NWFP) and the Interior Columbia Basin Ecosystem Management Project (ICBEMP), prohibited the construction of new roads on 58.5 million acres of inventoried roadless areas in the national forest system, which also meant closing those areas to most logging activities. For the Bush administration, this amounted to the creation of de facto wilderness areas and locked up resources that under the National Forest Management Act (NFMA) should be open to the full range of multiple uses.[3] (The administration received early support from the courts, which castigated the Forest Service for National Environmental Policy Act (NEPA) violations, and offered its own revised roadless rule; but after more than a decade of legal and political battles, the roadless rule survived.)

At the same time, the new Bush administration had learned from the Reagan administration's battles with environmentalists not to challenge the value of environmental protection directly.[4] The political strategist Frank Luntz warned the administration that "the environment is probably the single issue on which Republicans in general—and President Bush in particular—are most vulnerable. . . . Therefore, *any discussion of the environment has to be grounded in an effort to reassure a skeptical public that you care about the environment for its own sake*—that your intentions are strictly honorable."[5] Instead of challenging environmentalism directly, then, President Bush worked to define "a new environmentalism for the 21st century" built on "a new spirit of respect and cooperation."[6]

Under this banner, the Bush administration advanced a number of environmental initiatives, such as the Clear Skies Initiative, that environmental organizations said were just green-washed resource development projects. The initiative most relevant to ecosystem management was the president's Healthy Forest Initiative. In the wake of several dramatic fire seasons, the Bush administration argued that federal agencies needed to restore forest health throughout the country to increase forest productivity and to protect millions of Americans from catastrophic wildfires. He launched the Healthy Forest Initiative in 2002, which Congress endorsed in the Healthy Forest Restoration Act of 2003. In one sense, the Bush administration affirmed the previous administration's emphasis on forest health, but the Bush administration defined forest health quite differently. Rather than measuring forest health in terms of biodiversity or old-growth characteristics, the Bush administration measured it in terms of timber productivity and wildlands fire risk.[7]

For the Bush administration, the Healthy Forest Initiative provided language that was pro-environment while still challenging the power of environmental organizations. The administration argued that the greatest obstacles to forest health were environmental regulation and environmental organizations that used administrative appeals and litigation to obstruct forest improvement projects. To address this, the initiative included goals of reducing environmental assessment work, shortening planning processes, and shielding forest health projects from administrative appeals and litigation. These aspects of the Healthy Forest Initiative echoed the 1989 and 1995 salvage riders discussed in chapters 4 and 6.[8]

The Healthy Forest Initiative reflected the Bush administration's tendency to frame land and resource questions in political and economic terms rather than ecological terms and to focus federal land agencies on resource uses rather than on broader, and sometimes more abstract, ecosystem concerns. This was reinforced by the fact that while the Clinton administration had appointed a number of leaders with science backgrounds to oversee federal lands—Secretary Babbitt, Jack Ward Thomas, Mike Dombeck, etc.—the Bush administration appointed more leaders with backgrounds in business and law. Whereas the Clinton administration had tried to elevate science through the National Biological Survey, Rangeland Reform '94, and ecosystem management rooted in ecology and conservation biology, the Bush administration held certain areas of science at arm's length. This could be seen in the president's treatment of climate science and in the changing role of the president's science advisor.[9]

As with President George H. W. Bush's approach to wetlands delineation, this should not be interpreted as the triumph of politics over science in some simple or crass sense. As with wetlands, forest health was a broad concept without clear definition, and the Bush administration certainly drew on science to advance the Healthy Forest Initiative; and as chapter 6 shows, the Clinton administration hardly eschewed politics in ecosystem management, which was precisely Jack Ward Thomas's criticism of the final version of the final NWFP. The primary point here is simply that the two administrations framed forest health issues in very different terms, which privileged different points of view and guided land and resource management in different directions.

Clearly, the Bush administration rejected a number of ecosystem management principles advanced by the Clinton administration, particularly managing to protect ecological processes and functions more than resource outputs. In other words, the Bush administration rejected the substantive model of ecosystem management that came out of the Clinton administration's implementation of the Endangered Species Act (ESA). Yet the Bush administration continued to champion other ecosystem management principles, such as adaptive and collaborative management, adopting the rhetoric of "cooperative conservation." As mentioned in the introduction, Interior Secretary Gale Norton repeatedly pledged that the administration would use "Consultation, Communication, and Cooperation, all in the service of Conservation." She essentially promised an anti-regulatory model of conservation, which emphasized the importance of local participation in public lands issues, particularly participation from those who "live and work on the land."

In August 2004, President Bush formalized cooperative conservation in an executive order, directing federal agencies to carry out their legal mandates in a way that:

- (i) facilitates cooperative conservation;
- (ii) takes appropriate account of and respects the interests of persons with ownership or legally recognized interests in land and other natural resources;
- (iii) properly accommodates local participation in Federal decisionmaking; and
- (iv) provides that the programs, projects, and activities are consistent with protecting public health and safety.[10]

And the administration advanced its cooperative approach in a variety of other ways as well, so that ecosystem management projects gave way to collaborative conservation projects, and sustaining biodiversity and ecosystem integrity gave way to Sustaining Working Landscapes.[11] As with the Healthy Forest Initiative specifically, cooperative conservation was ecosystem management reframed in political and economic terms and stripped of many substantive ecological commitments.

So What Happened to Ecosystem Management?

In a recent essay entitled "Whatever Happened to Ecosystem Management and Federal Land Planning?" political scientist Martin Nie ponders whether he should write about ecosystem management "in the past tense, as a sort of obituary."[12] But, he notes, it is difficult to determine what happened to ecosystem management when it was never clearly defined, and he argues that certain principles advanced under ecosystem management—adaptive management, collaboration, and landscape-scale restoration—remain central in the rhetoric of federal land and resource management. The question left to ponder, then, is what happened to various models and principles of ecosystem management after the Clinton administration?

The real *ecosystem* management for environmentalists in the 1990s was built on the model of the ESA. It was a robust, substantive commitment to preserving biodiversity at the genetic, species, and ecosystem levels. This meant planning first to protect ecosystem integrity and then determining what land uses fit within that plan. As ecologists and conservation biologists made clear, this was not an easy commitment to keep, since it meant continually adjusting management plans in response to the dynamism, complexity, and uncertainty in ecological systems. Scientists and environmentalists, it should be noted, did not oppose all economic uses of the public lands; they just believed that these uses were subordinate to some threshold of ecosystem integrity.

The real ecosystem *management* to multiple-use agencies—the Forest Service and the BLM—as well as resource industries, was built on the model of NEPA. It was a commitment to gathering scientific information and using it in an open forum where various interest groups could deliberate on the appropriate goals for management, or at least have the opportunity to vie for priority in management decisions. Neither the agencies nor the resource industries generally argued that resource development should trump all environmental concerns in federal land and resource management, particularly concerns about future forest or range productivity. They just believed that national parks, wilderness areas, and other specially designated lands had already been withdrawn from resource development and committed to preservation and that the environmental protection afforded multiple-use lands needed to focus on ensuring a sustained yield of natural resources.

The Bush administration clearly rejected the substantive model of ecosystem management, arguing that it would run "willy-nilly all over the West," and it worked to constrain even the procedural model of ecosystem management by reducing scientific assessment and administrative appeals that slowed or blocked resource development projects. In both cases, the Bush administration downplayed the importance of scientific authority in legitimizing federal land and resource decisions and elevated the collaborative and participatory principles—the role of democratic authority—in ecosystem management. The fact that collaboration and participation in the Bush administration led to more resource development and fewer environmental restrictions points simply to the fact that the administration chose different collaboration partners and used collaborative processes to achieve different substantive ends. One is reminded of political scientist Theodore Lowi's warning at the end of the Clinton administration that one should "never fall in love with process. Americans love process, especially if it is grounded in competition, visibility, and democracy. I suppose this is because it helps us maintain the appearance of consensus. But process is a double-edged sword: eventually it will cut the other way, and then it will be called betrayal."[13]

So what did happen to ecosystem management? In essence, the Clinton administration never clarified the meaning of ecosystem management as an integrated and overarching paradigm for the federal estate; rather it gathered a variety of principles and tools under the ecosystem management banner and applied them when opportunities emerged. Without an administration actively advancing ecosystem management as a loosely defined paradigm, Martin Nie explains, it simply fragmented against the enduring realities of federal land law and politics.[14] In 1967, Lynton Caldwell complained that public land law provides "no means for taking the fundamental ecological context of land use into account."[15] And in 1991, Joseph Sax explained that "a fundamental purpose of the traditional system of property law has been to destroy the functioning of [ecosystems]."[16] In the twenty-first century, Caldwell's and Sax's assessments are still cogent. Federal land law and politics are powerful centrifugal forces that fragment federal land and resource management. Laws such as the ESA and NEPA provide countervailing centripetal forces that provide federal land agencies with some unified goals of environmental protection, which is why they served as the foundation for ecosystem management in the 1990s. In the twenty-first century, the Bush and the Obama administrations have emphasized different centrifugal and centripetal forces, prioritizing them in different ways and yielding different outcomes. Whether because of outright opposition or a lack of political energy, neither administration supported a paradigmatic interpretation of ecosystem management.

Yet ecosystem management remains instructive, because the basic resources and challenges that drove ecosystem management remain largely unchanged in the twenty-first century. Ecology continues to highlight the complexities and uncertainties of ecological systems, particularly in light of climate change, and it continues to highlight the need for management that protects broad ecological processes and functions that cross political boundaries. Agencies simply can't put resource management questions back into a simple formula of sustained yield or the rectilinear boundaries of national parks or national forests. Closely related to this, political science and public opinion continue to reject planning and management based primarily on administrative expertise and autonomy and continue to call for various forms of democratic accountability. Various schools of political thought continue to emphasize governance that is mutual, collaborative, and adaptable rather than governance administered primarily by governmental agencies with clear legal authority. Agencies such as the Forest Service will never regain the autonomy that they once enjoyed. And the basic paradoxes of federal land and resource management remain just as challenging as they were in the 1990s. The demands for environmental protection and use have increased rather than decreased, and changes in scientific knowledge, economic value, and political power ensure that balance between the two will remain elusive. The pressures for agencies to legitimize their decisions both scientifically and democratically remain equally powerful. These pressures fluctuate depending on the administration and the particular issues, but agencies must show that their decisions are supported by science and by some representative public.

So while "ecosystem management" as a rhetorical banner might not return and become a truly determinative paradigm for federal agencies, political leaders will have to gather many of the same principles under different banners in order to address the current challenges of land and resource management. It is worth returning, then, to Keiter's reflections on a legislative agenda that Congress could follow if it decided to mandate a more permanent paradigm informed by the principles of ecosystem management.[17] In the absence of national consensus, Congress could opt for a place-based approach, passing legislation that mandates ecosystem management for particular landscapes. "The place-based legislative approach," he writes, "is a particularly appropriate strategy for incorporating ecological principles into the American nature reserve system."[18] But if ecosystem management principles gain greater clarity and broader support, he suggests, Congress might opt to pass some overarching legislation for the public lands that would direct agencies to manage ecosystems "as dynamic entities to conserve and restore biodiversity levels and existing ecological processes in order to ensure a sustainable resource base."[19]

An ecosystem management statute—perhaps using different language such as adaptive or collaborative land management—would need to remain general, but according to Keiter it would need to impose at least two standards: "a nonimpairment standard that would establish a threshold basis for evaluating management proposals, and a biodiversity conservation standard that would impose an affirmative obligation on the agencies to protect and restore species diversity."[20] In other words, it would have to have legally actionable and substantive requirements. But with these thresholds in place, Keiter argues, the act would need to mandate collaborative and participatory planning, because agencies need to honor the nation's democratic heritage as well. In other words, the statute would need to differentiate between the substantive and procedural requirements, identifying areas where Congress has already made land-use decisions and space for truly open public deliberation.

With virtually no prospect of ecosystem management legislation on the horizon and a presidential administration that invests limited energy in the public lands arena, innovative use of ecosystem management principles has come from other quarters. The federal land agencies are contributing to this but not leading it in any clear sense. Consider, for example, the push for "large landscape conservation." In the absence of clear legal authority to made unilateral decisions, as is the case for most large-scale ecological challenges, networks, partnerships, and regional institutions have evolved to address particular landscapes. Co-authors of one report on large landscape conservation—Matthew McKinney, Lynn Scarlett, and Daniel Kemmis—argue for many of the basic principles of procedural ecosystem management, essentially packaging these principles in political rather than primarily ecological terms. While large landscape conservation does not have a precise definition, they argue that it is "multijurisdictional, multipurpose, and multistakeholder, and that [it operates] with various governance arrangements at diverse geographic scales. The common currency in large landscape conservation is regional collaboration—the ability to work across boundaries with people and organizations that have diverse interests yet share a common place and purpose."[21] These projects depend more on voluntary participation than on legal coercion and therefore promise to build what McKinney and Kemmis call collaborative democracy.

Looking around the country, there is a wide range of collaborative models that has evolved to address complex and often large-scale land-use conflicts. Independent and academic institutes, including the Lincoln Institute of Land Policy in Massachusetts, the Ecosystem Management Initiative at the University of Michigan, and the Center for Natural Resources and Environmental Policy at the University of Montana, maintain databases with hundreds of examples. The examples do not fit under a single banner, such as ecosystem management or large landscape conservation, but they all reflect some core elements of multiparty collaboration driven by shared values and concerns rather than simply by regulatory requirements. Said another way, they tend toward pragmatism, avoiding ideological debates over first principles and focusing on shared value commitments that can generate action. Learning from these projects may indeed prove valuable as federal, state, tribal, and private landowners negotiate over common land-use problems that cross jurisdictional boundaries.

Whether the various collaborative projects will lead to a unified paradigm for federal land and resource planning is unclear, but it seems unlikely. The collaborative governance movement begins with a robust commitment to democratic deliberation and a critique of rigid and coercive authority. It is pragmatic and open to compromise. It is grass-roots rather than hierarchical. These commitments may have been affirmed by the Clinton administration's rhetoric of government reinvention, the Bush administration's rhetoric of cooperative conservation, and a significant body of collaborative governance literature, but they are also challenged by the legal infrastructure of federal land and resource management, which emphasizes the importance of legal authority and accountability.

In the case of federal lands, one cannot have governance without the legal power and limitations of government. As legal scholar George Coggins puts it, "devolution as now advocated, is abdication of legal management responsibilities by federal land managers."[22] Federal employees are limited in the ways that they participate in collaborative projects precisely to ensure that they are held accountable to Congress. Federal agencies do not have complete freedom to negotiate compromises and make long-term promises. They have certain binding legal mandates that involve potentially coercive power, and they answer to congresses and presidents that change regularly. This is not to say that federal agencies can't participate in collaborative programs, it is simply to say that a future paradigm for federal land and resource management will need legislative support that defines the agencies' discretionary limits in collaborative governance and identifies ways to hold the agencies accountable.

In the meantime, federal agencies continue to pursue a variety of initiatives that began as ecosystem management projects. With congressional endorsement, the National Park Service and the US Fish and Wildlife Service continue to work with other federal agencies, state agencies, non-governmental organizations, and individuals to restore the Florida Everglades; a variety of federal agencies are engaged in a watershed restoration project for the Chesapeake Bay; and a number of federal agencies have played a role in the Klamath Basin Restoration Agreement that addresses contentious water rights and endangered species issues within the Klamath River Basin in California and Oregon. One could also look back at the Greater Yellowstone Ecosystem. Though the agencies abandoned the more ambitious vision document and with it hopes of integrated management of the ecosystem as a whole, the Park Service and Forest Service continue to work on ecosystem-wide issues outlined in chapter three, including bison and wolf management.[23] Or one could look within the boundaries of the Interior Columbia Basin Ecosystem Management Project where the BLM is playing a significant role in voluntary efforts to protect the sage grouse. It is true that one can find a regulatory mandate or threat behind most of these projects, whether it is the ESA or the Clean Water Act. Nonetheless, these projects involve real collaborative governance.

Perhaps in the end, then, "ecosystem management" as it developed during the Clinton administration did not endure as a controlling paradigm for federal land and resource management, but it nonetheless marked a meaningful turn in federal land and resource management. The new directions aren't necessarily predictable in a substantive sense of ensuring greater biodiversity protection or healthier public ranges, but they are predictable in a procedural sense of federal agencies engaging in more "multijurisdictional, multipurpose, and multistakeholder" negotiations at a wider range of geographic scales. What seems predictable, then, is process.

Notes

Introduction

1. Lynton K. Caldwell, "Environment: A New Focus for Public Policy?," *Public Administration Review* 23, no. 3 (1963): 136.
2. Andy Kerr, "Ecosystem Management Must Include the Most Human of Factors," *BioScience* 45, no. 6 (1995): 378.
3. R. Edward Grumbine, "What Is Ecosystem Management?," *Conservation Biology* 8, no. 1 (1994): 31. (Quote italicized in the original.)
4. Ibid., 34.
5. Hanna J. Cortner and Margaret A. Moote, *The Politics of Ecosystem Management* (Washington, DC: Island Press, 1999), 40.
6. IEMTF, "The Ecosystem Approach: Healthy Ecosystems *and* Sustainable Economies," (Washington, DC: GPO, 1995), 1.
7. Ibid.
8. Mike Dombeck, Christopher Wood, and Jack E. Williams, *From Conquest to Conservation: Our Public Lands Legacy* (Washington, DC: Island Press, 2003), 157.
9. Allan K. Fitzsimmons, *Defending Illusions: Federal Protection of Ecosystems* (New York: Rowman & Littlefield, 1999).
10. Martin Nie, "Whatever Happened to Ecosystem Management and Federal Land Planning?," in *The Laws of Nature: Reflections on the Evolution of Ecosystem Management Law and Policy*, ed. Kalyani Robbins (Akron, OH: University of Akron 2013), 67–94; G. Brown and C. C. Harris, "The US Forest Service: Whither the New Resource Management Paradigm?," *Journal of Environmental Management* 58, no. 1 (2000). Greg Brown and Charles C. Harris, "The U.S. Forest Service: Toward the New Resource Management Paradigm?," *Society and Natural Resources* 5 (1992): 231–45.
11. Robert B. Keiter, "Toward Legitimizing Ecosystem Management on the Public Domain," *Ecological Applications* 6, no. 3 (1996): 728. See also "Beyond the Boundary Line: Constructing a Law of Ecosystem Management," *Colorado Law Review* 1994 (Spring): 293–333; *Keeping Faith with Nature: Ecosystems, Democracy, and America's Public Lands* (New Haven, CT: Yale University Press, 2003).
12. Senator Mark Hatfield (R-OR) introduced an ecosystem management act in 1994, which would have amended BLM's Federal Land Policy and Management Act of 1976, but this never gained traction.
13. Bryan G. Norton, *Sustainability: A Philosophy of Adaptive Ecosystem Management* (Chicago: University of Chicago Press, 2005); Samuel P. Hays, *Wars in the Woods: The Rise of Ecological Forestry in America* (Pittsburgh: University of Pittsburgh Press, 2007); Daniel W. Schneider, "Environmental History and Ecosystem Management," *Environmental History* 10, no. 1 (2005): 89–90.

14. 42 USC §§4321-4347 (2012).

15. Thomas R. Dunlap, *Faith in Nature: Environmentalism as Religious Quest* (Seattle: University of Washington Press, 2004).

Chapter 1. The Intellectual Context of Ecosystem Management

1. Dean Bavington, "Managerial Ecology and Its Discontents: Exploring the Complexities of Control, Careful Use, and Coping in Resource and Environmental Management," *Environments* 30 (2002): 3–20.

2. "Plant community" refers simply to all of the plants in a given area that interact with one another. "Plant association" refers to a particular, recurring plant community, such as an oak-hickory forest or a short-grass prairie. William D. Rowley, "Historical Considerations in the Development of Range Science," in *Forest and Wildlife Science in America: A History*, ed. Harold K. Steen (Durham, NC: Forest History Society, 1999), 230–60.

3. Quoted in Robert P. McIntosh, "The Succession of Succession: A Lexical Chronology," *Bulletin of the Ecological Society of America* 80, no. 4 (1999): 257.

4. Donald Worster, *Nature's Economy: A History of Ecological Ideas*, 2nd ed. (New York: Cambridge University Press, 1994).

5. Rowley, "Historical Considerations in the Development of Range Science," 243.

6. Frederic E. Clements, *Research Methods in Ecology* (Lincoln, NE: University Publishing, 1905).

7. Quoted in Sharon E. Kingsland, *The Evolution of American Ecology, 1890–2000* (Baltimore, MD: Johns Hopkins University Press, 2005), 151.

8. Mark Westoby, Brian Walker, and Imanuel Noy-Meir, "Opportunistic Management for Rangelands Not at Equilibrium," *Journal of Range Management* 42, no. 4 (1989): 266.

9. One of Clements's students, Arthur Sampson, worked tirelessly to advance this perspective in the US Department of Agriculture and later at U.C. Berkeley: Arthur W. Sampson, "Plant Succession as a Factor in Range Management," *Journal of Forestry* 4 (May 1917): 593–96; "Plant Succession in Relation to Range Management" (Washington, DC: US Department of Agriculture, 1919).

10. Rowley, "Historical Considerations in the Development of Range Science," 241.

11. Stephen G. Boyce and Chadwick D. Oliver, "The History of Research in Forest Ecology and Silviculture," in *Forest and Wildlife Science in America*, ed. Harold K. Steen (Durham, NC: Forest History Society, 1999), 414–53.

12. A. G. Tansley, "The Use and Abuse of Vegetational Concepts and Terms," *Ecology* 16, no. 3 (1935): 293.

13. Ibid., 306.

14. Michael G. Barbour, "Ecological Fragmentation in the Fifties," in *Uncommon Ground: Toward Reinventing Nature*, ed. William Cronon (New York: W. W. Norton, 1995), 237; Malcolm Nicolson, Robert P. McIntosh, and M. Nicholson,

"H. A. Gleason and the Individualistic Hypothesis Revisited," *Bulletin of the Ecological Society of America* 83, no. 2 (2002): 133–42.

15. H. A. Gleason, "The Individualistic Concept of the Plant Association," *Bulletin of the Torrey Botanical Club* 53 (1926): 7–26; "The Individualistic Concept of the Plant Association," *American Midland Naturalist* 21 (1939): 92–110; "The Structure and Development of the Plant Association," *Bulletin of the Torrey Botanical Club* 44 (1917): 463–81.

16. In 1953, the Ecological Society of America granted its first distinguished ecologist award to Gleason in recognition of his pioneering work. Barbour, "Ecological Fragmentation in the Fifties," 238. James Lovelock's Gaia Hypothesis, later termed Gaia Theory, is a prominent example of continued organismic thinking in the second half of the twentieth century. James E. Lovelock and Lynn Margulis, "Atmospheric Homeostasis by and for the Biosphere: The Gaia Hypothesis," *Tellus* 26, no. 1-2 (1974): 2–10; James E. Lovelock, "Gaia as Seen through the Atmosphere," *Atmospheric Environment* 6 (1972): 579–80.

17. Kingsland, *The Evolution of American Ecology, 1890–2000.*

18. E. F. Haskell, quoted in Robert P. McIntosh, "The Background and Some Current Problems of Theoretical Ecology," *Synthese* 43, no. 2 (1980): 214.

19. Eugene P. Odum, "The New Ecology," *BioScience* 14, no. 7 (July 1964): 14.

20. Donato Bergandi, "'Reductionistic Holism': An Oxymoron or a Philosophical Chimera of Eugene Odum's Systems Ecology?," in *The Philosophy of Ecology: From Science to Synthesis*, ed. David R. Keller and Frank B. Golley (Athens: University of Georgia Press, 2000), 204–17.

21. Eugene P. Odum, *Fundamentals of Ecology* (Philadelphia: Saunders, 1953).

22. 42 USC 4321-4347.

23. Robert H. Whittaker, "A Criticism of the Plant Association and Climatic Climax Concept," *Northwest Science* 25, no. 1 (1951): 17–31.

24. Robert K. Colwell, "The Evolution of Ecology," *American Zoologist* 25, no. 3 (1985): 771–77.

25. Robert K. Colwell, "What's New? Community Ecology Discovers Biology," in *A New Ecology: Novel Approaches to Interactive Systems*, ed. Peter W. Price, C. N. Slobodchikoff, and William S. Gaud (New York: John Wiley & Sons, 1984), 393.

26. McIntosh, "The Background and Some Current Problems of Theoretical Ecology," 204; Peter W. Price, C. N. Slobodchikoff, and William S. Gaud, eds., *A New Ecology: Novel Approaches to Interactive Systems* (New York: John Wiley & Sons, 1984).

27. Colwell, "What's New? Community Ecology Discovers Biology"; "The Evolution of Ecology."

28. It should be noted that later editions of the Odums' text placed less emphasis on the balance of nature.

29. Colwell, "What's New? Community Ecology Discovers Biology," 392.

30. Frank E. Egler, *The Nature of Vegetation: Its Management and Mismanagement: An Introduction to Vegetation Science* (Norfolk, CT: Aton Forest, 1977).

31. Herbert A. Simon, "The Architecture of Complexity," *Proceedings of the American Philosophical Society* 106, no. 6 (December 1962): 468.

32. Ibid.

33. Herbert A. Simon, "The Organization of Complex Systems," in *Hierarchy Theory: The Challenge of Complex Systems*, ed. Howard H. Pattee (New York: George Braziller, 1973), 5.

34. T. F. R. Allen and Thomas B. Starr, *Hierarchy: Perspectives for Ecological Complexity* (Chicago: University of Chicago Press, 1982).

35. Richard Levins, "The Limits of Complexity," in *Hierarchy Theory: The Challenge of Complex Systems*, ed. Howard H. Pattee (New York: George Braziller, 1973), 113.

36. Simon Levin, "The Problem of Pattern and Scale in Ecology: The Robert H. MacArthur Award Lecture," *Ecology* 73, no. 6 (December 1992): 1959.

37. John Holland, *Hidden Order: How Adaptation Builds Complexity* (Reading, MA: Addison-Wesley, 1995).

38. C. S. Holling, "Resilience and Stability of Ecological Systems," *Annual Review of Ecological Systems* 4 (1973): 1–2.

39. It should be noted that ecologists did not always share clear definitions of terms like stability and resilience.

40. Craig R. Allen and C. S. Holling, eds., *Discontinuities in Ecosystems and Other Complex Systems* (New York: Columbia University Press, 2008); Lance H. Gunderson and C. S. Holling, eds., *Panarchy: Understanding Transformations in Human and Natural Systems* (Washington, DC: Island Press, 2002); Lance H. Gunderson, C. S. Holling, and Stephen S. Light, eds., *Barriers and Bridges to the Renewal of Ecosystems and Institutions* (New York: Columbia University Press, 1995).

41. C. S. Holling, "Understanding the Complexity of Economic, Ecological, and Social Systems," *Ecosystems* 4, no. 5 (August 2001): 391.

42. C. S. Holling, "Surprise for Science, Resilience for Ecosystems, and Incentives for People," *Ecological Applications* 6, no. 3 (August 1996): 733.

43. C. S. Holling and Gary K. Meffe, "Command and Control and the Pathology of Natural Resource Management," *Conservation Biology* 10, no. 2 (April 1996): 329.

44. Ibid., 330.

45. Holling, "Resilience and Stability of Ecological Systems," 21.

46. Holling and Meffe, "Command and Control and the Pathology of Natural Resource Management," 329, 30, and 34. Italics in the original.

47. Carl J. Walters and Ray Hilborn, "Ecological Optimization and Adaptive Management," *Annual Review of Ecology and Systematics* 9 (1978): 157.

48. Ibid.

49. C. Walters and C. S. Holling, "Large-Scale Management Experiments and Learning by Doing," *Ecology* 71, no. 6 (December 1990): 2060.

50. Elliot Norse, "A River That Flows to the Sea: The Marine Biological Diversity Movement," *Oceanography* 9, no. 1 (1996): 5.

51. Timothy J. Farnham, *Saving Nature's Legacy: Origins of the Idea of Biological Diversity* (New Haven, CT: Yale University Press, 2007), 12. Italics in the original.

52. See, for example, Edward O. Wilson and Frances M. Peter, eds., *Biodiversity* (Washington, DC: National Academies Press, 1988).

53. Paul R. Ehrlich and Anne H. Ehrlich, *Extinction: The Causes and Consequences of the Disappearance of Species* (New York: Random House, 1981).

54. Farnham, *Saving Nature's Legacy: Origins of the Idea of Biological Diversity*, 5.

55. Michael E. Soulé, "Conservation Biology and the 'Real World,'" in *Conservation Biology: The Science of Scarcity and Diversity*, ed. Michael E. Soulé (Sunderland, MA: Sinauer Associates, 1986), 6.

56. Gary K. Meffe, ed., *Ecosystem Management: Adaptive, Community-Based Conservation* (Washington, DC: Island Press, 2002), 84–85.

57. Stuart L. Pimm, *The Balance of Nature? Ecological Issues in the Conservation of Species and Communities* (Chicago: University of Chicago Press, 1991).

58. Mark L. Shaffer, "Minimum Population Sizes for Species Conservation," *BioScience* 31, no. 2 (February 1981): 132.

59. Michael E. Gilpin and Michael E. Soulé, "Minimum Viable Populations: Processes of Species Extinction," in *Conservation Biology: The Science of Scarcity and Diversity*, ed. Michael E. Soulé (Sunderland, MA: Sinauer Associates, 1986), 19–34; Shaffer, "Minimum Population Sizes for Species Conservation."

60. Gilpin and Soulé, "Minimum Viable Populations: Processes of Species Extinction."

61. Bruce G. Marcot and Dennis D. Murphy, "On Population Viability Analysis and Management," in *Biodiversity in Managed Landscapes: Theory and Practice*, ed. Robert C. Szaro and David W. Johnston (New York: Oxford University Press, 1996), 58–76.

62. 36 CFR 219.19 (1982).

63. Jack Ward Thomas and Hal Salwasser, "Bringing Conservation Biology into a Position of Influence in Natural Resource Management," *Conservation Biology* 3, no. 2 (June 1989): 123–27; Hal Salwasser, "Spotted Owls: Turning a Battleground into a Blueprint," *Ecology* 68, no. 4 (August 1987): 776–79; Salwasser, "Editorial," *Conservation Biology* 1, no. 4 (December 1987): 275–77.

64. Reed F. Noss, "Issues of Scale in Conservation Biology," in *Conservation Biology: The Theory and Practice of Nature Conservation, Preservation, and Management*, ed. Peggy L. Fiedler and Subodh K. Jain (New York: Chapman and Hill, 1992), 241.

65. Ibid., 248.

66. Jerry F. Franklin, "Preserving Biodiversity: Species, Ecosystems, or Landscapes?," *Ecological Applications* 3, no. 2 (1993): 202.

67. Peggy L. Fiedler et al., "The Contemporary Paradigm in Ecology and Its Implications for Endangered Species Conservation," *Endangered Species Update* 10, no. 3 & 4 (January/February 1993): 7.

68. Dean L. Urban, "Landscape Ecology and Ecosystem Management," in *Sustainable Ecological Systems: Implementing an Ecological Approach to Land Management*, ed. W. Wallace Covington and Leonard F. Debano (Flagstaff, AZ: USDA Forest Service Rocky Mountain Forest and Range Experiment Station, 1993), 128.

69. Reed F. Noss, "Some Principles of Conservation Biology, as They Apply to Environmental Law," *Chicago-Kent Law Review* 69, no. 4 (1994): 893–909.

70. In addition to the individual footnotes below, the author acknowledges a debt to Paul Culhane. Through conversations and unpublished writing, he contributed significantly to this section. He should not, however, be held accountable for any errors in the pages that follow.

71. Quoted in Leonard D. White, *The Republican Era: 1869–1901: A Study in Administrative History* (New York: MacMillan, 1958), 299.

72. Herbert Kaufman, "Emerging Conflicts in the Doctrines of Public Administration," *American Political Science Review* 50, no. 4 (December 1956): 1057–74.

73. Woodrow Wilson, "The Study of Administration," *Political Science Quarterly* 2, no. 2 (June 1887): 210.

74. Brian J. Cook, *Bureaucracy and Self-Government: Reconsidering the Role of Public Administration in American Politics* (Baltimore, MD: Johns Hopkins University Press, 1996).

75. Frank Goodnow, *Politics and Administration* (New York: Macmillan, 1900).

76. H. George Frederickson, "The Lineage of New Public Administration," *Administration and Society* 8, no. 2 (August 1976): 149–74.

77. Samuel P. Hays, *Conservation and the Gospel of Efficiency: The Progressive Conservation Movement, 1890–1920* (Cambridge, MA: Harvard University Press, 1959).

78. Frederick Taylor, *Principles of Scientific Management* (New York: Harper and Sons Publishers, 1911), 5.

79. Daniel Nelson, *Frederick W. Taylor and the Rise of Scientific Management* (Madison: University of Wisconsin Press, 1980).

80. Luther Gulick, "Politics, Administration, and the 'New Deal,'" *Annals of the American Academy of Political and Social Science* 169 (September 1933): 63.

81. Ibid., 64.

82. Vincent Ostrom, *The Intellectual Crisis in American Public Administration*, 3rd ed. (Tuscaloosa: University of Alabama Press, 2008), 36.

83. Luther Gulick and Lyndall Urwick, eds., *Papers on the Science of Administration* (New York: Columbia University, Institute of Public Administration, 1937).

84. Luther Gulick, *American Forest Policy: A Study of Government Administration and Economic Control* (New York: Duell, Sloan, and Pearce, 1951), 214.

85. Ibid., 231.

86. Frederickson, "The Lineage of New Public Administration," 153.

87. One can read Gulick's response in a number of publications, including "Management Is a Science," *Academy of Management Journal* 8, no. 1 (March 1965): 7–14.

88. Herbert A. Simon, "Proverbs of Administration," *Public Administration Review* 6, no. 1 (Winter 1946): 53.

89. James G. March and Herbert Simon, *Organizations* (New York: Wiley, 1958), 29.

90. Herbert Simon, *Administrative Behavior: A Study of Decision-Making Processes in Administrative Organization* (New York: Macmillan, 1947), 2.

91. March and Simon, *Organizations*, 84. Italics in the original.

92. Ostrom, *The Intellectual Crisis in American Public Administration*, 37.

93. March and Simon, *Organizations*, 140-41.

94. Herbert Kaufman, "Music of the Spheres: A Lifetime of Study of Public Administration," *Public Administration Review* 56, no. 2 (March-April 1996): 127–38.

95. Ostrom, *The Intellectual Crisis in American Public Administration,* 133.

96. Terence J. Tipple and J. Douglas Wellman, "Herbert Kaufman's Forest Ranger Thirty Years Later: From Simplicity and Homogeneity to Complexity and Diversity," *Public Administration Review* 51, no. 5 (September-October 1991): 422–24.

97. Herbert Kaufman, "Organizational Theory and Political Theory," *American Political Science Review* 58, no. 1 (March 1964): 7.

98. Ibid., 12.

99. Philip Selznick, *TVA and the Grass Roots* (Berkeley: University of California Press, 1949), 15.

100. Charles E. Lindblom, "The Science of Muddling Through," *Public Administration Review* 19, no. 2 (Spring 1959): 80.

101. Ibid., 80, 84.

102. Ibid., 80.

103. Phillip O. Foss, *Politics and Grass: The Administration of the Public Domain* (Seattle: University of Washington Press, 1960); Grant McConnell, *Private Power and American Democracy* (New York: Vintage Books, 1966).

104. David Frederick Paulsen, "An Approach to Organizational Analysis: A Case Study of the Bureau of Land Management" (PhD diss., University of Washington, 1966); George A. Shipman and David F. Paulsen, "The Bureau of Land Management: United States Department of the Interior: The Organization of the Washington Office" (Seattle: University of Washington: Institute for Administrative Research, 1963); Shipman and Paulsen, "The Bureau of Land Management: United States Department of the Interior: The Program Action System" (Seattle: University of Washington: Institute for Administrative Research, 1964).

105. Ostrom, *The Intellectual Crisis in American Public Administration,* 53.

106. Ibid., 58.

107. Ibid., 61.

108. Ibid., 63.

109. H. George Frederickson, *New Public Administration* (Tuscaloosa: University of Alabama Press, 1980), 5.

110. Laurent Dobuzinskis, "Historical and Epistemological Trends in Public Administration," *Journal of Management History* 3, no. 4 (1997): 298–316.

111. Frederickson, *New Public Administration,* 8.

112. Ibid., 69.

113. Ibid., 60.

114. Harold D. Lasswell, *Politics: Who Gets What, When, How* (New York: Peter Smith, 1950).

115. Theodore J. Lowi, "Four Systems of Policy, Politics and Choice," *Public Administration Review* 4 (July/August 1972): 298–310.

116. Theodore J. Lowi, *The End of Liberalism: The Second Republic of the United States,* 2nd ed. (New York: W. W. Norton, 1979), 62. Italics in the original.

117. Paul A. Sabatier, ed. *Theories of the Policy Process* (Boulder, CO: Westview Press, 1999).

118. Karl S. Zimmerer, "Human Geography and the 'New Ecology': Prospect and Promise of Integration," *Annals of the Association of American Geographers* 84, no. 1 (March 1994): 108–25.

119. Quoted in Dobuzinskis, "Historical and Epistemological Trends in Public Administration," 311.

120. James McCarthy, "Neoliberalism and the Politics of Alternatives: Community Forestry in British Columbia and the United States," *Annals of the Association of American Geographers* 96, no. 1 (2006): 100.

121. H. George Frederickson, "Whatever Happened to Public Administration? Governance, Governance Everywhere" (Belfast: Institute of Governance, Public Policy, and Social Research, 2004), 3.

122. Quoted in ibid.

123. Walter J. M. Kickert, "Public Governance in the Netherlands: An Alternative to Anglo-American 'Managerialism,'" *Public Administration* 75, no. 4 (Winter 1997): 735, 45.

124. Lester M. Salamon, ed. *The Tools of Government: A Guide to the New Governance* (New York: Oxford University Press, 2002), 41.

125. John DeWitt et al., "What Will New Governance Mean for the Federal Government," *Public Administration Review* 54, no. 2 (March-April 1994): 170–75.

126. Zhiyong Lan and David H. Rosenbloom, "Public Administration in Transition?," *Public Administration Review* 52, no. 6 (November-December 1992): 535–37; James R. Thompson, "Reinvention as Reform: Assessing the National Performance Review," *Public Administration Review* 60, no. 6 (November–December 2000): 508–21.

127. Walters and Holling, "Large-Scale Management Experiments and Learning by Doing."

128. Ronald J. Herring, "Authority and Scale in Political Ecology: Some Cautions on Localism," in *Biological Diversity: Balancing Interests through Adaptive Collaborative Management*, ed. Louise E. Buck et al. (New York: CRC Press, 2001), 187–204; Judith A. Layzer, *Natural Experiments: Ecosystem-Based Management and the Environment* (Cambridge, MA: MIT Press, 2008).

Chapter 2. The Policy Context of Ecosystem Management

1. Yellowstone Act of 1872, 16 USC §21 (2012).

2. Samuel Trask Dana and Sally K. Fairfax, *Forest and Range Policy: Its Development in the United States*, 2nd ed. (New York: McGraw-Hill, 1980); Robert L. Fischman, "The National Wildlife Refuge System and the Hallmarks of Modern Organic Legislation," *Ecological Law Quarterly* 29 (2002): 457–544; Alfred Runte, *National Parks: The American Experience* (Lincoln: University of Nebraska Press, 1979); James R. Skillen, *The Nation's Largest Landlord: The Bureau of Land Management in the American West* (Lawrence: University Press of Kansas, 2009).

3. Quoted in Gifford Pinchot, *The Fight for Conservation* (New York: Doubleday, Page, 1910), 233.

4. Brigitte W. Johnson, "The American Tree Farm System: Growing Stewardship from the Roots," *Forest History Today* 17, no. 1 (Spring 2012): 17.

5. 16 USC § 475 (2012).

6. Gifford Pinchot, "The Use of the National Forest Reserves: Regulations and Instructions" (Washington, DC: US Department of Agriculture, 1905), 11.

7. Pinchot, *The Fight for Conservation*, 42.

8. Timothy Egan, *The Big Burn: Teddy Roosevelt and the Fire That Saved America* (New York: Houghton Mifflin Harcourt, 2009); Stephen J. Pyne, *Year of the Fires: The Story of the Great Fires of 1910* (New York: Viking, 2001); Harold K. Steen, *The U.S. Forest Service: A History* (Seattle: University of Washington Press, 1976).

9. Paul W. Hirt, *A Conspiracy of Optimism: Management of the National Forests since World War Two* (Lincoln: University of Nebraska Press, 1994).

10. Public Law 86-517, 1960.

11. Hirt, *A Conspiracy of Optimism.*

12. Horace M. Albright and Robert Cahn, *The Birth of the National Park Service: The Founding Years, 1913–33* (Chicago: Howe Brothers, 1985), 32.

13. 16 USC §1 (2012).

14. Joseph Sax, *Mountains without Handrails: Reflections on the National Parks* (Ann Arbor: University of Michigan Press, 1980).

15. Hal Rothman, "'A Regular Ding-Dong Fight': Agency Culture and Evolution in the NPS-USFS Dispute, 1916–1937," *Western Historical Quarterly* 20, no. 2 (1989): 149.

16. Alfred Runte, *Trains of Discovery*, 4th ed. (Boulder, CO: Roberts Rinehart Publishers, 1998).

17. Steen, *The U.S. Forest Service: A History*; Albright and Cahn, *The Birth of the National Park Service*, 32–52.

18. Timothy Davis, Todd A. Croteau, and Christopher H. Marston, eds., *America's National Park Roads and Parkways: Drawings from the Historic American Engineering Record* (Baltimore, MD: Johns Hopkins University Press, 2004); Rothman, "'A Regular Ding-Dong Fight.'"

19. Robert L. Fischman, *The National Widlife Refuges: Coordinating a Conservation System through Law* (Washington, DC: Island Press, 2003), 213.

20. Anders Halverson, *An Entirely Synthetic Fish: How Rainbow Trout Beguiled America and Overran the World* (New Haven, CT: Yale University Press, 2011): 13–15.

21. Donald W. Hawthorne, "The History of Federal and Cooperative Animal Damage Control," *Sheep & Goat Research Journal* 19 (2004): 13–15.

22. Donald Worster, *Nature's Economy: A History of Ecological Ideas*, 2nd ed. (New York: Cambridge University Press, 1994), 258–90.

23. Hawthorne, "The History of Federal and Cooperative Animal Damage Control."

24. Worster, *Nature's Economy*, 260; See also Thomas R. Dunlap, *Saving America's Wildlife* (Princeton, NJ: Princeton University Press, 1988).

25. Fischman, *The National Wildlife Refuges*, 32–63.

26. As amended, 16 USC §742a (2012).

27. E. Louise Peffer, *The Closing of the Public Domain: Disposal and Reservation Policies 1900–50* (Stanford, CA: Stanford University Press, 1951).

28. Charles E. Winter, *Four Hundred Million Acres: The Public Lands and Resources* (New York: Arno Press, 1979), 187.

29. Karen R. Merrill, *Public Lands and Political Meaning: Ranchers, the Government, and the Property between Them* (Berkeley: University of California Press, 2002); Skillen, *The Nation's Largest Landlord*, 1–13.

30. Skillen, *The Nation's Largest Landlord*, 18.

31. Thomas R. Dunlap, *Faith in Nature: Environmentalism as Religious Quest* (Seattle: University of Washington Press, 2004), 95.

32. Samuel P. Hays, "From Conservation to Environment: Environmental Politics in the United States since World War II," *Environmental Review* 6 (Fall 1982): 21; *A History of Environmental Politics since 1945* (Pittsburgh: University of Pittsburgh Press, 2000).

33. Marion Clawson, "The Crisis in Outdoor Recreation," *American Forests* 65, no. 3 (1959): 22–31, 40–41; "The Crisis in Outdoor Recreation, Part II," *American Forests* 65, no. 4 (1959): 28–35, 61–62.

34. Marion Clawson and Carlton S. Van Doren, eds., *Statistics on Outdoor Recreation* (Washington, DC: Resources for the Future, 1984), 23, 42, 177, 97.

35. Dunlap, *Saving America's Wildlife*, xii.

36. Hays, "From Conservation to Environment."

37. 16 USC §1131.

38. Michael Egan, *Barry Commoner and the Science of Survival: The Remaking of American Environmentalism* (Cambridge, MA: MIT Press, 2007).

39. Samuel P. Hays, *Beauty, Health and Permanence: Environmental Politics in the United States, 1955–1985* (New York: Cambridge University Press, 1987), 171–206.

40. Rachel Carson, *Silent Spring* (New York: Houghton Mifflin, 1962).

41. Egan, *Barry Commoner and the Science of Survival.*

42. Quoted in Gerald Garvey, "False Promises: The NPR in Historical Perspective," in *Inside the Reinvention Machine: Appraising Governmental Reform*, ed. Donald F. Kettl and John J. DiIulio, Jr. (Washington, DC: Brookings Institution, 1995), 97.

43. Gail L. Achterman and Sally K. Fairfax, "The Public Participation Requirements of the Federal Land Policy and Management Act," *Arizona Law Review* 21, no. 2 (1979): 512.

44. Indeed, the commission only recommended that Congress repeal one major body of federal land law, namely the extensive nineteenth and early twentieth-century provisions that Congress had developed to dispose of unreserved public lands now managed by the BLM.

45. "One Third of the Nation's Land: A Report to the President and to Congress by the Public Land Law Review Commission" (Washington, DC: Public Land Law Review Commission, 1970), 2–3.

46. Ibid., 6.

47. Ibid., 2–3.
48. 16 USC §1604 (2012).
49. 43 USC §1701 (2012).
50. Steven C. Schulte, *Wayne Aspinall and the Shaping of the American West* (Boulder: University Press of Colorado, 2002), 241.
51. Lynton K. Caldwell, "Environment: A New Focus for Public Policy?," *Public Administration Review* 23, no. 3 (1963): 134–36.
52. Lynton K. Caldwell, "An Ecosystems Approach to Public Land Policy," in *Public Land Policy: Proceedings of the Western Resources Conference*, ed. Phillip O. Foss (Boulder: Colorado Associated University Press, 1968), 43, 45.
53. Lynton K. Caldwell, "Politics and Public Land Policy," in *America's Public Lands: Politics, Economics and Administration: Conference on the Public Land Law Review Commission Report*, ed. Harriet Nathan (San Francisco: Institute of Governmental Studies, University of California, Berkeley, 1970), 310.
54. The National Forest Management Act of 1976 also played a significant role in the rise of ecosystem management, but it only applies to the US Forest Service. I single out NEPA and ESA because they apply to all federal agencies.
55. Quoted in Linda Luther, "The National Environmental Policy Act: Background and Implementation" (Washington, DC: Congressional Research Service, 2005), 3.
56. Quoted in Matthew J. Lindstrom and Zachary A. Smith, *The National Environmental Policy Act: Judicial Misconstruction, Legislative Indifference, and Executive Neglect* (College Station: Texas A&M University Press, 2001), 24–25.
57. Quoted in Jayne E. Daly, "A Glimpse of the Past—A Vision for the Future: Senator Henry M. Jackson and National Land-Use Legislation," *Urban Lawyer* 28, no. 1 (Winter 1996): 7.
58. Lynton K. Caldwell, "The Ecosystem as a Criterion for Public Land Policy," *Natural Resources Journal* 10, no. 2 (April 1970): 208.
59. 42 USC §4321 (2012).
60. 42 USC §4332 (2012).
61. J. Brooks Flippen, *Nixon and the Environment* (Albuquerque: University of New Mexico Press, 2000), 50–51.
62. Lynton K. Caldwell, *The National Environmental Policy Act: An Agenda for the Future* (Bloomington: Indiana University Press, 1998); Lindstrom and Smith, *The National Environmental Policy Act.*
63. 449 F.2d 1112 (D.C. Cir. 1971).
64. Caldwell, *The National Environmental Policy Act*, xvi.
65. Shannon Peterson, *Acting for Endangered Species: The Statutory Ark* (Lawrence: University Press of Kansas, 2002), 22.
66. Brian Czech and Paul R. Krausman, *The Endangered Species Act: History, Conservation, Biology, and Public Policy* (Baltimore, MD: Johns Hopkins University Press, 2001), 21.
67. Quoted in Peterson, *Acting for Endangered Species*, 27.
68. 16 USC §1531.
69. The act exempts from protection "species of the Class Insecta determined by the Secretary to constitute a pest whose protection under the provisions of

this Act would present an overwhelming and overriding risk to man." 16 USC §1532.

70. 16 USC §1536.

71. 16 USC §1538.

72. Quoted in Peterson, *Acting for Endangered Species*, 30.

73. Ibid., 29–35.

74. Ibid., 33.

75. Ibid., 33–35.

76. Holly Doremus, "The Endangered Species Act: Static Law Meets Dynamic World," *Washington University Journal of Law & Policy* 32 (2010): 182; Peterson, *Acting for Endangered Species*, 34.

77. *TVA v. Hill*, 437 U.S. 153 (1978) and *Babbitt v. Sweet Home Chapter of Communities for a Greater Oregon*, 515 U.S. 687 (1995).

78. Peterson, *Acting for Endangered Species.*

79. Hirt, *A Conspiracy of Optimism*, 247.

80. Quoted in ibid., 249.

81. See *Izaak Walton League v. Butz* 522 F.2d 945 (4th Cir. 1975).

82. Samuel P. Hays, *Wars in the Woods: The Rise of Ecological Forestry in America* (Pittsburgh: University of Pittsburgh Press, 2007), 16–19.

83. 16 USC §1604.

84. 36 CFR §219.19 (1982).

85. A. S. Leopold et al., "Wildlife Management in the National Parks" (Washington, DC: Department of the Interior, 1963).

86. Ibid.

87. Quoted in James A. Pritchard, *Preserving Yellowstone's Natural Conditions: Science and the Perception of Nature* (Lincoln: University of Nebraska Press, 1999), 216.

88. Quoted in ibid., 218.

89. Quoted in ibid.

90. John C. Freemuth, *Islands under Siege: National Parks and the Politics of External Threats* (Lawrence: University Press of Kansas, 1991), 2.

91. Ibid.

92. James K. Agee and Darryll R. Johnson, eds., *Ecosystem Management for Parks and Wilderness* (Seattle: University of Washington Press, 1988).

93. Jeanne Nienaber Clarke and Daniel C. McCool, *Staking Out the Terrain: Power and Performance among Natural Resource Agencies* (Albany: State University of New York Press, 1996), 111; Fischman, "The National Wildlife Refuge System and the Hallmarks of Modern Organic Legislation," 477.

94. 16 USC 460a (2012).

95. P.L. 89-669, 1966.

96. Quoted in Dunlap, *Saving America's Wildlife*, 129.

97. Ibid., 127–41.

98. Executive Order 11643, February 8, 1972.

99. Clarke and McCool, *Staking Out the Terrain*, 116–17.

100. Ibid., 117.

101. Kristina Alexander, "Warranted but Precluded: What That Means under

the Endangered Species Act (ESA)" (Washington, DC: Congressional Research Service, 2010).

102. Clarke and McCool, *Staking Out the Terrain*, 119.

103. Skillen, *The Nation's Largest Landlord*, 38–71.

104. *Wilderness Society v. Hickel*, 325 F.Supp. 422 (D.C. 1970); *NRDC v. Morton* 458 F.2d 827 (D.C. Cir. 1972); and *NRDC v. Morton* 388 F.Supp. 829 (D.C. 1974).

105. R. McGreggor Cawley, *Federal Land, Western Anger: The Sagebrush Rebellion and Environmental Politics* (Lawrence: University Press of Kansas, 1993); Christopher James Klyza, *Who Controls Public Lands?: Mining, Forestry, and Grazing Policies, 1870–1990* (Chapel Hill: University of North Carolina Press, 1996); Robert F. Durant, *The Administrative Presidency Revisited: Public Lands, the BLM, and the Reagan Revolution* (Albany: State University of New York Press, 1992).

Chapter 3. Yellowstone: The Ecosystem Management Laboratory

1. Paul Schullery, *Searching for Yellowstone: Ecology and Wonder in the Last Wilderness* (New York: Houghton Mifflin, 1997), 2, 5.

2. Robert B. Keiter and Mark S. Boyce, eds., *The Greater Yellowstone Ecosystem: Redefining America's Wilderness Heritage* (New Haven, CT: Yale University Press, 1991).

3. Schullery, *Searching for Yellowstone*, 56; Aubrey Haines, *The Yellowstone Story: A History of Our First National Park, Vol. 1* (Boulder: University Press of Colorado, 1999).

4. Schullery, *Searching for Yellowstone*, 119.

5. Robert Pahre, "Political Opposition to Transboundary Cooperation in the Greater Yellowstone Area," *Journal of Tourism and Leisure Studies* 17, no. 2 (August 2011): 102.

6. Richard West Sellars, *Preserving Nature in the National Parks: A History* (New Haven, CT: Yale University Press, 1997), 70.

7. Ibid., 113–14.

8. Schullery, *Searching for Yellowstone*, 153.

9. Ibid., 154.

10. Sellars, *Preserving Nature in the National Parks*, 155–56.

11. Andrew C. Isenberg, *The Destruction of the Bison* (New York: Cambridge University Press, 2000); Margaret Mary Meagher, "The Bison of Yellowstone National Park" (Washington, DC: US National Park Service, 1973).

12. Isenberg, *The Destruction of the Bison*, 179–86.

13. Ibid., 81.

14. James A. Pritchard, *Preserving Yellowstone's Natural Conditions: Science and the Perception of Nature* (Lincoln: University of Nebraska Press, 1999), 175.

15. Quoted in ibid., 176.

16. Ibid., 182.

17. Quoted in Schullery, *Searching for Yellowstone*, 161.

18. Ibid.

19. Ibid.; Dunlap, *Saving America's Wildlife*, 112–23.

20. Quoted in Pritchard, *Preserving Yellowstone's Natural Conditions,* 120.

21. Ibid., 187.

22. Ibid., 190.

23. Sellars, *Preserving Nature in the National Parks,* 97.

24. A. S. Leopold et al., "Wildlife Management in the National Parks" (Washington, DC: Department of the Interior, 1963).

25. Pritchard, *Preserving Yellowstone's Natural Conditions,* 248.

26. Robert D. Barbee, Paul Schullery, and John D. Varley, "The Yellowstone Vision: An Experiment That Failed or a Vote for Posterity?," in *Partnerships in Parks and Preservation,* conference proceedings (Albany, NY, September 9–12, 1991).

27. Pritchard, *Preserving Yellowstone's Natural Conditions,* 252–53.

28. Cite 40 FR 31734-31736, July 28, 197.

29. S. P. Mealey, ed., *Guidelines for Management Involving Grizzly Bears in the Greater Yellowstone Area* (USDA Forest Service and USDI National Park Service, 1979).

30. Douglas B. Houston, "Ecosystems of National Parks," *Science* 172, no. 3984 (May 14, 1971): 648–51.

31. Pritchard, *Preserving Yellowstone's Natural Conditions,* 259.

32. GYCC, "The Greater Yellowstone Area: An Aggregation of National Park and National Forest Management Plans" (Greater Yellowstone Coordinating Committee, 1987).

33. J. Peterson, "Threats to Greater Yellowstone: Inventory of Management Issues and Proposed Projects in the GYE" (Bozeman, MT: Greater Yellowstone Coalition, 1984).

34. Subcommittee on Public Lands and the Subcommittee on National Parks and Recreation, *Greater Yellowstone Ecosystem,* 99th Congress, 1st Session, October 24, 1985, 3.

35. Ibid., 16, 47.

36. Ibid., 40.

37. Ibid., 35.

38. Ibid., 57, 17.

39. M. Lynne Corn and Ross W. Gorte, "Yellowstone: Ecosystem, Resources, and Management" (Washington, DC: Congressional Research Service, 1986), 30.

40. Ibid., v.

41. GYCC, "The Greater Yellowstone Area."

42. Bruce Goldstein, "The Struggle over Ecosystem Management at Yellowstone," *BioScience* 42, no. 3 (March 1992): 185. This supports Robert Keiter's interpretation of the conflict as involving three distinct parties: environmentalists concerned with biodiversity, groups concerned with resource development, and agencies concerned with protecting their autonomy. Robert B. Keiter, "Taking Account of the Ecosystem on the Public Domain: Law and Ecology in the Greater Yellowstone Region," *Colorado Law Review* 60, no. 4 (1989): 924–1007.

43. GYCC, "Vision for the Future: A Framework for Coordination in the Greater Yellowstone Area" (Washington, DC: GPO, 1990), 1-6.

44. Ibid., 1-6, 3-1. See also goal number two: "Encourage Opportunities That Are Biologically and Economically Sustainable."

45. Ibid., 3-7, 4-2.

46. Ibid., 3-7.

47. Barbee, Schullery, and Varley, "The Yellowstone Vision," 82.

48. Karen J. Budd, "Ecosystem Management: Will National Forests Be 'Managed' into National Parks?," in *The Greater Yellowstone Ecosystem: Redefining America's Wilderness Heritage*, ed. Robert B. Keiter and Mark S. Boyce (New Haven, CT: Yale University Press, 1991), 67.

49. Letter from David B. Rovig of Billings, MT, to Interior Secretary Lujan, August 3, 1990. In "Interference in Environmental Programs by Political Appointees: The Improper Treatment of a Senior Executive Service Official: A Report" (Washington, DC: GPO, 1993), 149.

50. Letter from Senators Alan Wallop and Al Simpson and Congressman Craig Thomas to Interior Secretary Lujan, August 22, 1990. Ibid.

51. Pamela Litchman and Tim W. Clark, "Rethinking the 'Vision' Exercise in the Greater Yellowstone Ecosystem," *Society and Natural Resources* 7, no. 5 (1994): 459–78.

52. Allan K. Fitzsimmons, *Defending Illusions: Federal Protection of Ecosystems* (New York: Rowman & Littlefield, 1999), 81.

53. David Garber, "History of the Greater Yellowstone Coordinating Committee" (Bozeman, MT: Greater Yellowstone Coordinating Committee, 1994).

54. "A Framework for Coordination of National Parks and National Forests in the Greater Yellowstone Area" (Billings, MT: Greater Yellowstone Coordinating Committee, 1991), 4.

55. Barbee, Schullery, and Varley, "The Yellowstone Vision," 82.

56. "Interference in Environmental Programs by Political Appointees," 103.

57. Ibid., 2.

58. Pritchard, *Preserving Yellowstone's Natural Conditions*, 294.

59. Pahre, "Political Opposition to Transboundary Cooperation in the Greater Yellowstone Area," 101.

60. Matthew A. Wilson, "The Wolf in Yellowstone: Science, Symbol, or Politics?," *Society and Natural Resources* 10, no. 5 (2008): 454.

61. Robert B. Keiter, *Keeping Faith with Nature: Ecosystems, Democracy, and America's Public Lands* (New Haven, CT: Yale University Press, 2003), 128–36.

62. Dunlap, *Saving America's Wildlife*.

63. Alistair J. Bath, "Public Attitudes about Wolf Restoration in Yellowstone," in *The Greater Yellowstone Ecosystem: Redefining America's Wilderness Heritage*, ed. Robert B. Keiter and Mark S. Boyce (New Haven, CT: Yale University Press, 1991), 367–76.

64. L. David Mech, "Returning the Wolf to Yellowstone," in *The Greater Yellowstone Ecosystem: Redefining America's Wilderness Heritage*, ed. Robert B. Keiter and Mark S. Boyce (New Haven, CT: Yale University Press, 1991), 309–22.

65. Robert J. Noecker, "Reintroduction of Wolves" (Washington, DC: Congressional Research Service, 1997).

66. Roberta A. Klein, "Wolf Recovery in the Northern Rockies," in *Finding*

Common Ground: Governance and Natural Resources in the American West, ed. Ronald D. Brunner et al. (New Haven, CT: Yale University Press, 2002), 88–125.

67. U.S. Fish and Wildlife Service, "Wolves for Yellowstone? A Report to the United States Congress, Volume I and II" (Billings, MT: US Fish and Wildlife Service, 1990).

68. Noecker, "Reintroduction of Wolves"; Klein, "Wolf Recovery in the Northern Rockies."

69. Subcommittee on Public Lands, National Parks and Forests, *Northern Rocky Mountain Gray Wolf Recovery Act of 1990*, 2nd Session, September 19, 1990.

70. Ibid., 111.

71. Ibid., 147.

72. Ibid.

73. Klein, "Wolf Recovery in the Northern Rockies," 97–98.

74. Ibid., 98.

75. Ibid., 100–103.

76. Noecker, "Reintroduction of Wolves."

77. Defenders of Wildlife website, http://www.defenders.org/publications/statistics_on_payments_from_the_defenders_wildlife_foundation_wolf_compensation_trust.pdf.

78. US Fish and Wildlife Service, "Service Declares Wyoming Gray Wolf Recovered under the Endangered Species Act and Returns Management Authority to the State," August 31, 2012 (press release).

79. Robert B. Keiter, "Greater Yellowstone's Bison: Unraveling of an Early American Wildlife Conservation Achievement," *Journal of Wildlife Management* 61, no. 1 (1997): 1–11.

80. Schullery, *Searching for Yellowstone*, 232.

81. James Duffus III, "Wildlife Management: Many Issues Unresolved in Yellowstone Bison-Cattle Brucellosis Conflict" (Washington, DC: GAO, 1992), 3.

82. Robert B. Keiter and Peter H. Froelicher, "Bison, Brucellosis, and Law in the Greater Yellowstone Ecosystem," *Land and Water Law Review* 28, no. 1 (1993): 1–75.

83. Christina M. Cromley, "Bison Management in Greater Yellowstone," in *Finding Common Ground: Governance and Natural Resources in the American West*, ed. Ronald D. Brunner et al. (New Haven, CT: Yale University Press, 2002), 132.

84. Ibid., 135–36.

85. Ibid.; see also Norman F. Cheville, Dale R. McCullough, and Lee R. Paulson, "Brucellosis in the Greater Yellowstone Area" (Washington, DC: National Research Council, 1998).

86. Bruce Babbitt and Daniel Glickman, "Record of Decision for Final Environmental Impact Statement and Bison Management Plan for the State of Montana and Yellowstone National Park" (Washington, DC: Departments of Interior and Agriculture, 2000).

87. Robin M. Nazzaro, "Preliminary Observations on the Implementation of the Interagency Bison Management Plan" (Washington, DC: GAO, 2007).

88. Robin M. Nazzaro, "Yellowstone Bison: Interagency Plan and Agencies'

Management Need Improvement to Better Address Bison-Cattle Brucellosis Controversy" (Washington, DC: GAO, 2008).

89. Ibid., 7.

90. John C. Freemuth, *Islands under Siege: National Parks and the Politics of External Threats* (Lawrence: University Press of Kansas, 1991).

Chapter 4. Resistance and Acceptance: Ecosystem Management in the Bush Administration

1. "A Year of Change: To Restore America's Greatness: Department of the Interior, January 20, 1981–January 20, 1982" (Washington, DC: Department of the Interior, 1982).

2. Norman J. Vig and Michael E. Kraft, eds., *Environmental Policy: New Directions for the Twenty-First Century*, 6th ed. (Washington, DC: CQ Press, 2006), 104–5.

3. Quoted in Eliot Marshall, "Campaigning on the Environment," *Science* 241, no. 4873 (September 23, 1988): 1589.

4. Norman J. Vig, "Presidential Leadership and the Environment: From Reagan to Clinton," in *Environmental Policy in the 1990s: Reform or Reaction?*, ed. Norman J. Vig and Michael E. Kraft (Washington, DC: CQ Press, 1997), 95–118.

5. Marshall, "Campaigning on the Environment," 1589.

6. Quoted in Mary H. Cooper, "Jobs vs. Environment: Should Jobs Be Sacrificed for the Sake of Environmental Protection?," *Congressional Researcher* 2, no. 18 (May 15, 1992): 409–32.

7. Ibid.

8. Quoted in Edward Schiappa, "Towards a Pragmatic Approach to Definition: 'Wetlands' and the Politics of Meaning," in *Environmental Pragmatism*, ed. Andrew Light and Eric Katz (New York: Routledge, 1996), 215.

9. Quoted in John Lancaster, "Environmentalists Hit 'Shocking' Appointment: Political Battle Looms over USDA Nominee," *Washington Post*, July 11, 1989; Lancaster, "New Look, Old Mission at Lujan's Department: Interior Still Bound to the West," *Washington Post*, January 19 1990.

10. Thomas R. Dunlap, *Faith in Nature: Environmentalism as Religious Quest* (Seattle: University of Washington Press, 2004), 95.

11. Quoted in John Lancaster, "Lujan: Endangered Species Act 'Too Tough,' Needs Change," *Washington Post*, May 12, 1990.

12. Remarks to Vaagen Brothers Lumber Employees in Colville, Washington, September 14, 1992.

13. Office of the Secretary, DOI, to Duncan Hunter, undated, FGNR002, WHORM: Subject File—Oversized Attachments, Bush Presidential Records, George Bush Presidential Library (hereafter cited as GBPL).

14. GAO, "Ecosystem Management: Additional Actions Needed to Adequately Test a Promising Approach" (Washington, DC: GAO, 1994), 24.

15. Quoted in Edward Schiappa, *Defining Reality: Definitions and the Politics of Meaning* (Carbondale: Southern Illinois University Press, 2003), 77.

16. Quoted in ibid.

17. Ibid., 78.

18. W. R. Walker and S. C. Richardson, "The Federal Wetlands Manual: Swamped by Controversy" (Blacksburg: Virginia Water Resources Research Center, 1991).

19. Alyson C. Flournoy, "Preserving Dynamic Systems: Wetlands, Ecology, and Law" *Duke Environmental Law and Policy Forum* 7 (Winter 1996): 111.

20. William M. Lewis, Jr., "Wetlands: Characteristics and Boundaries" (Washington, DC: National Research Council, 1995), 59.

21. Jon Kusler, "Wetlands Delineation: An Issue of Science or Politics," *Environment* 34, no. 2 (March 1992): 7–11, 29–37; Lewis, "Wetlands: Characteristics and Boundaries."

22. Lewis, "Wetlands: Characteristics and Boundaries," 63.

23. Quoted in ibid., 50.

24. "Federal Manual for Identifying and Delineating Jurisdictional Wetlands" (Washington, DC: Federal Interagency Committee for Wetland Delineation, 1989), 9.

25. Lewis, "Wetlands: Characteristics and Boundaries."

26. Ibid., 73.

27. Schiappa, *Defining Reality*, 86.

28. Ibid., 79.

29. Lewis, "Wetlands: Characteristics and Boundaries," 74–88.

30. Kusler, "Wetlands Delineation," 34.

31. Quoted in Schiappa, *Defining Reality*, 87.

32. "Great Dismal Swamp May Lose Wetlands Protection," *Christian Science Monitor*, October 12, 1991.

33. Schiappa, *Defining Reality*, 81.

34. Flournoy, "Preserving Dynamic Systems."

35. Oliver Wendell Holmes, "The Path of the Law," *Harvard Law Review* 10, no. 8 (1897): 457–78.

36. Flournoy, "Preserving Dynamic Systems: Wetlands, Ecology, and Law," 112.

37. Ibid.

38. Steven Lewis Yaffee, *The Wisdom of the Spotted Owl: Policy Lessons for a New Century* (Washington, DC: Island Press, 1994), 3–35.

39. James D. Proctor, "Whose Nature? The Contested Moral Terrain of Ancient Forests," in *Uncommon Ground: Toward Reinventing Nature*, ed. William Cronon (New York: W. W. Norton & Company, 1995), 269–97.

40. *Northern Spotted Owl Briefing Book*, US Fish and Wildlife Service, May 1, 1990, FGNR002, WHORM: Subject File—Oversized Attachments, Bush Presidential Records, GBPL.

41. *Northern Spotted Owl v. Hodel*, 716 F.Supp. 479 (W.D. 1988).

42. See, for example, Victor M. Sher, "Travels with Strix: The Spotted Owl's Journey through the Federal Courts," *Public Land & Resources Law Review* 14 (1993): 41–79; Brendon Swedlow, "Scientists, Judges, and Spotted Owls:

Policymakers in the Pacific Northwest," *Duke Environmental Law and Policy Forum* 13 (Spring 2003): 187–278; Yaffee, *The Wisdom of the Spotted Owl.*

43. *The Wisdom of the Spotted Owl,* 16.

44. Ibid., 23.

45. Ibid., 23–24.

46. Ibid., 34–35.

47. Quoted in *Portland Audubon Soc. v. Lujan,* 712 F.Supp. 1456, 1459 (D. Ore. 1989).

48. Quoted in *Portland Audubon Society v. Lujan,* 712 F.Supp. 1456, 1473 (D. Ore. 1989).

49. Ibid., 1477.

50. Ibid., 1481.

51. Ibid., 1484.

52. *Portland Audubon Society v. Lujan,* 795 F.Supp. 1489 (D. Or. 1992).

53. Quoted in Yaffee, *The Wisdom of the Spotted Owl,* 77.

54. Ibid., 95–96.

55. Ibid., 100–101.

56. Quoted in ibid., 105.

57. Andrea L. Hungerford, "Changing the Management of Public Land Forests: The Role of the Spotted Owl Injunctions," *Environmental Law* 24 (1994): 1407.

58. Ibid.

59. The appropriations rider led to another interesting legal twist. The Ninth Circuit Court of Appeals reversed Judge Dwyer's dismissal in 1991, ruling that the Section 318 rider violated the constitutional separation of powers by prohibiting the courts from reviewing the Forest Service's plans. The Supreme Court reversed in 1992, ruling that the rider was essentially amending legislation to exempt Forest Service timber plans from review rather than prohibiting the courts from reviewing congressional legislation.

60. Jack Ward Thomas et al., "A Conservation Strategy for the Northern Spotted Owl: Report of the Interagency Scientific Committee to Address the Conservation of the Northern Spotted Owl" (Portland, OR: USDA Forest Service, USDI Bureau of Land Management, USDI Fish and Wildlife Service, USDI National Park Service, 1990), 1, 39.

61. Ibid., 1.

62. *Seattle Audubon Society v. Evans,* No. C89-169WD (W.D. Wash. March 7, 1991).

63. James Duffus III, "Endangered Species: Spotted Owl Petition Evaluation Beset by Problems" (Washington, DC: GAO, 1989).

64. Ibid.

65. Ibid.

66. Ibid., 8.; *Northern Spotted Owl Briefing Book,* US Fish and Wildlife Service, May 1, 1990, FGNR002, WHORM: Subject File—Oversized Attachments, Bush Presidential Records, GBPL, 4.

67. David S. Cloud, "Environmental Record Puts Nominee under Attack,"

Congressional Quarterly Weekly Report (October 7, 1989), *Northern Spotted Owl Briefing Book,* US Fish and Wildlife Service, May 1, 1990, FGNR002, WHORM: Subject File—Oversized Attachments, Bush Presidential Records, GBPL, 4.

68. Manuel Lujan, Jr., "Inside Interior Interview," by Charles F. Wilkinson, March 17, 2004 (Boulder: Center of the American West, University of Colorado).

69. "Interagency Agreement between the Fish and Wildlife Service, Bureau of Land Management, National Parks Service, USDOI, and the Forest Service, USDA," June 1989, FGNR002, WHORM: Subject File—Oversized Attachments, Bush Presidential Records, GBPL.

70. Yaffee, *The Wisdom of the Spotted Owl,* 127.

71. Bureau of Land Management, *Northern Spotted Owl: The Jamison Plan Detailed Management Strategy,* September 24, 1990, FGNR002, WHORM: Subject File—Oversized Attachments, Bush Presidential Records, GBPL; Thomas et al., "A Conservation Strategy for the Northern Spotted Owl."

72. *Lane County Audubon Society v. Jamison,* 958 F.2d. 290 (9th Cir. 1992).

73. *Northern Spotted Owl v. Lujan,* 758 F.Supp. 621 (W.D. Wash. 1991).

74. Yaffee, *The Wisdom of the Spotted Owl,* 132.

75. Shannon Peterson, *Acting for Endangered Species: The Statutory Ark* (Lawrence: University Press of Kansas, 2002), 99.

76. Bruce G. Marcot and Jack Ward Thomas, "Of Spotted Owls, Old Growth, and New Policies: A History since the Interagency Scientific Committee Report" (Washington, DC: USDA Forest Service, 1997), 5.

77. *Northern Spotted Owl Briefing Book,* US Fish and Wildlife Service, May 1, 1990, FGNR002, WHORM: Subject File—Oversized Attachments, Bush Presidential Records, GBPL, 1.

78. Ibid., 5.

79. Quoted in Yaffee, *The Wisdom of the Spotted Owl,* 128.

80. Department of the Interior Press Release, "Yeutter and Lujan Announce Five-Point Plan to Preserve Owl and Protect Jobs," June 26, 1990.

81. "Administration Vows to Protect Spotted Owl, Logging Jobs," *Cox News Service,* June 27, 1990.

82. Scott Sonner, "Scientists Favored Owl Plan Bush Administration Resists," *Seattle Times,* August 29, 1990.

83. "Interior Secretary Sees Less Impact from Owl Plan," *Seattle Times,* August 27, 1990.

84. Alyson Pytte, "Bush's Modest Proposal," *CQ Weekly,* September 29, 1990.

85. Memorandum from Bill Hecht to Jim Ciconni, May 18, 1990, FGNR002, WHORM: Subject File—Oversized Attachments, Bush Presidential Records, GBPL, 3 (emphasis in the original).

86. Memorandum from Michael Deland to Governor Sununu, Andy Card, Ede Holiday, C. Boyden Gray, Roger Porter, Richard Darman, and Bob Grady, July 5, 1990, FGNR002, WHORM: Subject File—Oversized Attachments, Bush Presidential Records, GBPL.

87. Quoted in Yaffee, *The Wisdom of the Spotted Owl,* 135–36.

88. Memorandum, "Directive to Northern Spotted Owl Recovery Team,"

December 5, 1990, FGNR002, WHORM: Subject File—Oversized Attachments, Bush Presidential Records, GBPL.

89. Yaffee, *The Wisdom of the Spotted Owl*, 137.

90. Slade Gorton to John Sununu, June 26, 1991, FGNR002, WHORM: Subject File—Oversized Attachments, Bush Presidential Records, GBPL.

91. *Lane County Audubon Society v. Jamison*, 958 F.2d. 290, 293 (9th Cir. 1992).

92. Meeting of the Endangered Species Committee, May 14, 1992, FGNR002, WHORM: Subject File—Oversized Attachments, Bush Presidential Records, GBPL.

93. Ibid.

94. Marcot and Thomas, "Of Spotted Owls, Old Growth, and New Policies," 7.

95. Memorandum, "Work Group on Northern Spotted Owl Management Options," March 17, 1992, FGNR002, WHORM: Subject File—Oversized Attachments, Bush Presidential Records, GBPL.

96. Memorandum, "Outline the Responsibilities of the Work Group on Spotted Owl Management Options," February 14, 1992, FGNR002, WHORM: Subject File—Oversized Attachments, Bush Presidential Records, GBPL.

97. Secretary Lujan to Governor Roberts, February 21, 1992, FGNR002, WHORM: Subject File—Oversized Attachments, Bush Presidential Records, GBPL.

98. Peterson, *Acting for Endangered Species*, 106.

99. Robert M. Teeter to John Sununu, August 5, 1991, marked "shared with POTUS," FGNR002, WHORM: Subject File—Oversized Attachments, Bush Presidential Records, GBPL.

100. Bill Bradbury to President George Bush, February 24, 1992, FGNR002, WHORM: Subject File—Oversized Attachments, Bush Presidential Records, GBPL.

101. Remarks to Vaagen Brothers Lumber Employees in Colville, Washington, September 14, 1992.

102. Ibid.

103. Ibid.

104. Mark O. Hatfield, "Can't See the Forest for the Endangered Species," *Washington Post*, June 12, 1992.

105. Memorandum, "Ecosystem Management of the National Forests and Grasslands," June 3, 1992, Teresa Gorman Files, UNCED Follow Up, Bush Presidential Records, GBPL.

106. Richard Freeman, "The EcoFactory: The United States Forest Service and the Political Construction of Ecosystem Management," *Environmental History* 7, no. 4 (October 2002): 463–65.

107. F. Dale Robertson, Interview by Harold K. Steen, August 12–14, 1999 (Durham, NC: Forest History Society), 86.

108. F. Dale Robertson, "A History of New Perspectives and Ecosystem Management," in *Ouachita and Ozark Mountains Symposium: Ecosystem Management Research*, ed. James M. Guldin (Hot Springs, AR: USDA Forest Service, 1999), 3.

109. Richard W. Behan, "The Irony of the Multiple Use/Sustained Yield Concept: Nothing Is So Powerful as an Idea Whose Time Has Passed" (paper

presented at Multiple Use and Sustained Yield: Changing Philosophies for Federal Land Management?, Washington, DC, March 5–6, 1992), 96.

110. Ibid., 104.

111. Chris Maser, "Do We Owe Anything to the Future?" (paper presented at Multiple Use and Sustained Yield: Changing Philosophies for Federal Land Management?, Washington, DC, March 5–6, 1992), 206.

112. Quoted in *Seattle Audubon Society v. Evans*, 952 F.2d. 297, 302 (9th Cir. 1991).

113. Robertson, Interview by Harold K. Steen, 82.

114. Gerald W. Williams, "New Perspectives, New Forestry, or Biological Diversity: How Did We Get Here?" Paper presented at the Biological Diversity/Silviculture Workshop, Gleneden Beach, OR, May 6, 1991.

115. Quoted in Anna Maria Gillis, "The New Forestry," *BioScience* 40, no. 8 (September 1990): 558.

116. F. J. Swanson and Jerry F. Franklin, "New Forestry Principles from Ecosystem Analysis of Pacific Northwest Forests," *Ecological Applications* 2, no. 3 (August 1992): 262–74.

117. Robertson, "A History of New Perspectives and Ecosystem Management," 4.

118. Swanson and Franklin, "New Forestry Principles from Ecosystem Analysis of Pacific Northwest Forests," 271.

119. Robertson, "A History of New Perspectives and Ecosystem Management," 4.

120. William E. Shands, Anne Black, and Jim Giltmier, "From New Perspectives to Ecosystem Management," *George Wright Forum* 11, no. 1 (1994): 35–47.

121. Ibid., 36–37.

122. Christopher A. Frissell, Richard K. Nawa, and Reed Noss, "Is There Any Conservation Biology in 'New Perspectives'?: A Response to Salwasser," *Conservation Biology* 6, no. 3 (September 1992): 461–64; Nathaniel Lawrence and Dennis Murphy, "New Perspectives or Old Priorities?," *Conservation Biology* 6, no. 3 (September 1992): 465–68.

123. Lawrence and Murphy, "New Perspectives or Old Priorities?," 467.

124. Rick Freeman, Oral History with Hal Salwasser, May 18, 1997 (OH 370-01, Archives and Special Collections, Maureen and Mike Mansfield Library, University of Montana, Missoula), 10.

125. See, for example, George M. Leonard, Interview by Harold K. Steen (Durham, NC: Forest History Society, 1999).

126. Merrill R. Kaufmann et al., "An Ecological Basis for Ecosystem Management" (Fort Collins, CO: US Forest Service, 1994), 1.

127. Ibid.

Chapter 5. Putting Ecosystem Management to the Test in the Clinton Administration

1. Bruce Babbitt, "Inside Interior: Bruce Babbitt," interview by Charles Wilkinson for the Center of the American West, University of Colorado, Boulder, CO, April 20, 2004.

2. W. Wallace Covington, and Leonard F. DeBona, eds., *Sustainable Ecological Systems: Implementing an Ecological Approach to Land Management* (Flagstaff, AZ: USDA Forest Service: Rocky Mountain Forest and Range Experiment Station, 1993); L. H. Foley, ed., *Silviculture: From the Cradle of Forestry to Ecosystem Management* (Asheville, NC: USDA Forest Service: Southeastern Forest Experiment Station, 1993).

3. Hal Salwasser and Robert D. Pfister, "Ecosystem Management: From Theory to Practice," in *Sustainable Ecological Systems: Implementing an Ecological Approach to Land Management,* ed. W. Wallace Covington and Leonard F. DeBona (Flagstaff, AZ: Forest Service Rocky Mountain Forest and Range Experiment Station, 1993), 150–61.

4. Ann M. Bartuska, "Ecosystem Management in the Forest Service," in *Silviculture: From the Cradle of Forestry to Ecosystem Management,* ed. L. H. Foley (Asheville, NC: Forest Service Southeastern Forest Experiment Station, 1993), 13.

5. Dean L. Urban, "Landscape Ecology and Ecosystem Management," in *Sustainable Ecological Systems: Implementing an Ecological Approach to Land Management,* ed. W. Wallace Covington and Leonard F. Debano (Flagstaff, AZ: USDA Forest Service Rocky Mountain Forest and Range Experiment Station, 1993), 128.

6. Joseph F. McNeel and Ron L. Copstead, "Operational Factors Affecting the Implementation of Ecosystem Management Objectives," in *Silviculture: From the Cradle of Forestry to Ecosystem Management,* ed. L. H. Foley (Asheville, NC: USDA Forest Service Southeastern Forest Experiment Station, 1993), 167–77.

7. Patricia N. Manley, ed., "Draft Region 5 Ecosystem Management Guidebook" (Vallejo, CA: US Forest Service Region 5, 1994).

8. Ibid., 216.

9. Ibid., 220.

10. Robert Costanza, Bryan G. Norton, and Benjamin Haskell, eds., *Ecosystem Health: New Goals for Environmental Management* (Washington, DC: Island Press, 1992).

11. Again, this does not include the Department of Defense and the Department of Energy, which both have agencies that manage federally owned land.

12. Tom Kenworthy, "Pragmatic Critic Is Set to Be Interior's Next Landlord," *Washington Post,* January 19, 1993.

13. Ibid.

14. Bruce Babbitt, "Science: Opening the Next Century of Conservation History," *Science* 267, no. 5206 (March 31, 1995): 1954.

15. Robert Mason and Bruce Babbitt, "Interview: Bruce Babbitt," *Natural Resources and Environment* 11, no. 1 (Summer 1996): 35–41.

16. Bruce Babbitt, *Cities in the Wilderness: A New Vision of Land Use in America* (Washington, DC: Island Press, 2005), 60.

17. Ibid.

18. Babbitt, "Science"; Bruce Babbitt, "The Future Environmental Agenda for the United States," October 29, 1992, Box 11-2, Clinton History Project, William J. Clinton Library.

19. N. L. Christensen et al., "The Report of the Ecological Society of America

Committee on the Scientific Basis for Ecosystem Management," *Ecological Applications* 6, no. 3 (1996): 668–69.

20. Ibid., 669.

21. Quoted in Peter Raven, "A Biological Survey for the Nation: Committee on the Formation of the National Biological Survey and Commission on the Formation of the National Biological Survey" (Washington, DC: National Academy of Sciences, 1993), vii.

22. Quoted in Diane Krahe, "The Ill-Fated NBS: A Historical Analysis of Bruce Babbitt's Vision to Overhaul Interior Science" (paper presented at Rethinking Protected Areas in a Changing World: Proceedings of the 2011 George Wright Society Biennial Conference on Parks, Protected Areas, and Cultural Sites, Hancock, MI, 2012), 161.

23. Richard West Sellars, *Preserving Nature in the National Parks: A History* (New Haven, CT: Yale University Press, 1997).

24. Statement of Bruce Babbitt, Secretary of the Interior before the Senate Appropriations Subcommittee on Interior and Related Agencies, June 8, 1993, Clinton History Project, Box 10-6, William J. Clinton Presidential Library.

25. Raven, "A Biological Survey for the Nation," 158–59.

26. Secretarial Order No. 3165: Organizational Preparations for the Creation of the National Biological Survey, May 17, 1993.

27. House Committee on Merchant Marine and Fisheries, Subcommittee on National Parks, Forests, and Public Lands, *H.R. 1845: A Bill to Establish the Biological Survey in the Department of the Interior*, July 15, 1993, 6–7.

28. Ibid., 28.

29. Ibid., 152.

30. Catalina Camia, "Interior: Opponents Fear Ramifications of Plant and Animal Survey," *CQ Weekly*, July 31, 1993.

31. Ibid.

32. Andrew Taylor, "Biological Survey Bill Halted after Heated House Debate," *CQ Weekly*, October 9, 1993, 2730.

33. Jeffrey P. Cohn, "The National Biological Survey," *BioScience* 43, no. 8 (September 1993): 521–22.

34. M. Lynne Corn, "The National Biological Service" (Washington, DC: Congressional Research Service, 1995).

35. H. Ronald Pulliam, "The Political Education of a Biologist: Part 1," *Wildlife Society Bulletin* 26, no. 2 (Summer 1998): 199.

36. Ibid., 200.

37. Ibid., 202.

38. Krahe, "The Ill-Fated NBS," 163.

39. Robert T. Lackey, "Is Science Biased toward Natural?," *Northwest Science Forum* 83, no. 3 (2009): 277–79.

40. James R. Skillen, *The Nation's Largest Landlord: The Bureau of Land Management in the American West* (Lawrence: University Press of Kansas, 2009), 124–28.

41. Senate Committee on Energy and Natural Resources, *Grazing*, 103rd Cong., 2nd sess., April 20, May 14, 1994.

42. Skillen, *The Nation's Largest Landlord*, 145–47.

43. Will Stelle, "Executive Branch Policies and Activities in Ecosystem Management," in *Ecosystem Management: Status and Potential* (Washington, DC: GPO, 1994), 269.
44. Ibid.
45. Jeffrey Zinn and M. Lynne Corn, "Executive Summary," in *Ecosystem Management: Status and Potential* (Washington, DC: GPO, 1994), 5.
46. Ibid.
47. Ibid., 8.
48. Ibid., 13.
49. Bill Clinton and Al Gore, *Putting People First* (New York: Times Books, 1992).
50. Quoted in William A. Galston and Geoffrey L. Tibbetts, "Reinventing Federalism: The Clinton/Gore Program for a New Partnership among the Federal, State, Local, and Tribal Governments," *Publius* 24, no. 3 (Summer 1994): 25.
51. Al Gore, "From Red Tape to Results: Creating a Government That Works Better and Costs Less" (Washington, DC: National Performance Review, 1993), i.
52. Ibid., 52.
53. Ibid.
54. IEMTF, "The Ecosystem Approach: Healthy Ecosystems *and* Sustainable Economies" (Washington, DC: GPO, 1995).
55. Ibid., 17.
56. Ibid., 19.
57. Ibid., 50.
58. Memorandum of Understanding to Foster the Ecosystem Approach, December 15, 1995, 1.
59. Ibid., 2.
60. Allan K. Fitzsimmons, "Federal Ecosystem Management: A 'Train Wreck' in the Making" (Washington, DC: Cato Institute, 1994).
61. Allan K. Fitzsimmons, *Defending Illusions: Federal Protection of Ecosystems* (New York: Rowman & Littlefield, 1999), 288.
62. Alston Chase, *In a Dark Wood: The Fight over Forests and the Rising Tyranny of Ecology* (New York: Houghton Mifflin, 1995), 388–89.
63. Ibid., 391.
64. Ronald N. Johnson, "Ecosystem Management and Reinventing Government," in *Breaking the Environmental Policy Gridlock*, ed. Terry L. Anderson (Stanford, CA: Hoover Institution Press, 1997), 43.
65. CEQ, "Incorporating Biodiversity into Environmental Impact Analysis under the National Environmental Policy Act" (Washington, DC: Council on Environmental Quality, 1993), 5.
66. Ibid., 7–8.
67. Ibid., 23.
68. Jon Healy, "Ten Environmental Nettles That Await the 103rd Congress," *Congressional Quarterly Weekly Report* 51, no. 7 (February 13, 1992): 310–11.
69. *The Fund for Animals, et al. v. Turner*, Civ. No. 92-800.
70. Eric R. Glitzenstein, "On the USFWS Settlement Regarding Federal Listing of Endangered Species," *Endangered Species Update* 10, no. 5 (March 1993):

1–3.

71. J. B. Ruhl, "The Endangered Species Act and Private Property: A Matter of Timing and Location" *Cornell Journal of Law and Policy* 8 (1999): 37–53.

72. *Babbitt v. Sweet Home Chapter of Communities for a Great Oregon*, 515 U.S. 687 (1995).

73. John Copeland Nagle, *Law's Environment: How the Law Shapes the Places We Live* (New Haven, CT: Yale University Press, 2010), 50–92.

74. "Statement of Secretary of the Interior Bruce Babbitt: Senate Committee on Environment and Public Works Oversight Hearing on the Endangered Species Act," June 15, 1994, Box 10, Clinton Administrative History Project, William J. Clinton Presidential Library.

75. Bruce Babbitt, "To Reauthorize the Endangered Species Act: Why, Where and How We Should Translate Our Success Stories into Law," Speech to the National Press Club, Washington DC, July 17, 1997, Box 11, Clinton Administrative History Project, William J. Clinton Presidential Library.

76. Randy T. Simmons, "Fixing the Endangered Species Act," in *Breaking the Environmental Policy Gridlock*, ed. Terry L. Anderson (Stanford, CA: Hoover Institution Press, 1997), 82.

77. Quoted in Glitzenstein, "On the USFWS Settlement," 3.

78. Edward T. LaRoe, "Implementation of an Ecosystem Approach to Endangered Species Conservation," *Endangered Species Update* 10, no. 3 & 4 (January/February 1993): 3–4.

79. "An Ecosystem Approach to Fish and Wildlife Conservation: An Approach to More Effectively Conserve the Nation's Biodiversity" (Washington, DC: US Fish and Wildlife Service, 1994).

80. Ibid., 4.

81. "Endangered and Threatened Wildlife and Plants: Notice of Interagency Cooperative Policy for the Ecosystem Approach to the Endangered Species Act," *Federal Register* 59, no. 126 (July 1, 1994): 34273.

82. "Memorandum of Understanding between Federal Agencies on Implementation of the Endangered Species Act," September 28, 1994. Quoted in J. B. Ruhl, "Who Needs Congress? An Agenda for Administrative Reform of the Endangered Species Act," *New York University Environmental Law Journal* 6 (1998): 378.

83. Press Release, "Administrative Changes Will Make Endangered Species Act User-Friendly and Improve Benefits to At-Risk Species," June 14, 1994, Box 1, Clinton Administration History Project Files, RG 48, National Archives, College Park, MD.

84. Robert D. Thornton, "Habitat Conservation Plans: Frayed Safety Nets or Creative Partnerships?," *Natural Resources and Environment* 16, no. 2 (2001): 94.

85. Kirk Emerson and Charles R. Wise, "Statutory Approaches to Regulatory Takings: State Property Rights Legislation Issues and Implications for Public Administration," *Public Administration Review* 57, no. 5 (September-October 1997): 412. For a broader discussion, see Bruce Yandle, ed. *Land Rights: The 1990s Property Rights Rebellion* (Lanham, MD: Rowman and Littlefield, 1995).

86. Harold K. Steen, *The US Forest Service: A History* (Seattle: University of

Washington Press, 1976), 222–45.

87. John R. Nolon, "The National Land Use Policy Act," *Pace Environmental Law Review* 13 (1996): 519–23.

88. Bruce Babbitt, "ESA and Private Property: The Endangered Species Act and 'Takings': A Call for Innovation within the Terms of the Act," *Environmental Law* 24 (Spring 1994): 362.

89. Ruhl, "Who Needs Congress?," 370.

90. Ed Marston, "Interior View: Bruce Babbitt Took the Real West to Washington: A High Country News Interview," *High Country News* 33, no. 3 (2001): 8–11.

91. Babbitt, "ESA and Private Property," 361.

92. Ibid.

93. Thornton, "Habitat Conservation Plans."

94. 16 USC 1539(a)(1)(B).

95. 16 USC 1539.

96. Jennifer Jester, "Habitat Conservation Plans under Section 10 of the Endangered Species Act: The Alabama Beach Mouse and the Unfulfilled Mandate of Species Recovery," *Boston College Environmental Affairs Law Review* 26, no. 1 (1998): 133.

97. Babbitt, *Cities in the Wilderness*, 63; Robert D. Thornton, "The Search for a Conservation Planning Paradigm: Section 10 of the ESA," *Natural Resources and Environment* 8, no. 1 (Summer 1993): 21–23, 65–67.

98. Press Release, "Administration Announces Special Rule on Timber Harvest in Spotted Owl Habitat on Non-Federal Lands," December 10, 1993, Box 2, Clinton Administration History Project Files, RG 48, National Archives, College Park, MD.

99. "Protecting America's Heritage: A Fair, Cooperative and Scientifically Sound Approach to Improving the Endangered Species Act" (Washington, DC: US Fish and Wildlife Service, 1995).

100. Ibid.

101. Ibid.

102. "Endangered and Threatened Wildlife and Plants: Notice of Interagency Cooperative Policy for Peer Review in Endangered Species Act Activities," *Federal Register* 59, no. 126 (July 1, 1994): 34270.

103. "Endangered and Threatened Wildlife and Plants: Notice of Interagency Cooperative Policy for Endangered Species Act Section 9 Prohibitions," *Federal Register* 59, no. 126 (July 1, 1994): 34272.

104. Quoted in Lynn E. Dwyer, Dennis D. Murphy, and Paul R. Ehrlich, "Property Rights Case Law and the Challenge to the Endangered Species Act," *Conservation Biology* 9, no. 4 (August 1995): 730.

105. Press Release, "Administration's New Assurance Policy Tells Landowners: 'No Surprises' in Endangered Species Planning," August 11, 1994, Box 1, Clinton Administration History Project Files, RG 48, National Archives, College Park, MD.

106. Dwyer, Murphy, and Ehrlich, "Property Rights Case Law and the Challenge to the Endangered Species Act," 730.

107. Press Release, "Administration's New Assurance Policy Tells Landowners:

'No Surprises' in Endangered Species Planning," August 11, 1994, Box 1, Clinton Administration History Project Files, RG 48, National Archives, College Park, MD.

108. Quoted in Ruhl, "Who Needs Congress?," 383.

109. Ibid., 385–87.

110. "Safe Harbor Agreements and Candidate Conservation Agreements," *Federal Register* 62, no. 113 (June 12, 1997): 32189–94.

111. "Proposed Rule Exempting Certain Small Landowners and Low-Impact Activities from Endangered Species Act Requirements for Threatened Species," *Federal Register* 60, no. 139 (July 20, 1995): 37419.

112. Ruhl, "Who Needs Congress?," 396.

113. Thornton, "Habitat Conservation Plans," 95.

114. Ibid.

115. J. B. Ruhl, "Past, Present, and Future Trends of the Endangered Species Act," *Public Land and Resources Law Review* 25 (2004): 15–37.

116. Alexander Cockburn, "'Win-Win' with Bruce Babbitt: The Clinton Administration Meets the Environment," *New Left Review* 201 (September/October 1993): 46–59.

117. Babbitt, *Cities in the Wilderness*, 90–100.

118. Robert B. Keiter, *Keeping Faith with Nature: Ecosystems, Democracy, and America's Public Lands* (New Haven, CT: Yale University Press, 2003); Judith A. Layzer, *Natural Experiments: Ecosystem-Based Management and the Environment* (Cambridge, MA: MIT Press, 2008); Steven L. Yaffee et al., eds., *Ecosystem Management in the United States: An Assessment of Current Experience* (Washington, DC: Island Press, 1996).

Chapter 6. The Northwest Forest Plan: Substantive Ecosystem Management

1. James Pipkin, "The Northwest Forest Plan Revisited" (Washington, DC: Council on Environmental Quality, 1998), 13–14.

2. *Seattle Audubon Soc. v. Robertson*, No. 89-160 (WD Wash., Mar. 24, 1989).

3. Jack Ward Thomas et al., "The Northwest Forest Plan: Origins, Components, Implementation Experience, and Suggestions for Change," *Conservation Biology* 20, no. 2 (April 2006): 277–87.

4. *Seattle Audubon Society v. Moseley*, 798 F.Supp. 1473 (W.D. Wash. 1992).

5. *Portland Audubon Society v. Lujan*, 795 F.Supp. (D. Or. 1992).

6. *Seattle Audubon Society v. Espy*, 998 F.2d 699 (9th Circ. 1993).

7. *Portland Audubon Society v. Babbitt*, 998 F.2d 705 (9th Cir. 1993).

8. Paul Bedard, "Jobs Come before Owls, Bush Says in Northwest; Warns of Veto for Species Act," *Washington Times*, September 15, 1992.

9. "Campaign '92: Transcript of the Third Presidential Debate (Part 1)," *Washington Post*, October 20, 1992.

10. John Dillin, "Presidential Candidates Spar over US Environmental

Issues," *Christian Science Monitor,* April 28, 1992.

11. Bedard, "Jobs Come before Owls, Bush Says in Northwest."

12. E. Thomas Tuchmann et al., "The Northwest Forest Plan: A Report to the President and Congress" (Washington, DC: USDA Office of Forestry and Economic Assistance, 1996), 28.

13. K. Norman Johnson, "Science-Based Assessment of the Forests of the Pacific Northwest," in *Creating a Forestry for the 21st Century,* ed. Kathryn A. Kohm and Jerry F. Franklin (Washington, DC: Island Press, 1997), 397–410.

14. Tuchmann et al., "The Northwest Forest Plan," 28.

15. Quoted in FEMAT, "Forest Ecosystem Management: An Ecological, Economic, and Social Assessment" (Washington, DC : GPO, 1993), VII-A, 81.

16. Quoted in ibid., 88–89.

17. Quoted in ibid., 38.

18. Quoted in ibid., ii.

19. Lynton K. Caldwell, "An Ecosystems Approach to Public Land Policy," in *Public Land Policy: Proceedings of the Western Resources Conference,* ed. Phillip O. Foss (Boulder: Colorado Associated University Press, 1968), 44.

20. Ibid.

21. FEMAT, "Forest Ecosystem Management," ii.

22. Quoted in ibid., VII-A, 87.

23. Ibid., 14.

24. Memorandum, Forest Conference Executive Committee to Forest Conference Inter-Agency Working Groups, May 7, 1993. Found in *Final Supplemental Environmental Impact Statement on Management of Habitat for Late-Successional and Old-Growth Forest Related Species within the Range of the Northern Spotted Owl,* Volume 2 (Washington, DC: US Department of Agriculture and US Department of the Interior, February 1994), Appendix C.

25. Gerald W. Williams, "FEMAT to SEIS: Transforming the Social Assessment" (paper presented at Integrating Social Science in Ecosystem Management: A National Challenge, Helen, GA, December 12–14, 1994).

26. Ibid.

27. FEMAT, "Forest Ecosystem Management," III-1.

28. Ibid.

29. Bruce G. Marcot and Jack Ward Thomas, "Of Spotted Owls, Old Growth, and New Policies: A History since the Interagency Scientific Committee Report" (Washington, DC: USDA Forest Service, 1997), 11.

30. Logan A. Norris, "Science Review," in *Bioregional Assessments: Science at the Crossroads of Management and Policy,* ed. K. Norman Johnson et al. (Washington, DC: Island Press, 1999), 118.

31. Johnathan B. Haufler, Carolyn A. Mehl, and Gary J. Roloff, "Using a Coarse-Filter Approach with Species Assessment for Ecosystem Management" *Wildlife Society Bulletin* 24, no. 2 (Summer 1996): 200–208.

32. Thomas R. Wellock, "The Dickey Bird Scientists Take Charge: Science, Policy, and the Spotted Owl," *Environmental History* 15 (July 2010): 391.

33. FEMAT, "Forest Ecosystem Management," II-26.

34. "Draft Supplemental Environmental Impact Statement on Management of Habitat for Late-Successional and Old-Growth Forest Related Species within the Range of the Northern Spotted Owl" (Washington, DC: Department of Agriculture and Department of the Interior, 1993), V29.
35. Ibid.
36. Ibid.
37. FEMAT, "Forest Ecosystem Management," II-26.
38. Ibid., II-24.
39. Ibid., III-24.
40. Ibid., VII-217.
41. Ibid., II-84–85.
42. Ibid., II-86–88.
43. Ibid., II-89.
44. *Northwest Forest Resource Council v. Espy*, 846 F.Supp. 1009 (D.D.C. 1994).
45. Ross W. Gorte, "The Clinton Administration's Forest Plan for the Pacific Northwest" (Washington, DC: Congressional Research Service, 1993).
46. Susan Charnley, "The Northwest Forest Plan as a Model for Broad-Scale Ecosystem Management: A Social Perspective," *Conservation Biology* 20, no. 2 (April 2006): 330–40; Thomas Michael Power, "Public Timber Supply, Market Adjustments, and Local Economies: Economic Assumptions of the Northwest Forest Plan," *Conservation Biology* 20, no. 2 (April 2006): 341–50.
47. Tuchmann et al., "The Northwest Forest Plan," 44.
48. *Public Papers of the Presidents of the United States, William J. Clinton, 1993.* (Washington, DC: GPO, 1994), 388.
49. Jack Ward Thomas, Interview by Harold K. Steen, May 10–12, 2001 (Durham, NC: Forest History Society).
50. William J. Clinton and Albert Gore, Jr., "The Forest Plan: For a Sustainable Economy and a Sustainable Environment" (Washington, DC.: The White House, 1993).
51. *Public Papers of the Presidents of the United States, William J. Clinton, 1993*, 965.
52. Ibid.
53. Clinton and Gore, Jr., "The Forest Plan."
54. Senator Mark O. Hatfield, speaking on the replacement of Dale Robertson as Forest Service chief, 103rd Cong., 1st sess., *Congressional Record* 139 (November 4, 1993).
55. Jack Ward Thomas, *The Journals of a Forest Service Chief*, ed. Harold K. Steen (Seattle: Forest History Society and University of Washington Press, 2004), 60–74.
56. Tuchmann et al., "The Northwest Forest Plan," 46.
57. Ibid., 49.
58. Thomas, *The Journals of a Forest Service Chief*, 75–80.
59. *Final Supplemental Environmental Impact Statement on Management of Habitat*, Appendix B11; Randy Molina, Bruce G. Marcot, and Robin Lesher, "Protecting Rare, Old-Growth, Forest-Associated Species under the Survey and Manage Program Guidelines of the Northwest Forest Plan," *Conservation Biology* 20, no. 2 (April 2006): 306–18.
60. *Final Supplemental Environmental Impact Statement on Management of Habitat.*

61. Thomas, "An Interview with Jack Ward Thomas," 18.
62. Thomas, *The Journals of a Forest Service Chief*, 77.
63. "Record of Decision: Amendments to Forest Service and Bureau of Land Management Planning Documents within the Range of the Northern Spotted Owl" (Washington, DC: Department of Agriculture and Department of the Interior, 1994).
64. *Seattle Audubon Society v. Lyons*, 871 F.Supp. 1291, 1311–1321 (W. D. W. 1994).
65. 109 Stat. 194 (1995).
66. Paul Larmer, "Cut to the Past: Logging Wars Resume," *High Country News* 45 (October 16, 1995).
67. Press Release, "Remarks by the President on Safe and Drug-Free Schools," June 7, 1995.
68. Memorandum, President Clinton to Secretary of the Interior, Secretary of Agriculture, Secretary of Commerce, and Administrator of the Environmental Protection Agency, "Implementing Timber-Related Provisions to Public Law 104-19," August 1, 1995, Box 90, History of USDA, Clinton Library.
69. Pre-Work Documents, RIEC Meeting, November 18, 1993. On file with the author.
70. Ibid.
71. Ibid.
72. George H. Stankey, Roger N. Clark, and Bernard T. Bormann, "Adaptive Management of Natural Resources: Theory, Concepts, and Management Institutions" (Portland, OR: USDA Forest Service, Pacific Northwest Research Station, 2005).
73. Bernard T. Bormann, Richard W. Haynes, and Jon R. Martin, "Adaptive Management of Forest Ecosystems: Did Some Rubber Hit the Road?," *BioScience* 57, no. 2 (February 2007): 187, 186–91.
74. Tuchmann et al., "The Northwest Forest Plan," 52.
75. George H. Stankey et al., "The Limits of Acceptable Change (LAC) System for Wilderness Planning" (Ogden, UT: Forest Service Intermountain Forest and Range Experiment Station, 1985).
76. Bruce Shindler, Peter List, and Brent Steel, "Managing Federal Forests: Public Attitudes in Oregon and Nationwide" *Journal of Forestry* 91, no. 7 (July 1993): 36–42; Bruce Shindler, Brent Steel, and Peter List, "Public Judgments of Adaptive Management: A Response from Forest Communities," *Journal of Forestry* 94, no. 6 (June 1996): 4–12; Bruce A. Shindler, Mark Brunson, and George H. Stankey, "Social Acceptability of Forest Conditions and Management Practices: A Problem Analysis" (Portland, OR: USDA Forest Service, 2002).
77. Bernard T. Bormann et al., "Adaptive Ecosystem Management in the Pacific Northwest," (Portland, OR: USDA Forest Service, 1994).
78. "Interagency Ecosystem Management: Information Resource Management: Status Report," February 1, 1994, on file with the author.
79. Meeting Notes, Regional Interagency Executive Committee, December 16 and 17, 1993, on file with the author.
80. Tier 1 Key Watersheds "contribute directly to the conservation of at-risk

anadromous salmonids, bull trout, and resident fish species. They also have a high potential of being restored as part of a watershed restoration program." Tier 2 Key Watersheds "may not contain at-risk fish stocks, [but] they are important sources of high quality water." "Record of Decision: Amendments to Forest Service and Bureau of Land Management Planning Documents within the Range of the Northern Spotted Owl," Standards and Guidelines, B-18.

81. Ibid., C-30.

82. Ibid., B-13.

83. Tuchmann et al., "The Northwest Forest Plan."

84. "Ecosystem Analysis at the Watershed Scale: Federal Guide for Watershed Analysis" (Portland, OR: Regional Ecosystem Office, 1995).

85. Ibid.

86. Leslie M. Reid, Robert R. Ziemer, and Michael J. Furniss, "Watershed Analysis on Federal Lands of the Pacific Northwest" (Berkeley, CA: Pacific Southwest Research Station, 1996).

87. Ibid.

88. Gordon E. Grant and Frederick J. Swanson, "Watershed Analysis on the Public Lands: Where Are We Now and Where Should We Be Going?," in *Watersheds '94: Respect, Rethink, Restore,* ed. Richard R. Harris et al. (Davis: University of California, Water Resources Center, 1994).

89. Dave Baker et al., "Implementation Monitoring: Summary of Regional Interagency Monitoring Results" (Portland, OR: Forest Service, Pacific Northwest Research Station, 2005), 12.

90. Gordon H. Reeves et al., "The Aquatic Conservation Strategy of the Northwest Forest Plan," *Conservation Biology* 20, no. 2 (2006).

91. Leslie M. Reid, Robert R. Ziemer, and Thomas E. Lisle, "What a Long Strange Trip It's Been—or—Who Took the Synthesis out of Analysis?," *Watershed Management Council Network* 6, no. 4 (1996): 6–7.

92. Tuchmann et al., "The Northwest Forest Plan," 92–97.

93. Baker et al., "Implementation Monitoring," 10.

94. Barry S. Mulder et al., "The Strategy and Design of the Effectiveness Monitoring Program for the Northwest Forest Plan" (Portland, OR: Forest Service, Pacific Northwest Research Station, 1999), 49.

95. Ibid.

96. Reeves et al., "The Aquatic Conservation Strategy of the Northwest Forest Plan"; Jack Ward Thomas, "Sustainability of the Northwest Forest Plan: Still to Be Tested," in *Forest Futures: Science, Politics, and Policy for the Next Century,* ed. Karen Arabas and Joe Bowersox (Boulder, CO: Rowman & Littlefield, 2004), 3–22.

97. *Final Supplemental Environmental Impact Statement on Management of Habitat.*

98. Molina, Marcot, and Lesher, "Protecting Rare, Old-Growth, Forest-Associated Species."

99. *Final Supplemental Environmental Impact Statement on Management of Habitat,* Standards and Guidelines, C-5.

100. Pre-Work, RIEC Meeting, November 16, 1994. On file with the author.

101. Pipkin, "The Northwest Forest Plan Revisited."

102. Bruce G. Marcot and Randy Molina, "Considerations of Other Species Associated with Older Forest Conditions," in *Northwest Forest Plan—The First 10 Years (1994–2003): Synthesis of Monitoring and Research Results*, ed. Richard W. Haynes et al. (Portland, OR: Pacific Northwest Research Station, 2006), 150, 53.

103. *Oregon Natural Resources Council Action v. United States Forest Service*, 59 F. Supp. 2d 1085 (W.D. Wash. 1999).

104. For details of the modified S&M standards, see Randy Molina et al., "Strategic Survey Framework for the Northwest Forest Plan Survey and Manage Program" (Portland, OR: Forest Service Pacific Northwest Research Station, 2003). For an outline of the litigation, see background information in *Northwest Ecosystem Alliance v. Rey*, 380 F.Supp. 2d 1175 (W.D. Wash. 2006) and *Conservation Northwest v. Sherman*, 715 F.3d 1181 (9th Cir. 2013).

105. *Final Supplemental Environmental Impact Statement on Management of Habitat*, Standards and Guidelines, A-4, D-5.

106. Ibid., D-4. Emphasis added.

107. George H. Stankey and Bruce Shindler, "Adaptive Management Areas: Achieving the Promise, Avoiding the Peril" (Pacific Northwest Research Station: US Forest Service, 1997), 3–4.

108. Quoted in Sally Duncan, "Too Early to Tell, or Too Late to Rescue?: Adaptive Management under Scrutiny," *Science Findings*, no. 33 (2001): 3.

109. Thomas, "Sustainability of the Northwest Forest Plan," 14.

110. Richard W. Haynes et al., *Northwest Forest Plan—The First 10 Years (1994–2003): Synthesis of Monitoring and Research Results* (Portland, OR: Forest Service Pacific Northwest Research Station, 2006), 117–43.

111. Thomas, "Sustainability of the Northwest Forest Plan," 14.

112. G. H. Stankey et al., "Adaptive Management and the Northwest Forest Plan: Rhetoric and Reality," *Journal of Forestry* 101 (2003): 40–46; Bernard T. Bormann and George H. Stankey, "Crisis as a Positive Role in Implementing Adaptive Management after the Biscuit Fire, Pacific Northwest, U.S.A," in *Adaptive Environmental Management: A Practitioner's Guide*, ed. C. Allan and G. H. Stankey (New York: Springer, 2009).

113. Bernard T. Bormann and A. Ross Kiester, "Options Forestry: Acting on Uncertainty," *Journal of Forestry* 102 (June 2004): 156.

114. Martin Nie and Courtney Schultz, "Decision Making Triggers in Adaptive Management" (Portland, OR: Forest Service, Pacific Northwest Research Station, 2011).

115. Judith A. Layzer, *Natural Experiments: Ecosystem-Based Management and the Environment* (Cambridge, MA: MIT Press, 2008), 289.

116. Ronald J. Herring, "Authority and Scale in Political Ecology: Some Cautions on Localism," in *Biological Diversity: Balancing Interests through Adaptive Collaborative Management*, ed. Louise E. Buck et al. (New York: CRC Press, 2001), 200.

Chapter 7. ICBEMP: Procedural Ecosystem Management

1. For background information, see Kai N. Lee, *Compass and Gyroscope: Integrating Science and Politics for the Environment* (Washington, DC: Island Press, 1993).

2. See, for example, analysis of the Vision process's failure in Susan G. Clark, *Ensuring Greater Yellowstone's Future: Choices for Leaders and Citizens* (New Haven, CT: Yale University Press, 2008).

3. Thomas Quigley, Russell T. Graham, and Richard W. Haynes, "Interior Columbia Basin Ecosystem Management Project: Case Study," in *Bioregional Assessments: Science at the Crossroads of Management and Policy*, ed. K. Norman Johnson et al. (Washington, DC: Island Press, 1999), 271.

4. Nancy Langston, *Forest Dreams, Forest Nightmares* (Seattle: University of Washington Press, 1995), 265.

5. Quigley, Graham, and Haynes, "Interior Columbia Basin Ecosystem Management Project," 271.

6. Langston, *Forest Dreams, Forest Nightmares*, 265.

7. William R. Gast, Jr., et al., "Blue Mountains Forest Health Report" (Portland, OR: USDA Forest Service, Pacific Northwest Region, 1991), i.

8. Langston, *Forest Dreams, Forest Nightmares*, 265.

9. Gast et al., "Blue Mountains Forest Health Report," V-1.

10. Ibid., VI-11.

11. Mark G. Henjum et al., *Interim Protection for Late-Successional Forests, Fisheries, and Watersheds: National Forests East of the Cascade Crest, Oregon and Washington*, Wildlife Society Technical Review; 94-2 (Bethesda, MD: The Wildlife Society, 1994), 1.

12. Ibid., 6.

13. Richard Everett et al., *Eastside Forest Ecosystem Health Assessment*, Volume 1 (Portland, OR: USDA Forest Service Pacific Northwest Research Station, 1994).

14. Ibid., 3.

15. Ibid., 16.

16. Ibid., 6.

17. Ibid., 8.

18. Bernard T. Bormann et al., *Eastside Forest Ecosystem Health Assessment:* Volume 5 (Portland, OR: USDA Forest Service, Pacific Northwest Research Station, 1994).

19. Everett et al., *Eastside Forest Ecosystem Health Assessment*, Volume 1, 6.

20. Bormann et al., *Eastside Forest Ecosystem Health Assessment*, Volume 5.

21. Everett et al., *Eastside Forest Ecosystem Health Assessment*, Volume 1, 36.

22. Ibid., 37.

23. Michael C. Blumm, "The Amphibious Salmon: The Evolution of Ecosystem Management in the Columbia River Basin," *Ecology Law Quarterly* 24, no. 653 (1997): 653–76.

24. Quoted in *Pacific Rivers Council v. Robertson*, 845 F.Supp. 713 (D. Or. 1993).

25. Quigley, Graham, and Haynes, "Interior Columbia Basin Ecosystem Management Project."

26. 958 F.2D 290 (9th Cir. 1992).

27. Geoff Middaugh, "ICBEMP: A Health Appreciation for the Bizarre," Unpublished manuscript, March 9, 2011, on file with the author; Jack Ward Thomas and Jim Baca, "Interior Columbia Basin Ecosystem Management Project Charter" (Portland, OR: US Forest Service and Bureau of Land Management, 1994).

28. Quigley, Graham, and Haynes, "Interior Columbia Basin Ecosystem Management Project," 273.

29. Susan Giannettine et al., "Broadscale Assessments and Ecosystem Management: Four Perspectives," *Journal of Land, Resources, and Environmental Law* 21 (2001): 331–48; Martha G. Hahn, "Interior Columbia Basin Ecosystem Management Project: Management Review," in *Bioregional Assessments: Science at the Crossroads of Management and Policy*, ed. K. Norman Johnson et al. (Washington, DC: Island Press, 1999).

30. Senate Committee on Energy and Natural Resources, *Federal Forest Management*, 1st Session, April 5, 1995, 58.

31. Mike Dombeck, Christopher Wood, and Jack E. Williams, *From Conquest to Conservation: Our Public Lands Legacy* (Washington, DC: Island Press, 2003), 157.

32. ICBEMP, Executive Steering Committee Minutes, June 23, 1995, 4 and 6.

33. Jack Ward Thomas and Michael P. Dombeck, "Decision Notice/Decision Record for the Interim Strategies for Managing Anadromous Fish-Producing Watersheds in Eastern Oregon and Washington, Idaho, and Portions of California" (Washington, DC: Bureau of Land Management and US Forest Service, 1995).

34. *Friends of the Wild Swan, Inc. v. United States Forest Service*, 966 F.Supp. 1002 (D.Or 1997).

35. Thomas and Baca, "Interior Columbia Basin Ecosystem Management Project Charter."

36. ICBEMP, Executive Steering Committee Minutes, October 19, 1995, and December 19, 1995.

37. Richard W. Haynes, Russell T. Graham, and Thomas Quigley, "A Framework for Ecosystem Management in the Interior Columbia Basin" (Portland, OR: USDA Forest Service, 1996).

38. Ibid., 10.

39. Ibid., 11.

40. Ibid., 57.

41. Ibid., 18.

42. Ibid., 24–27.

43. Ibid., 30–31.

44. Ibid., 37.

45. "Status of the Interior Columbia Basin: Summary of Scientific Findings" (Portland, OR: USDA Forest Service, 1996), 18.

46. Ibid., 100.

47. James K. Agee, "Interior Columbia Ecosystem Management Project: Science Review," in *Bioregional Assessments: Science at the Crossroads of Management and Policy*, ed. K. Norman Johnson et al. (Washington, DC: Island Press, 1999).

48. "Status of the Interior Columbia Basin," 115.

49. Ibid., 133–42.

50. Ibid.; Thomas Quigley and Heidi Bigler Cole, "Highlighted Scientific Findings of the Interior Columbia Basin Ecosystem Management Project" (Portland, OR: USDA Forest Service, 1997).

51. ICBEMP Project History, http://www.icbemp.gov/html/icbhome.html.

52. Agee, "Interior Columbia Ecosystem Management Project: Science Review."

53. ICBEMP, Executive Steering Committee Minutes, January 16, 1996, 3.

54. ICBEMP, Executive Steering Committee Minutes, April 22, May 10, and May 31–June 1, 1996.

55. Jack Ward Thomas, *The Journals of a Forest Service Chief*, ed. Harold K. Steen (Seattle: Forest History Society and University of Washington Press, 2004), 374.

56. Ibid.

57. Ibid., 207.

58. "Eastside Environmental Impact Statement" (Portland, OR: Bureau of Land Management and US Forest Service, 1997); "Upper Columbia River Basin Draft Environmental Impact Statement" (Portland, OR: Bureau of Land Management and US Forest Service, 1997).

59. William S. Keeton and Gregory H. Aplet, "Ecosystem Management in the Interior Columbia River Basin" (Washington, DC: The Wilderness Society, 1997), 1.

60. Thomas M. Quigley, Kristine M. Lee, and Sylvia J. Arbelbide, "Evaluation of EIS Alternatives by the Science Integration Team" (Portland, OR: USDA Forst Service, Pacific Northwest Research Station, 1997).

61. ICBEMP, Executive Steering Committee Minutes, January 8, 1997.

62. ICBEMP, Executive Steering Committee Minutes, February 10, 1997.

63. ICBEMP, Executive Steering Committee Minutes, April 16, 1997.

64. Senate Committee on Energy and Natural Resources and the House Committee on Resources, *Interior Columbia Basin Ecosystem Management Plan*, 1st Session, May 15, 1997, 34.

65. Middaugh, "ICBEMP," 5.

66. ICBEMP, Executive Steering Committee Minutes, December 18, 1997.

67. House Subcommittee on Forest and Forest Health of the Committee on Resources, *Oversight Hearings on Clinton-Gore Administration's Forest Service Roadless Area Moratorium*, 2nd Session, February 25 and March 17, 1998, 61.

68. ICBEMP Project History, http://www.icbemp.gov/html/icbhome.html.

69. House Subcommittee on Forest and Forest Health of the Committee on Resources, *Hearings on the Interior Columbia Basin Ecosystem Management Project*, 2nd Session, March 10 and April 14, 1998, 1–2.

70. Senate Subcommittee on the Department of the Interior and Related Agencies of the Committee on Appropriations and the Subcommittee on Forests and Public Lands Management of the Committee on Energy and Natural Resources, *Columbia Basin Ecosystem Management Plan*, 2nd Session, April 29; May 1, 4, and 28; October 15; December 7, 1998, 31.

71. House Subcommittee on Forest and Forest Health, *Hearings on the Interior Columbia Basin Ecosystem Management Project*, 4.

72. Michael P. Dombeck, Interview by Harold K. Steen, May 15–18, 2003 (Durham, NC: Forest History Society).

73. Middaugh, "ICBEMP," 5.

74. ICBEMP, Executive Steering Committee Minutes, November 30, 1998.

75. ICBEMP, Executive Steering Committee Minutes, June 23, 1998.

76. "Interior Columbia Basin Ecosystem Management Project: Letter to Congress from the Secretaries of Agriculture and Interior," November 17, 1998, briefing paper on file with the author.

77. Middaugh, "ICBEMP."

78. Quoted in ICBEMP, Executive Steering Committee Minutes, May 26, 1998.

79. Frank Walker to William M. Daley, August 3, 1999, on file with the author.

80. Quoted in ICBEMP, Executive Steering Committee Minutes, May 26, 1998.

81. ICBEMP, Executive Steering Committee Minutes, October 7, 1998.

82. Quoted in ICBEMP, Executive Steering Committee Minutes, May 26, 1998.

83. "Interior Columbia Basin Supplemental Draft Environmental Impact Statement" (Portland, OR: Bureau of Land Management and the US Forest Service, 2000), 3.

84. ICBEMP, Executive Steering Committee Minutes, December 2, 1998.

85. Ibid.

86. Senate Committee on Energy and Natural Resources and the House Committee on Resources, *Interior Columbia Basin Ecosystem Management Plan*, 44.

87. "Preliminary Summary of Predominant Public Comments on ICBEMP Supplemental Draft EIS," Executive Steering Committee Minutes, September 11, 2000.

88. "Preliminary Summary of Comments on the ICBEMP SDEIS," Executive Steering Committee Minutes, September 2000, 4.

89. "90-Day Finding and Commencement of Status Review for a Petition to List the Western Sage Grouse in Washington as Threatened or Endangered," *Federal Register* 65, no. 165 (August 24, 2000): 51578–84.

90. "Preliminary Summary of Comments on the ICBEMP SDEIS."

91. Michael R. Crouse to Susan Giannettino, "Interior Columbia Basin Ecosystem Management Project Supplemental Draft Environmental Impact Statement," July 6, 2000, on file with the author.

92. "NMFS Comments on ICBEMP March 2000 SDEIS," July 6, 2000, on file with the author, 2.

93. *Friends of the Wild Swan and Prairie Wood Products v. Glickman*, 971 F.Supp. 457 (D.Or. 1997).

94. "Consultation on ICBEMP," Consultation discussion paper for the Executive Steering Committee meeting in October 2000, August 29, 2000, on file with the author.

95. Sherm Karl to Geoff Middaugh, "Consistency—Lack of Broad-Scale Component," September 9, 2000, on file with the author.

96. "Decision Points: Funding as a Transition Issue," Discussion document

for the Executive Steering Committee, November 29, 2000, on file with the author.

97. Ibid.

98. "Implications of the NMFS Supplemental Biological Opinion for OR/WA BLM," January 26, 2001.

99. "ICBEMP White Paper—BLM OR/WA Perspective," February 7, 2001, on file with the author.

100. Middaugh, "ICBEMP," 9.

101. "ICBEMP White Paper—BLM OR/WA Perspective."

102. Press Release, "Federal Land Agencies Adopt Science Strategy for Interior Columbia River Basin," Portland, OR, February 19, 2003.

103. "The Interior Columbia Basin Strategy: A Strategy for Applying the Knowledge Gained by the Interior Columbia Basin Ecosystem Management Project to the Revision of Forest and Resource Management Plans and Project Implementation" (Portland, OR: Bureau of Land Management and US Forest Service, 2003).

104. Agee, "Interior Columbia Ecosystem Management Project," 292.

Conclusion

1. B. Drummond Ayres, Jr., "The 2000 Campaign: Campaign Briefing," *New York Times*, August 25, 2000; Joel Connelly, "Bush Would Seize Reins on Lands Policy," *Seattle Post-Intelligencer*, June 26, 2000.

2. James R. Skillen, *The Nation's Largest Landlord: The Bureau of Land Management in the American West* (Lawrence: University Press of Kansas, 2009), 164–89.

3. Martin Nie, *The Governance of Western Public Lands: Mapping Its Present and Future* (Lawrence: University Press of Kansas, 2008), 87–124.

4. Judith A. Layzer, *Open for Business: Conservatives' Opposition to Environmental Regulation* (Cambridge, MA: MIT Press, 2012).

5. "The Environment: A Cleaner, Safer, Healthier America" (Alexandria, VA: The Luntz Research Companies, 2002).

6. "President Bush on Environmental Agenda," *Washington Post*, May 30, 2001.

7. Jacqueline Vaughn and Hanna J. Cortner, *George W. Bush's Healthy Forests: Reframing the Environmental Debate* (Boulder: University Press of Colorado, 2005).

8. Ibid.

9. Skillen, *The Nation's Largest Landlord*, 164–89; Roger Pielke, Jr., and Roberta Klein, "The Rise and Fall of the Science Advisor to the President of the United States," *Minerva* 47, no. 1 (2009): 7–29.

10. "Executive Order 13352: Facilitation of Cooperative Conservation," *Federal Register* 69, no. 137 (August 26, 2004).

11. http://georgewbush-whitehouse.archives.gov/government/norton-bio.html

12. Martin Nie, "Whatever Happened to Ecosystem Management and Federal

Land Planning?," in *The Laws of Nature: Reflections on the Evolution of Ecosystem Management Law and Policy*, ed. Kalyani Robbins (Akron, OH: University of Akron, 2013), 67.

13. Theodore J. Lowi, "Frontyard Propaganda," *Boston Review* (October/November 1999).

14. Nie, "Whatever Happened to Ecosystem Management and Federal Land Planning?"

15. Lynton K. Caldwell, "An Ecosystems Approach to Public Land Policy," in *Public Land Policy: Proceedings of the Western Resources Conference*, ed. Phillip O. Foss (Boulder: Colorado Associated University Press, 1968), 43.

16. Joseph L. Sax, "Ecosystems and Property Rights in Greater Yellowstone: The Legal System in Transition," in *The Greater Yellowstone Ecosystem: Redefining America's Wilderness Heritage*, ed. Robert B. Keiter and Mark S. Boyce (New Haven, CT: Yale University Press, 1991), 77.

17. Robert B. Keiter, "Beyond the Boundary Line: Constructing a Law of Ecosystem Management," *Colorado Law Review* 1994 (Spring): 293–333; Keiter, "Toward Legitimizing Ecosystem Management on the Public Domain," *Ecological Applications* 6, no. 3 (August 1996): 727–30; Keiter, "Ecosystems and the Law: Toward an Integrated Approach," *Ecological Applications* 8, no. 2 (May 1998): 332–41.

18. *Keeping Faith with Nature: Ecosystems, Democracy, and America's Public Lands* (New Haven, CT: Yale University Press, 2003), 307–8.

19. Ibid., 308.

20. Ibid., 309.

21. Matthew McKinney, Lynn Scarlett, and Daniel Kemmis, "Large Landscape Conservation: A Strategic Framework for Policy and Action" (Cambridge, MA: Lincoln Institute of Land Policy, 2010), 20.

22. George Cameron Coggins, "'Devolution' in Federal Land Law: Abdication by Any Other Name . . . ," *West-Northwest Journal of Environmental Law and Policy* 14 (2008): 211–18.

23. Leslie Bienen and Gary Tabor, "Applying an Ecosystem Approach to Brucellosis Control: Can an Old Conflict between Wildlife and Agriculture Be Successfully Managed?," *Frontiers in Ecology and the Environment* 4, no. 6 (August 2006): 319–27.

Bibliography

Achterman, Gail L., and Sally K. Fairfax. "The Public Participation Requirements of the Federal Land Policy and Management Act." *Arizona Law Review* 21, no. 2 (1979): 501–39.

"Administration Vows to Protect Spotted Owl, Logging Jobs." *Cox News Service*, June 27, 1990.

Agee, James K. "Interior Columbia Ecosystem Management Project: Science Review." In *Bioregional Assessments: Science at the Crossroads of Management and Policy*, edited by K. Norman Johnson, Frederick Swanson, Margaret Herring, and Sarah Greene, 288–92. Washington, DC: Island Press, 1999.

Agee, James K., and Darryll R. Johnson, eds. *Ecosystem Management for Parks and Wilderness.* Seattle: University of Washington Press, 1988.

Albright, Horace M., and Robert Cahn. *The Birth of the National Park Service: The Founding Years, 1913–33.* Chicago: Howe Brothers, 1985.

Alexander, Kristina. "Warranted but Precluded: What That Means under the Endangered Species Act (ESA)." Washington, DC: Congressional Research Service, 2010.

Allen, Craig R., and C. S. Holling, eds. *Discontinuities in Ecosystems and Other Complex Systems.* New York: Columbia University Press, 2008.

Allen, T. F. R., and Thomas B. Starr. *Hierarchy: Perspectives for Ecological Complexity.* Chicago: University of Chicago Press, 1982.

Anderson, Terry L., ed. *Breaking the Environmental Policy Gridlock.* Stanford, CA: Hoover Institution Press, 1997.

Ayres, B. Drummond, Jr. "The 2000 Campaign: Campaign Briefing." *New York Times*, August 25, 2000.

Babbitt, Bruce. *Cities in the Wilderness: A New Vision of Land Use in America.* Washington, DC: Island Press, 2005.

———. "ESA and Private Property: The Endangered Species Act and 'Takings': A Call for Innovation within the Terms of the Act." *Environmental Law* 24 (Spring 1994): 355–67.

———. "Inside Interior: Bruce Babbitt." Interview by Charles Wilkinson for the Center of the American West, University of Colorado, Boulder, CO, April 20, 2004.

———. "Science: Opening the Next Century of Conservation History." *Science* 267, no. 5206 (March 31, 1995): 1954–55.

Babbitt, Bruce, and Daniel Glickman. "Record of Decision for Final Environmental Impact Statement and Bison Management Plan for the State of Montana and Yellowstone National Park." Washington, DC: Departments of Interior and Agriculture, 2000.

Baker, Dave, Gery Ferguson, Craig Palmer, and Tim Tolle. "Implementation

Monitoring: Summary of Regional Interagency Monitoring Results." Portland, OR: Forest Service, Pacific Northwest Research Station, 2005.

Barbee, Robert D., Paul Schullery, and John D. Varley. "The Yellowstone Vision: An Experiment that Failed or a Vote for Posterity?" In *Partnerships in Parks and Preservation.* Conference proceedings. Albany, NY, September 9–12, 1991.

Barbour, Michael G. "Ecological Fragmentation in the Fifties." In *Uncommon Ground: Toward Reinventing Nature,* edited by William Cronon, 233–55. New York: W. W. Norton, 1995.

Bartuska, Ann M. "Ecosystem Management in the Forest Service." In *Silviculture: From the Cradle of Forestry to Ecosystem Management,* edited by L. H. Foley, 12–15. Asheville, NC: Forest Service Southeastern Forest Experiment Station, 1993.

Bath, Alistair J. "Public Attitudes about Wolf Restoration in Yellowstone." In *The Greater Yellowstone Ecosystem: Redefining America's Wilderness Heritage,* edited by Robert B. Keiter and Mark S. Boyce, 367–76. New Haven, CT: Yale University Press, 1991.

Bavington, Dean. "Managerial Ecology and Its Discontents: Exploring the Complexities of Control, Careful Use, and Coping in Resource and Environmental Management." *Environments* 30 (2002): 3–20.

Bedard, Paul. "Jobs Come before Owls, Bush Says in Northwest; Warns of Veto for Species Act." *Washington Times,* September 15, 1992.

Behan, Richard W. "The Irony of the Multiple Use/Sustained Yield Concept: Nothing Is So Powerful as an Idea Whose Time Has Passed." Paper presented at the Multiple Use and Sustained Yield: Changing Philosophies for Federal Land Management?, Washington, DC, March 5–6, 1992.

Bergandi, Donato. "'Reductionistic Holism': An Oxymoron or a Philosophical Chimera of Eugene Odum's Systems Ecology?" In *The Philosophy of Ecology: From Science to Synthesis,* edited by David R. Keller and Frank B. Golley, 204–17. Athens: University of Georgia Press, 2000.

Bienen, Leslie, and Gary Tabor. "Applying an Ecosystem Approach to Brucellosis Control: Can an Old Conflict between Wildlife and Agriculture Be Successfully Managed?" *Frontiers in Ecology and the Environment* 4, no. 6 (August 2006): 319–27.

Blumm, Michael C. "The Amphibious Salmon: The Evolution of Ecosystem Management in the Columbia River Basin." *Ecology Law Quarterly* 24, no. 653 (1997): 653–76.

Bormann, Bernard T., Martha H. Brookes, E. David Ford, A. Ross Kiester, Chadwick D. Oliver, and James F. Weigand. "Eastside Forest Ecosystem Health Assessment: Volume 5: A Framework for Sustainable-Ecosystem Management." Portland, OR: USDA Forest Service, Pacific Northwest Research Station, 1994.

Bormann, Bernard T., Patrick G. Cunningham, Martha H. Brookes, Van W. Manning, and Michael W. Collopy. "Adaptive Ecosystem Management in the Pacific Northwest." Portland, OR: USDA Forest Service, 1994.

Bormann, Bernard T., Richard W. Haynes, and Jon R. Martin. "Adaptive

Management of Forest Ecosystems: Did Some Rubber Hit the Road?" *BioScience* 57, no. 2 (February 2007): 186–91.
Bormann, Bernard T., and A. Ross Kiester. "Options Forestry: Acting on Uncertainty." *Journal of Forestry* 102 (June 2004): 22–27.
Bormann, Bernard T., and George H. Stankey. "Crisis as a Positive Role in Implementing Adaptive Management after the Biscuit Fire, Pacific Northwest, U.S.A." In *Adaptive Environmental Management: A Practitioner's Guide*, edited by C. Allan and G. H. Stankey, 143–67. New York: Springer, 2009.
Boyce, Stephen G., and Chadwick D. Oliver. "The History of Research in Forest Ecology and Silviculture." In *Forest and Wildlife Science in America*, edited by Harold K. Steen, 414–53. Durham, NC: Forest History Society, 1999.
Brown, G., and C. C. Harris. "The US Forest Service: Whither the New Resource Management Paradigm?" [In English]. *Journal of Environmental Management* 58, no. 1 (2000): 1–19.
Brown, Greg, and Charles C. Harris. "The US Forest Service: Toward the New Resource Management Paradigm?" *Society and Natural Resources* 5 (1992): 231–45.
Budd, Karen J. "Ecosystem Management: Will National Forests Be 'Managed' into National Parks?" In *The Greater Yellowstone Ecosystem: Redefining America's Wilderness Heritage*, edited by Robert B. Keiter and Mark S. Boyce, 65–76. New Haven, CT: Yale University Press, 1991.
Caldwell, Lynton K. "The Ecosystem as a Criterion for Public Land Policy." *Natural Resources Journal* 10, no. 2 (April 1970): 203–21.
———. "An Ecosystems Approach to Public Land Policy." In *Public Land Policy: Proceedings of the Western Resources Conference*, edited by Phillip O. Foss, 43–56. Boulder: Colorado Associated University Press, 1968.
———. "Environment: A New Focus for Public Policy?" *Public Administration Review* 23, no. 3 (September 1963): 132–39.
———. *The National Environmental Policy Act: An Agenda for the Future*. Bloomington: Indiana University Press, 1998.
———. "Politics and Public Land Policy." In *America's Public Lands: Politics, Economics and Administration: Conference on the Public Land Law Review Commission Report*, edited by Harriet Nathan, 297–329. San Francisco: Institute of Governmental Studies, University of California, Berkeley, 1970.
Camia, Catalina. "Interior: Opponents Fear Ramifications of Plant and Animal Survey." *CQ Weekly*, July 31, 1993, 1.
"Campaign '92: Transcript of the Third Presidential Debate (Part 1)." *Washington Post*, October 20, 1992.
Carson, Rachel. *Silent Spring*. New York: Houghton Mifflin, 1962.
Cawley, R. McGreggor. *Federal Land, Western Anger: The Sagebrush Rebellion and Environmental Politics*. Lawrence: University Press of Kansas, 1993.
CEQ. "Incorporating Biodiversity into Environmental Impact Analysis under the National Environmental Policy Act." Washington, DC: Council on Environmental Quality, 1993.
Charnley, Susan. "The Northwest Forest Plan as a Model for Broad-Scale

Ecosystem Management: A Social Perspective." *Conservation Biology* 20, no. 2 (April 2006): 330–40.

Chase, Alston. *In a Dark Wood: The Fight over Forests and the Rising Tyranny of Ecology*. New York: Houghton Mifflin, 1995.

Cheville, Norman F., Dale R. McCullough, and Lee R. Paulson. "Brucellosis in the Greater Yellowstone Area." Washington, DC: National Research Council, 1998.

Christensen, N. L., Ann M. Bartuska, James H. Brown, Stephen Carpenter, Carla D'Antonio, Robert Francis, Jerry F. Franklin, et al. "The Report of the Ecological Society of America Committee on the Scientific Basis for Ecosystem Management." *Ecological Applications* 6, no. 3 (1996): 665–91.

Clark, Susan G. *Ensuring Greater Yellowstone's Future: Choices for Leaders and Citizens.* New Haven, CT: Yale University Press, 2008.

Clarke, Jeanne Nienaber, and Daniel C. McCool. *Staking Out the Terrain: Power and Performance among Natural Resource Agencies.* Albany: State University of New York Press, 1996.

Clawson, Marion. "The Crisis in Outdoor Recreation." *American Forests* 65, no. 3 (March 1959): 22–31, 40–41.

———. "The Crisis in Outdoor Recreation, Part II." *American Forests* 65, no. 4 (April 1959): 28–35, 61–62.

Clawson, Marion, and Carlton S. Van Doren, eds. *Statistics on Outdoor Recreation.* Washington, DC: Resources for the Future, 1984.

Clements, Frederic E. *Research Methods in Ecology.* Lincoln, NE: University Publishing, 1905.

Clinton, Bill, and Al Gore. *Putting People First.* New York: Times Books, 1992.

Clinton, William J., and Albert Gore, Jr. "The Forest Plan: For a Sustainable Economy and a Sustainable Environment." Washington, DC: The White House, 1993.

Cloud, David S. "Environmental Record Puts Nominee under Attack." *Congressional Quarterly Weekly Report* (October 7, 1989): 1.

Cockburn, Alexander. "'Win-Win' with Bruce Babbitt: The Clinton Administration Meets the Environment." *New Left Review* 201 (September/October 1993): 46–59.

Coggins, George Cameron. "'Devolution' in Federal Land Law: Abdication by Any Other Name . . . ." *West-Northwest Journal of Environmental Law and Policy* 14 (2008): 211–18.

———. "Regulating Federal Natural Resources: A Summary Case against Devolved Collaboration." *Ecological Law Quarterly* 25 (1999): 602–10.

Cohn, Jeffrey P. "The National Biological Survey." *BioScience* 43, no. 8 (September 1993): 521–22.

Colwell, Robert K. "The Evolution of Ecology." *American Zoologist* 25, no. 3 (1985): 771–77.

———. "What's New? Community Ecology Discovers Biology." In *A New Ecology: Novel Approaches to Interactive Systems,* edited by Peter W. Price, C. N. Slobodchikoff, and William S. Gaud, 387–97. New York: John Wiley & Sons, 1984.

Connelly, Joel. "Bush Would Seize Reins on Lands Policy." *Seattle Post-Intelligencer*, June 26, 2000.

Cook, Brian J. *Bureaucracy and Self-Government: Reconsidering the Role of Public Administration in American Politics.* Baltimore, MD: Johns Hopkins University Press, 1996.

Cooper, Mary H. "Jobs vs. Environment: Should Jobs Be Sacrificed for the Sake of Environmental Protection?" *Congressional Researcher* 2, no. 18 (May 15, 1992): 409–32.

Corn, M. Lynne. "The National Biological Service." Washington, DC: Congressional Research Service, 1995.

Corn, M. Lynne, and Ross W. Gorte. "Yellowstone: Ecosystem, Resources, and Management." Washington, DC: Congressional Research Service, 1986.

Cortner, Hanna J., and Margaret A. Moote. *The Politics of Ecosystem Management.* Washington, DC: Island Press, 1999.

Costanza, Robert, Bryan G. Norton, and Benjamin Haskell, eds. *Ecosystem Health: New Goals for Environmental Management.* Washington, DC: Island Press, 1992.

Covington, W. Wallace, and Leonard F. DeBona, eds. *Sustainable Ecological Systems: Implementing an Ecological Approach to Land Management.* Flagstaff, AZ: USDA Forest Service: Rocky Mountain Forest and Range Experiment Station, 1993.

Cromley, Christina M. "Bison Management in Greater Yellowstone." In *Finding Common Ground: Governance and Natural Resources in the American West*, edited by Ronald D. Brunner, Christine H. Colburn, Christina M. Cromley, Roberta A. Klein, and Elizabeth A. Olson, 126–58. New Haven, CT: Yale University Press, 2002.

Czech, Brian, and Paul R. Krausman. *The Endangered Species Act: History, Conservation, Biology, and Public Policy.* Baltimore, MD: Johns Hopkins University Press, 2001.

Daly, Jayne E. "A Glimpse of the Past—A Vision for the Future: Senator Henry M. Jackson and National Land-Use Legislation." *Urban Lawyer* 28, no. 1 (Winter 1996): 7–39.

Dana, Samuel Trask, and Sally K. Fairfax. *Forest and Range Policy: Its Development in the United States.* 2nd ed. New York: McGraw-Hill, 1980.

Davis, Timothy, Todd A. Croteau, and Christopher H. Marston, eds. *America's National Park Roads and Parkways: Drawings from the Historic American Engineering Record.* Baltimore, MD: Johns Hopkins University Press, 2004.

DeWitt, John, Donald F. Kettl, Barbara Dyer, and W. Robert Lovan. "What Will New Governance Mean for the Federal Government." *Public Administration Review* 54, no. 2 (March-April 1994): 170–75.

Dillin, John. "Presidential Candidates Spar over US Environmental Issues." *Christian Science Monitor*, April 28, 1992.

Dobuzinskis, Laurent. "Historical and Epistemological Trends in Public Administration." *Journal of Management History* 3, no. 4 (1997): 298–316.

Dombeck, Michael P. Interview by Harold K. Steen, May 15–18, 2003. Durham, NC: Forest History Society.

Dombeck, Mike, Christopher Wood, and Jack E. Williams. *From Conquest to Conservation: Our Public Lands Legacy.* Washington, DC: Island Press, 2003.

Doremus, Holly. "The Endangered Species Act: Static Law Meets Dynamic World." *Washington University Journal of Law & Policy* 32 (2010): 175–235.

"Draft Supplemental Environmental Impact Statement on Management of Habitat for Late-Successional and Old-Growth Forest Related Species within the Range of the Northern Spotted Owl." Washington, DC: Department of Agriculture and Department of the Interior, 1993.

Duffus, James, III. "Endangered Species: Spotted Owl Petition Evaluation Beset by Problems." Washington, DC: GAO, 1989.

———. "Wildlife Management: Many Issues Unresolved in Yellowstone Bison-Cattle Brucellosis Conflict." Washington, DC: GAO, 1992.

Duncan, Sally. "Too Early to Tell, or Too Late to Rescue?: Adaptive Management under Scrutiny." *Science Findings,* no. 33 (2001): 1–5.

Dunlap, Thomas R. *Faith in Nature: Environmentalism as Religious Quest.* Seattle: University of Washington Press, 2004.

———. *Saving America's Wildlife.* Princeton, NJ: Princeton University Press, 1988.

Durant, Robert F. *The Administrative Presidency Revisited: Public Lands, the BLM, and the Reagan Revolution.* Albany: State University of New York Press, 1992.

Dwyer, Lynn E., Dennis D. Murphy, and Paul R. Ehrlich. "Property Rights Case Law and the Challenge to the Endangered Species Act." *Conservation Biology* 9, no. 4 (August 1995): 725–41.

"Eastside Environmental Impact Statement." Portland, OR: Bureau of Land Management and US Forest Service, 1997.

"Ecosystem Analysis at the Watershed Scale: Federal Guide for Watershed Analysis." Portland, OR: Regional Ecosystem Office, 1995.

"An Ecosystem Approach to Fish and Wildlife Conservation: An Approach to More Effectively Conserve the Nation's Biodiversity." Washington, DC: US Fish and Wildlife Service, 1994.

Egan, Michael. *Barry Commoner and the Science of Survival: The Remaking of American Environmentalism.* Cambridge, MA: MIT Press, 2007.

Egan, Timothy. *The Big Burn: Teddy Roosevelt and the Fire That Saved America.* New York: Houghton Mifflin Harcourt, 2009.

Egler, Frank E. *The Nature of Vegetation: Its Management and Mismanagement: An Introduction to Vegetation Science.* Norfolk, CT: Aton Forest, 1977.

Ehrlich, Paul R., and Anne H. Ehrlich. *Extinction: The Causes and Consequences of the Disappearance of Species.* New York: Random House, 1981.

Emerson, Kirk, and Charles R. Wise. "Statutory Approaches to Regulatory Takings: State Property Rights Legislation Issues and Implications for Public Administration." *Public Administration Review* 57, no. 5 (September–October 1997): 411–22.

"Endangered and Threatened Wildlife and Plants: Notice of Interagency Cooperative Policy for Endangered Species Act Section 9 Prohibitions." *Federal Register* 59, no. 126 (July 1, 1994): 34272.

"Endangered and Threatened Wildlife and Plants: Notice of Interagency

Cooperative Policy for Peer Review in Endangered Species Act Activities." *Federal Register* 59, no. 126 (July 1, 1994): 34270.
"Endangered and Threatened Wildlife and Plants: Notice of Interagency Cooperative Policy for the Ecosystem Approach to the Endangered Species Act." *Federal Register* 59, no. 126 (July 1, 1994): 34273.
"The Environment: A Cleaner, Safer, Healthier America." Alexandria, VA: The Luntz Research Companies, 2002.
Everett, Richard, Paul Hessburg, Mark Jensen, and Bernard Bormann. "Eastside Forest Ecosystem Health Assessment, Volume 1." Portland, OR: USDA Forest Service Pacific Northwest Research Station, 1994.
"Executive Order 13352: Facilitation of Cooperative Conservation." *Federal Register* 69, no. 137 (August 26, 2004): 52989–90.
Farnham, Timothy J. *Saving Nature's Legacy: Origins of the Idea of Biological Diversity*. New Haven, CT: Yale University Press, 2007.
"Federal Manual for Identifying and Delineating Jurisdictional Wetlands." Washington, DC: Federal Interagency Committee for Wetland Delineation, 1989.
FEMAT. "Forest Ecosystem Management: An Ecological, Economic, and Social Assessment." Washington, DC: GPO, 1993.
Fiedler, Peggy L., Robert A. Leidy, Richard D. Laven, Norman Gershenz, and Leslie Saul. "The Contemporary Paradigm in Ecology and Its Implications for Endangered Species Conservation." *Endangered Species Update* 10, no. 3 & 4 (January/February 1993): 7–12.
Final Supplemental Environmental Impact Statement on Management of Habitat for Late-Successional and Old-Growth Forest Related Species within the Range of the Northern Spotted Owl. Washington, DC: US Department of Agriculture and US Department of the Interior, February 1994.
Fischman, Robert L. *The National Widlife Refuges: Coordinating a Conservation System through Law*. Washington, DC: Island Press, 2003.
———. "The National Wildlife Refuge System and the Hallmarks of Modern Organic Legislation." *Ecological Law Quarterly* 29 (2002): 457–544.
Fitzsimmons, Allan K. *Defending Illusions: Federal Protection of Ecosystems*. New York: Rowman & Littlefield, 1999.
———. "Federal Ecosystem Management: A 'Train Wreck' in the Making." Washington, DC: Cato Institute, 1994.
Flippen, J. Brooks. *Nixon and the Environment*. Albuquerque: University of New Mexico Press, 2000.
Flournoy, Alyson C. "Preserving Dynamic Systems: Wetlands, Ecology, and Law." *Duke Environmental Law and Policy Forum* 7 (Winter 1996): 105–32.
Foley, L. H., ed. *Silviculture: From the Cradle of Forestry to Ecosystem Management*. Asheville, NC: USDA Forest Service: Southeastern Forest Experiment Station, 1993.
Foss, Phillip O. *Politics and Grass: The Administration of the Public Domain*. Seattle: University of Washington Press, 1960.
"A Framework for Coordination of National Parks and National Forests in the

Greater Yellowstone Area." Billings, MT: Greater Yellowstone Coordinating Committee, 1991.

Franklin, Jerry F. "Preserving Biodiversity: Species, Ecosystems, or Landscapes?" *Ecological Applications* 3, no. 2 (1993): 202–5.

Frederickson, H. George. "The Lineage of New Public Administration." *Administration and Society* 8, no. 2 (August 1976): 149–74.

———. *New Public Administration.* Tuscaloosa: University of Alabama Press, 1980.

———. "Whatever Happened to Public Administration? Governance, Governance Everywhere." Belfast: Institute of Governance, Public Policy, and Social Research, 2004.

Freeman, Richard. "The EcoFactory: The United States Forest Service and the Political Construction of Ecosystem Management." *Environmental History* 7, no. 4 (October 2002): 632–58.

Freemuth, John C. "Ecosystem Management and Its Place in the National Park Service." *University of Denver Law Review* 74 (1997): 697–727.

———. "The Emergence of Ecosystem Management: Reinterpreting the Gospel?" *Society and Natural Resources* 9, no. 4 (1996): 411–17.

———. *Islands under Siege: National Parks and the Politics of External Threats.* Lawrence: University Press of Kansas, 1991.

Frissell, Christopher A., Richard K. Nawa, and Reed Noss. "Is There Any Conservation Biology in 'New Perspectives'?: A Response to Salwasser." *Conservation Biology* 6, no. 3 (September 1992): 461–64.

Galston, William A., and Geoffrey L. Tibbetts. "Reinventing Federalism: The Clinton/Gore Program for a New Partnership among the Federal, State, Local, and Tribal Governments." *Publius* 24, no. 3 (Summer 1994): 23–48.

GAO. "Ecosystem Management: Additional Actions Needed to Adequately Test a Promising Approach." Washington, DC: GAO, 1994.

Garber, David. "History of the Greater Yellowstone Coordinating Committee." Bozeman, MT: Greater Yellowstone Coordinating Committee, 1994.

Garvey, Gerald. "False Promises: The NPR in Historical Perspective." In *Inside the Reinvention Machine: Appraising Governmental Reform,* edited by Donald F. Kettl and John J. DiIulio, Jr., 87–106. Washington, DC: Brookings Institution, 1995.

Gast, William R., Jr., Donald W. Scott, Craig Schmitt, David Clemens, Steven Howes, Charles G. Johnson, Jr., Robert Mason, Francis Mohr, and Robert A. Clapp, Jr. "Blue Mountains Forest Health Report." Portland, OR: USDA Forest Service, Pacific Northwest Region, 1991.

Giannettine, Susan, Kent Connaughton, Lance R. Clark, and Joan Frey. "Broadscale Assessments and Ecosystem Management: Four Perspectives." *Journal of Land, Resources, and Environmental Law* 21 (2001): 331–48.

Gillis, Anna Maria. "The New Forestry." *BioScience* 40, no. 8 (September 1990): 558–62.

Gilpin, Michael E., and Michael E. Soulé. "Minimum Viable Populations: Processes of Species Extinction." In *Conservation Biology: The Science of Scarcity and Diversity,* edited by Michael E. Soulé, 19–34. Sunderland, MA: Sinauer Associates, 1986.

Gleason, H. A. "The Individualistic Concept of the Plant Association." *Bulletin of the Torrey Botanical Club* 53 (1926): 7–26.

———. "The Individualistic Concept of the Plant Association." *American Midland Naturalist* 21 (1939): 92–110.

———. "The Structure and Development of the Plant Association." *Bulletin of the Torrey Botanical Club* 44 (1917): 463–81.

Glitzenstein, Eric R. "On the USFWS Settlement Regarding Federal Listing of Endangered Species." *Endangered Species Update* 10, no. 5 (March 1993): 1–3.

Goldstein, Bruce. "The Struggle over Ecosystem Management at Yellowstone." *BioScience* 42, no. 3 (March 1992): 183–87.

Goodnow, Frank. *Politics and Administration.* New York: Macmillan, 1900.

Gore, Al. "From Red Tape to Results: Creating a Government That Works Better and Costs Less." Washington, DC: National Performance Review, 1993.

Gorte, Ross W. "The Clinton Administration's Forest Plan for the Pacific Northwest." Washington, DC: Congressional Research Service, 1993.

Grant, Gordon E., and Frederick J. Swanson. "Watershed Analysis on the Public Lands: Where Are We Now and Where Should We Be Going?" In *Watersheds '94: Respect, Rethink, Restore,* edited by Richard R. Harris, Rick Kattelmann, Hannah Kerner, and Jeff Woled. Oakland, CA: University of California, Water Resources Center, 1994.

"Great Dismal Swamp May Lose Wetlands Protection." *Christian Science Monitor,* October 12, 1991.

Grumbine, R. Edward. "Reflections on 'What is Ecosystem Management'." *Conservation Biology* 11, no. 1 (1997): 41–47.

———. "What Is Ecosystem Management?" *Conservation Biology* 8, no. 1 (1994): 27–38.

Gulick, Luther. *American Forest Policy: A Study of Government Administration and Economic Control.* New York: Duell, Sloan, and Pearce, 1951.

———. "Management Is a Science." *Academy of Management Journal* 8, no. 1 (March 1965): 7–14.

———. "Politics, Administration, and the 'New Deal.'" *Annals of the American Academy of Political and Social Science* 169 (September 1933): 55–66.

Gulick, Luther, and Lyndall Urwick, eds. *Papers on the Science of Administration.* New York: Columbia University, Institute of Public Administration, 1937.

Gunderson, Lance H., and C. S. Holling, eds. *Panarchy: Understanding Transformations in Human and Natural Systems.* Washington, DC: Island Press, 2002.

Gunderson, Lance H., C. S. Holling, and Stephen S. Light, eds. *Barriers and Bridges to the Renewal of Ecosystems and Institutions.* New York: Columbia University Press, 1995.

GYCC. "The Greater Yellowstone Area: An Aggregation of National Park and National Forest Management Plans." Greater Yellowstone Coordinating Committee, 1987.

GYCC. "Vision for the Future: A Framework for Coordination in the Greater Yellowstone Area." Washington, DC: GPO, 1990.

Hahn, Martha G. "Interior Columbia Basin Ecosystem Management Project: Management Review." In *Bioregional Assessments: Science at the Crossroads of*

Management and Policy, edited by K. Norman Johnson, Frederick Swanson, Margaret Herring, and Sarah Greene, 293–98. Washington, DC: Island Press, 1999.

Haines, Aubrey. *The Yellowstone Story: A History of Our First National Park, Vol. 1.* Boulder: University Press of Colorado, 1999.

Halverson, Anders. *An Entirely Synthetic Fish: How Rainbow Trout Beguiled America and Overran the World.* New Haven, CT: Yale University Press, 2011.

Hatfield, Mark O. "Can't See the Forest for the Endangered Species." *Washington Post*, June 12, 1992.

Haufler, Johnathan B., Carolyn A. Mehl, and Gary J. Roloff. "Using a Coarse-Filter Approach with Species Assessment for Ecosystem Management." *Wildlife Society Bulletin* 24, no. 2 (Summer 1996): 200–208.

Hawthorne, Donald W. "The History of Federal and Cooperative Animal Damage Control." *Sheep & Goat Research Journal* 19 (2004): 13–15.

Haynes, Richard W., Bernard T. Bormann, Danny C. Lee, and Jon R. Martin. "Northwest Forest Plan—The First 10 Years (1994–2003): Synthesis of Monitoring and Research Results." Portland, OR: Forest Service Pacific Northwest Research Station, 2006.

Haynes, Richard W., Russell T. Graham, and Thomas Quigley. "A Framework for Ecosystem Management in the Interior Columbia Basin." Portland, OR: USDA Forest Service, 1996.

Hays, Samuel P. *Beauty, Health and Permanence: Environmental Politics in the United States, 1955–1985.* New York: Cambridge University Press, 1987.

———. *Conservation and the Gospel of Efficiency: The Progressive Conservation Movement, 1890–1920.* Cambridge, MA: Harvard University Press, 1959.

———. "From Conservation to Environment: Environmental Politics in the United States since World War II." *Environmental Review* 6 (Fall 1982): 14–41.

———. *A History of Environmental Politics since 1945.* Pittsburgh, PA: University of Pittsburgh Press, 2000.

———. *Wars in the Woods: The Rise of Ecological Forestry in America.* Pittsburgh, PA: University of Pittsburgh Press, 2007.

Healy, Jon. "Ten Environmental Nettles That Await the 103rd Congress." *Congressional Quarterly Weekly Report* 51, no. 7 (February 13, 1992): 310–11.

Henjum, Mark G., James R. Karr, Ellen W. Chu, and Eastside Forests Scientific Society Panel. *Interim Protection for Late-Successional Forests, Fisheries, and Watersheds: National Forests East of the Cascade Crest, Oregon and Washington.* Wildlife Society technical review; 94–92. Bethesda, MD: The Wildlife Society, 1994.

Herring, Ronald J. "Authority and Scale in Political Ecology: Some Cautions on Localism." In *Biological Diversity: Balancing Interests through Adaptive Collaborative Management*, edited by Louise E. Buck, Charles C. Geisler, John Schelhas, and Eva Wollenberg, 187–204. New York: CRC Press, 2001.

Hirt, Paul W. *A Conspiracy of Optimism: Management of the National Forests since World War Two.* Lincoln: University of Nebraska Press, 1994.

Holland, John. *Hidden Order: How Adaptation Builds Complexity.* Reading, MA: Addison-Wesley, 1995.

Holling, C. S. *Adaptive Environmental Assessment and Management.* New York: John Wiley, 1978.
———. "Resilience and Stability of Ecological Systems." *Annual Review of Ecological Systems* 4 (1973): 1–23.
———. "Surprise for Science, Resilience for Ecosystems, and Incentives for People." *Ecological Applications* 6, no. 3 (August 1996): 733–35.
———. "Understanding the Complexity of Economic, Ecological, and Social Systems." *Ecosystems* 4, no. 5 (August 2001): 390–405.
Holling, C. S., and Gary K. Meffe. "Command and Control and the Pathology of Natural Resource Management." *Conservation Biology* 10, no. 2 (April 1996): 328–37.
Holmes, Oliver Wendell. "The Path of the Law." *Harvard Law Review* 10, no. 8 (1897).
House Committee on Merchant Marine and Fisheries, Subcommittee on National Parks, Forests, and Public Lands. *H.R. 1845: A Bill to Establish the Biological Survey in the Department of the Interior,* July 15, 1993.
House Subcommittee on Forest and Forest Health of the Committee on Resources. *Hearings on the Interior Columbia Basin Ecosystem Management Project,* 2nd Session, March 10 and April 14, 1998.
House Subcommittee on Forest and Forest Health of the Committee on Resources. *Oversight Hearings on Clinton-Gore Administration's Forest Service Roadless Area Moratorium,* 2nd Session, February 25 and March 17, 1998.
Houston, Douglas B. "Ecosystems of National Parks." *Science* 172, no. 3984 (May 14, 1971): 648–51.
Hungerford, Andrea L. "Changing the Management of Public Land Forests: The Role of the Spotted Owl Injunctions." *Environmental Law* 24 (1994): 1395–434.
IEMTF. "The Ecosystem Approach: Healthy Ecosystems *and* Sustainable Economies." Washington, DC: GPO, 1995.
"Interference in Environmental Programs by Political Appointees: The Improper Treatment of a Senior Executive Service Official: A Report." Washington, DC: GPO, 1993.
"The Interior Columbia Basin Strategy: A Strategy for Applying the Knowledge Gained by the Interior Columbia Basin Ecosystem Management Project to the Revision of Forest and Resource Management Plans and Project Implementation." Portland, OR: Bureau of Land Management and US Forest Service, 2003.
"Interior Columbia Basin Supplemental Draft Environmental Impact Statement." Portland, OR: Bureau of Land Management and the US Forest Service, 2000.
"Interior Secretary Sees Less Impact from Owl Plan." *Seattle Times,* August 27, 1990.
Isenberg, Andrew C. *The Destruction of the Bison.* New York: Cambridge University Press, 2000.
Jester, Jennifer. "Habitat Conservation Plans under Section 10 of the Endangered Species Act: The Alabama Beach Mouse and the Unfulfilled Mandate

of Species Recovery." *Boston College Environmental Affairs Law Review* 26, no. 1 (1998): 131–87.

Johnson, Brigitte W. "The American Tree Farm System: Growing Stewardship from the Roots." *Forest History Today* 17, no. 1 (Spring 2012): 16–21.

Johnson, K. Norman. "Science-Based Assessment of the Forests of the Pacific Northwest." In *Creating a Forestry for the 21st Century*, edited by Kathryn A. Kohm and Jerry F. Franklin, 397–410. Washington, DC: Island Press, 1997.

Johnson, Ronald N. "Ecosystem Management and Reinventing Government." In *Breaking the Environmental Policy Gridlock*, edited by Terry L. Anderson, 22–52. Stanford, CA: Hoover Institution Press, 1997.

Kaufman, Herbert. "Emerging Conflicts in the Doctrines of Public Administration." *American Political Science Review* 50, no. 4 (December 1956): 1057–74.

———. "Music of the Spheres: A Lifetime of Study of Public Administration." *Public Administration Review* 56, no. 2 (March-April 1996): 127–38.

———. "Organizational Theory and Political Theory." *American Political Science Review* 58, no. 1 (March 1964): 5–14.

Kaufmann, Merrill R., Russell T. Graham, Brian J. Boyle, William H. Moir, Lee Perry, Richard T. Reynolds, Richard L. Bassett, et al. "An Ecological Basis for Ecosystem Management." Fort Collins, CO: US Forest Service, 1994.

Keeton, William S., and Gregory H. Aplet. "Ecosystem Management in the Interior Columbia River Basin." Washington, DC: The Wilderness Society, 1997.

Keiter, Robert B. "Beyond the Boundary Line: Constructing a Law of Ecosystem Management." *Colorado Law Review* 1994 (Spring): 293–333.

———. "Ecological Concepts, Legal Standards, and Public Land Law: An Analysis and Assessment." *Natural Resources Journal* 44 (Fall 2004): 943–88.

———. "Ecosystems and the Law: Toward an Integrated Approach." *Ecological Applications* 8, no. 2 (May 1998): 332–41.

———. "Greater Yellowstone's Bison: Unraveling of an Early American Wildlife Conservation Achievement." *Journal of Wildlife Management* 61, no. 1 (1997): 1–11.

———. *Keeping Faith with Nature: Ecosystems, Democracy, and America's Public Lands.* New Haven, CT: Yale University Press, 2003.

———. "Taking Account of the Ecosystem on the Public Domain: Law and Ecology in the Greater Yellowstone Region." *Colorado Law Review* 60, no. 4 (1989): 924–1007.

———. "Toward Legitimizing Ecosystem Management on the Public Domain." *Ecological Applications* 6, no. 3 (August 1996): 727–30.

Keiter, Robert B., and Mark S. Boyce, eds. *The Greater Yellowstone Ecosystem: Redefining America's Wilderness Heritage.* New Haven, CT: Yale University Press, 1991.

Keiter, Robert B., and Peter H. Froelicher. "Bison, Brucellosis, and Law in the Greater Yellowstone Ecosystem." *Land and Water Law Review* 28, no. 1 (1993): 1–75.

Kenworthy, Tom. "Pragmatic Critic Is Set to Be Interior's Next Landlord." *Washington Post*, January 19, 1993, 9.

Kerr, Andy. "Ecosystem Management Must Include the Most Human of Factors." *BioScience* 45, no. 6 (June 1995): 378.

Kickert, Walter J. M. "Public Governance in the Netherlands: An Alternative to Anglo-American 'Managerialism.'" *Public Administration* 75, no. 4 (Winter 1997): 731–52.

Kingsland, Sharon E. *The Evolution of American Ecology, 1890–2000.* Baltimore, MD: Johns Hopkins University Press, 2005.

Klein, Roberta A. "Wolf Recovery in the Northern Rockies." In *Finding Common Ground: Governance and Natural Resources in the American West,* edited by Ronald D. Brunner, Christine H. Colburn, Christina M. Cromley, Roberta A. Klein, and Elizabeth A. Olson, 88–125. New Haven, CT: Yale University Press, 2002.

Klyza, Christopher James. *Who Controls Public Lands?: Mining, Forestry, and Grazing Policies, 1870–1990.* Chapel Hill: University of North Carolina Press, 1996.

Krahe, Diane. "The Ill-Fated NBS: A Historical Analysis of Bruce Babbitt's Vision to Overhaul Interior Science." Paper presented at Rethinking Protected Areas in a Changing World: Proceedings of the 2011 George Wright Society Biennial Conference on Parks, Protected Areas, and Cultural Sites, Hancock, MI, 2012.

Kusler, Jon. "Wetlands Delineation: An Issue of Science or Politics." *Environment* 34, no. 2 (March 1992): 7–11, 29–37.

Lackey, Robert T. "Is Science Biased toward Natural?" *Northwest Science Forum* 83, no. 3 (2009): 277–79.

———. "Seven Pillars of Ecosystem Management." *Landscape and Urban Planning* 40, no. 1/3 (1998): 21–30.

Lan, Zhiyong, and David H. Rosenbloom. "Public Administration in Transition?" *Public Administration Review* 52, no. 6 (November-December 1992): 535–37.

Lancaster, John. "Environmentalists Hit 'Shocking' Appointment: Political Battle Looms over USDA Nominee." *Washington Post,* July 11, 1989.

———. "Lujan: Endangered Species Act 'Too Tough,' Needs Change." *Washington Post,* May 12, 1990.

———. "New Look, Old Mission at Lujan's Department: Interior Still Bound to the West." *Washington Post,* January 19, 1990.

Langston, Nancy. *Forest Dreams, Forest Nightmares.* Seattle: University of Washington Press, 1995.

Larmer, Paul. "Cut to the Past: Logging Wars Resume." *High Country News* 45 (October 16, 1995).

LaRoe, Edward T. "Implementation of an Ecosystem Approach to Endangered Species Conservation." *Endangered Species Update* 10, no. 3 & 4 (January/February 1993): 3–6.

Lasswell, Harold D. *Politics: Who Gets What, When, How.* New York: Peter Smith, 1950.

Lavigne, Jean. "Where the Buffalo Roam: Boundaries and Politics of Scale in the Yellowstone Region." *GeoJournal* 58, no. 4 (2002): 285–92.

Lawrence, Nathaniel, and Dennis Murphy. "New Perspectives or Old Priorities?" *Conservation Biology* 6, no. 3 (September 1992): 465–68.

Layzer, Judith A. *Natural Experiments: Ecosystem-Based Management and the Environment.* Cambridge, MA: MIT Press, 2008.

———. *Open for Business: Conservatives' Opposition to Environmental Regulation.* Cambridge, MA: MIT Press, 2012.

Lee, Kai N. *Compass and Gyroscope: Integrating Science and Politics for the Environment.* Washington, DC: Island Press, 1993.

Leonard, George M. Interview by Harold K. Steen. Durham, NC: Forest History Society, 1999.

Leopold, Aldo. *A Sand County Almanac.* New York: Oxford University Press, 1966.

Leopold, A. S. , S. A. Cain, C. M. Cottam, I. N. Gabrielson, and T. L. Kimball. "Wildlife Management in the National Parks." Washington, DC: Department of the Interior, 1963.

Leshy, John D. "Is the Multiple Use/Sustained Yield Management Philosophy Still Applicable Today?" Paper presented at the Multiple Use and Sustained Yield: Changing Philosophies for Federal Land Management?, Washington, March 5 & 6, 1992.

Levin, Simon. "The Problem of Pattern and Scale in Ecology: The Robert H. MacArthur Award Lecture." *Ecology* 73, no. 6 (December 1992): 1943–67.

Levins, Richard. "The Limits of Complexity." In *Hierarchy Theory: The Challenge of Complex Systems,* edited by Howard H. Pattee, 111–27. New York: George Braziller, 1973.

Lewis, William M., Jr. "Wetlands: Characteristics and Boundaries." Washington, DC: National Research Council, 1995.

Lindblom, Charles E. "The Science of Muddling Through." *Public Administration Review* 19, no. 2 (Spring 1959): 79–88.

Lindstrom, Matthew J., and Zachary A. Smith. *The National Environmental Policy Act: Judicial Misconstruction, Legislative Indifference, and Executive Neglect.* College Station: Texas A&M University Press, 2001.

Litchman, Pamela, and Tim W. Clark. "Rethinking the 'Vision' Exercise in the Greater Yellowstone Ecosystem." *Society and Natural Resources* 7, no. 5 (1994): 459–78.

Lovelock, James E. "Gaia as Seen through the Atmosphere." *Atmospheric Environment* 6 (1972): 579–80.

Lovelock, James E., and Lynn Margulis. "Atmospheric Homeostasis by and for the Biosphere: The Gaia Hypothesis." *Tellus* 26, no. 1-2 (1974): 2–10.

Lowi, Theodore J. *The End of Liberalism: The Second Republic of the United States.* 2nd ed. New York: W. W. Norton, 1979.

———. "Four Systems of Policy, Politics and Choice." *Public Administration Review* 4 (July/August 1972): 298–310.

———. "Frontyard Propaganda." *Boston Review* (October/November 1999).

Lujan, Manuel, Jr. "Inside Interior Interview." By Charles F. Wilkinson, March 17, 2004. Boulder, CO: Center of the American West, University of Colorado.

Luther, Linda. "The National Environmental Policy Act: Background and Implementation." Washington, DC: Congressional Research Service, 2005.

Manley, Patricia N., ed. "Draft Region 5 Ecosystem Management Guidebook." Vallejo, CA: US Forest Service Region 5, 1994.

March, James G., and Herbert Simon. *Organizations.* New York: Wiley, 1958.
Marcot, Bruce G., and Randy Molina. "Conservation of Other Species Associated with Older Forest Conditions." In *Northwest Forest Plan—The First 10 Years (1994–2003): Synthesis of Monitoring and Research Results,* edited by Richard W. Haynes, Bernard T. Bormann, Danny C. Lee, and Jon R. Martin, 145–49. Portland, OR: Pacific Northwest Research Station, 2006.
Marcot, Bruce G., and Dennis D. Murphy. "On Population Viability Analysis and Management." In *Biodiversity in Managed Landscapes: Theory and Practice,* edited by Robert C. Szaro and David W. Johnston, 58–76. New York: Oxford University Press, 1996.
Marcot, Bruce G., and Jack Ward Thomas. "Of Spotted Owls, Old Growth, and New Policies: A History since the Interagency Scientific Committee Report." Washington, DC: USDA Forest Service, 1997.
Marshall, Eliot. "Campaigning on the Environment." *Science* 241, no. 4873 (September 23, 1988): 1589–90.
Marston, Ed. "Interior View: Bruce Babbitt Took the Real West to Washington: A High Country News Interview." *High Country News* 33, no. 3 (February 12, 2001): 8–11.
Maser, Chris. "Do We Owe Anything to the Future?" Paper presented at Multiple Use and Sustained Yield: Changing Philosophies for Federal Land Management?, Washington, DC, March 5–6, 1992.
Mason, Robert, and Bruce Babbitt. "Interview: Bruce Babbitt." *Natural Resources and Environment* 11, no. 1 (Summer 1996): 35–41.
McCarthy, James. "Neoliberalism and the Politics of Alternatives: Community Forestry in British Columbia and the United States." *Annals of the Association of American Geographers* 96, no. 1 (2006): 84–104.
McConnell, Grant. *Private Power and American Democracy.* New York: Vintage Books, 1966.
McIntosh, Robert P. "The Background and Some Current Problems of Theoretical Ecology." *Synthese* 43, no. 2 (February 1980): 195–255.
———. "The Succession of Succession: A Lexical Chronology." *Bulletin of the Ecological Society of America* 80, no. 4 (October 1999): 256–65.
McKinney, Matthew, Lynn Scarlett, and Daniel Kemmis. "Large Landscape Conservation: A Strategic Framework for Policy and Action." Cambridge, MA: Lincoln Institute of Land Policy, 2010.
McNeel, Joseph F., and Ron L. Copstead. "Operational Factors Affecting the Implementation of Ecosystem Management Objectives." In *Silviculture: From the Cradle of Forestry to Ecosystem Management,* edited by L. H. Foley, 167–77. Asheville, NC: USDA Forest Service Southeastern Forest Experiment Station, 1993.
Meagher, Margaret Mary. "The Bison of Yellowstone National Park." Washington, DC: US National Park Service, 1973.
Mealey, S. P., ed. *Guidelines for Management Involving Grizzly Bears in the Greater Yellowstone Area.* USDA Forest Service and USDI National Park Service, 1979.
Mech, L. David. "Returning the Wolf to Yellowstone." In *The Greater Yellowstone*

Ecosystem: Redefining America's Wilderness Heritage, edited by Robert B. Keiter and Mark S. Boyce, 309–322. New Haven, CT: Yale University Press, 1991.

Meffe, Gary K., ed. *Ecosystem Management: Adaptive, Community-Based Conservation.* Washington, DC: Island Press, 2002.

Merrill, Karen R. *Public Lands and Political Meaning: Ranchers, the Government, and the Property between Them.* Berkeley: University of California Press, 2002.

Molina, Randy, Bruce G. Marcot, and Robin Lesher. "Protecting Rare, Old-Growth, Forest-Associated Species under the Survey and Manage Program Guidelines of the Northwest Forest Plan." *Conservation Biology* 20, no. 2 (April 2006): 306–18.

Molina, Randy, Dan McKenzie, Robin Lesher, Jan Ford, Jim Alegria, and Richard Cutler. "Strategic Survey Framework for the Northwest Forest Plan Survey and Manage Program." Portland, OR: Forest Service Pacific Northwest Research Station, 2003.

Morrissey, Wayne A., Jeffrey A. Zinn, and Lynne M. Corn. "Ecosystem Management: Federal Agency Activities." Washington, DC: Congressional Research Service, 1994.

Mulder, Barry S., Barry R. Noon, Thomas A. Spies, Martin G. Raphael, Craig J. Palmer, Anthony R. Olsen, Gordon H. Reeves, and Hartwell H. Welsh. "The Strategy and Design of the Effectiveness Monitoring Program for the Northwest Forest Plan." Portland, OR: Forest Service, Pacific Northwest Research Station 1999.

Nagle, John Copeland. *Law's Environment: How the Law Shapes the Places We Live.* New Haven, CT: Yale University Press, 2010.

Nazzaro, Robin M. "Preliminary Observations on the Implementation of the Interagency Bison Management Plan." Washington, DC: GAO, 2007.

———. "Yellowstone Bison: Interagency Plan and Agencies' Management Need Improvement to Better Address Bison-Cattle Brucellosis Controversy." Washington, DC: GAO, 2008.

Nelson, Daniel. *Frederick W. Taylor and the Rise of Scientific Management.* Madison: University of Wisconsin Press, 1980.

Nicolson, Malcolm, Robert P. McIntosh, and M. Nicholson. "H. A. Gleason and the Individualistic Hypothesis Revisited." *Bulletin of the Ecological Society of America* 83, no. 2 (April 2002): 133–42.

Nie, Martin. *The Governance of Western Public Lands: Mapping Its Present and Future.* Lawrence: University Press of Kansas, 2008.

———. "The Underappreciated Role of Regulatory Enforcement in Natural Resource Conservation." *Policy Sciences* 41, no. 2 (June 2008): 139–64.

———. "Whatever Happened to Ecosystem Management and Federal Land Planning?" In *The Laws of Nature: Reflections on the Evolution of Ecosystem Management Law and Policy*, edited by Kalyani Robbins, 67–94. Akron, OH: University of Akron, 2013.

Nie, Martin, and Courtney Schultz. "Decision Making Triggers in Adaptive Management." Portland, OR: Forest Service, Pacific Northwest Research Station, 2011.

"90-Day Finding and Commencement of Status Review for a Petition to List the

Western Sage Grouse in Washington as Threatened or Endangered." *Federal Register* 65, no. 165 (August 24, 2000): 51578–84.
Noecker, Robert J. "Reintroduction of Wolves." Washington, DC: Congressional Research Service, 1997.
Nolon, John R. "The National Land Use Policy Act." *Pace Environmental Law Review* 13 (1996): 519–23.
Norris, Logan A. "Science Review." In *Bioregional Assessments: Science at the Crossroads of Management and Policy*, edited by K. Norman Johnson, Frederick Swanson, Margaret Herring, and Sarah Greene, 117–20. Washington, DC: Island Press, 1999.
Norse, Elliot. "A River That Flows to the Sea: The Marine Biological Diversity Movement." *Oceanography* 9, no. 1 (1996): 5–9.
Norton, Bryan G. *Sustainability: A Philosophy of Adaptive Ecosystem Management.* Chicago: University of Chicago Press, 2005.
Noss, Reed F. "Issues of Scale in Conservation Biology." In *Conservation Biology: The Theory and Practice of Nature Conservation, Preservation, and Management*, edited by Peggy L. Fiedler and Subodh K. Jain, 239–52. New York: Chapman and Hill, 1992.
———. "Some Principles of Conservation Biology, as They Apply to Environmental Law." *Chicago-Kent Law Review* 69, no. 4 (1994): 893–909.
Odum, Eugene P. *Fundamentals of Ecology*. Philadelphia: Saunders, 1953.
———. "The New Ecology." *BioScience* 14, no. 7 (July 1964): 14–16.
"One Third of the Nation's Land: A Report to the President and to Congress by the Public Land Law Review Commission." Washington, DC: Public Land Law Review Commission, 1970.
Ostrom, Vincent. *The Intellectual Crisis in American Public Administration.* 3rd ed. Tuscaloosa: University of Alabama Press, 2008.
Pahre, Robert. "Political Opposition to Transboundary Cooperation in the Greater Yellowstone Area." *Journal of Tourism and Leisure Studies* 17, no. 2 (August 2011): 99–128.
Paulsen, David Frederick. "An Approach to Organizational Analysis: A Case Study of the Bureau of Land Management." PhD diss., University of Washington, 1966.
Peffer, E. Louise. *The Closing of the Public Domain: Disposal and Reservation Policies 1900–50.* Stanford, CA: Stanford University Press, 1951.
Peterson, J. "Threats to Greater Yellowstone: Inventory of Management Issues and Proposed Projects in the GYE." Bozeman, MT: Greater Yellowstone Coalition, 1984.
Peterson, Shannon. *Acting for Endangered Species: The Statutory Ark.* Lawrence: University Press of Kansas, 2002.
Pielke, Roger, Jr., and Roberta Klein. "The Rise and Fall of the Science Advisor to the President of the United States." *Minerva* 47, no. 1 (March 2009): 7–29.
Pimm, Stuart L. *The Balance of Nature? Ecological Issues in the Conservation of Species and Communities.* Chicago: University of Chicago Press, 1991.
Pinchot, Gifford. *The Fight for Conservation.* New York: Doubleday, Page, 1910.

———. "The Use of the National Forest Reserves: Regulations and Instructions." Washington, DC: US Department of Agriculture, 1905.

Pipkin, James. "The Northwest Forest Plan Revisited." Washington, DC: Council on Environmental Quality, 1998.

Power, Thomas Michael. "Public Timber Supply, Market Adjustments, and Local Economies: Economic Assumptions of the Northwest Forest Plan." *Conservation Biology* 20, no. 2 (April 2006): 341–50.

"President Bush on Environmental Agenda." *Washington Post*, May 30, 2001.

Price, Peter W., C. N. Slobodchikoff, and William S. Gaud, eds. *A New Ecology: Novel Approaches to Interactive Systems*. New York: John Wiley & Sons, 1984.

Pritchard, James A. *Preserving Yellowstone's Natural Conditions: Science and the Perception of Nature*. Lincoln: University of Nebraska Press, 1999.

Proctor, James D. "Whose Nature? The Contested Moral Terrain of Ancient Forests." In *Uncommon Ground: Toward Reinventing Nature*, edited by William Cronon, 269–97. New York: W. W. Norton, 1995.

"Proposed Rule Exempting Certain Small Landowners and Low-Impact Activities from Endangered Species Act Requirements for Threatened Species." *Federal Register* 60, no. 139 (July 20, 1995): 37419–23.

"Protecting America's Heritage: A Fair, Cooperative and Scientifically Sound Approach to Improving the Endangered Species Act." Washington, DC: US Fish and Wildlife Service, 1995.

Public Papers of the Presidents of the United States, William J. Clinton, 1993. Washington, DC: GPO, 1994.

Pulliam, H. Ronald. "The Political Education of a Biologist: Part 1." *Wildlife Society Bulletin* 26, no. 2 (Summer 1998): 199–202.

Pyne, Stephen J. *Fire in America: A Cultural History of Wildland and Rural Fire*. Seattle: University of Washington Press, 1997.

———. *Year of the Fires: The Story of the Great Fires of 1910*. New York: Viking, 2001.

Pytte, Alyson. "Bush's Modest Proposal." *CQ Weekly*, September 29, 1990, 3105.

Quigley, Thomas, and Heidi Bigler Cole. "Highlighted Scientific Findings of the Interior Columbia Basin Ecosystem Management Project." Portland, OR: USDA Forest Service, 1997.

Quigley, Thomas, Russell T. Graham, and Richard W. Haynes. "Interior Columbia Basin Ecosystem Management Project: Case Study." In *Bioregional Assessments: Science at the Crossroads of Management and Policy*, edited by K. Norman Johnson, Frederick Swanson, Margaret Herring, and Sarah Greene. Washington, DC: Island Press, 1999.

Quigley, Thomas M., Kristine M. Lee, and Sylvia J. Arbelbide. "Evaluation of EIS Alternatives by the Science Integration Team." Portland, OR: USDA Forst Service, Pacific Northwest Research Station, 1997.

Raven, Peter. "A Biological Survey for the Nation: Committee on the Formation of the National Biological Survey and Commission on the Formation of the National Biological Survey." Washington, DC: National Academy of Sciences, 1993.

"Record of Decision: Amendments to Forest Service and Bureau of Land Management Planning Documents within the Range of the Northern Spotted

Owl." Washington, DC: Department of Agriculture and Department of the Interior, 1994.

Reeves, Gordon H., Jack E. Williams, Kelly M. Burnett, and Kirsten Gallo. "The Aquatic Conservation Strategy of the Northwest Forest Plan." *Conservation Biology* 20, no. 2 (2006): 319–29.

Reid, Leslie M., Robert R. Ziemer, and Michael J. Furniss. "Watershed Analysis on Federal Lands of the Pacific Northwest." Berkeley, CA: Pacific Southwest Research Station, 1996.

Reid, Leslie M., Robert R. Ziemer, and Thomas E. Lisle. "What a Long Strange Trip It's Been—or—Who Took the Synthesis out of Analysis?" *Watershed Management Council Network* 6, no. 4 (1996): 6–7.

Robertson, F. Dale. "A History of New Perspectives and Ecosystem Management." In *Ouachita and Ozark Mountains Symposium: Ecosystem Management Research*, edited by James M. Guldin, 3–7. Hot Springs, AR: USDA Forest Service, 1999.

———. Interview by Harold K. Steen, August 12–14, 1999. Durham, NC: Forest History Society.

Rothman, Hal. "'A Regular Ding-Dong Fight': Agency Culture and Evolution in the NPS-USFS Dispute, 1916–1937." *Western Historical Quarterly* 20, no. 2 (May 1989): 141–61.

Rowley, William D. "Historical Considerations in the Development of Range Science." In *Forest and Wildlife Science in America: A History*, edited by Harold K. Steen, 230–60. Durham, NC: Forest History Society, 1999.

Ruhl, J. B. "The Endangered Species Act and Private Property: A Matter of Timing and Location." *Cornell Journal of Law and Policy* 8 (1999): 37–53.

———. "Past, Present, and Future Trends of the Endangered Species Act." *Public Land and Resources Law Review* 25 (2004): 15–37.

———. "Who Needs Congress? An Agenda for Administrative Reform of the Endangered Species Act." *New York University Environmental Law Journal* 6 (1998): 367–410.

Runte, Alfred. *National Parks: The American Experience.* Lincoln: University of Nebraska Press, 1979.

———. *Trains of Discovery.* 4th ed. Boulder, CO: Roberts Rinehart Publishers, 1998.

Sabatier, Paul A., ed. *Theories of the Policy Process.* Boulder, CO: Westview Press, 1999.

"Safe Harbor Agreements and Candidate Conservation Agreements." *Federal Register* 62, no. 113 (June 12, 1997): 32189–94.

Salamon, Lester M., ed. *The Tools of Government: A Guide to the New Governance.* New York: Oxford University Press, 2002.

Salwasser, Hal. "Editorial." *Conservation Biology* 1, no. 4 (December 1987): 275–77.

———. Oral History by Rick Freeman. May 18, 1997. OH 370-01, Archives and Special Collections, Maureen and Mike Mansfield Library, University of Montana, Missoula.

———. "Spotted Owls: Turning a Battleground into a Blueprint." *Ecology* 68, no. 4 (August 1987): 776–79.

Salwasser, Hal, Douglas W. MacCleery, and Thomas A. Snellgrove. "An Ecosystem Perspective on Sustainable Forestry and New Directions for the US National Forest System." In *Defining Sustainable Forestry*, edited by Gregory H. Aplet, Nels Johnson, Jeffrey T. Olson, and V. Alaric Sample. Washington, DC: Island Press, 1993.

Salwasser, Hal, and Robert D. Pfister. "Ecosystem Management: From Theory to Practice." In *Sustainable Ecological Systems: Implementing an Ecological Approach to Land Management*, edited by W. Wallace Covington and Leonard F. DeBona, 150–61. Flagstaff, AZ: Forest Service Rocky Mountain Forest and Range Experiment Station, 1993.

Sampson, Arthur W. "Plant Succession as a Factor in Range Management." *Journal of Forestry* 4 (May 1917): 593–96.

———. "Plant Succession in Relation to Range Management." Washington, DC: US Department of Agriculture, 1919.

Sax, Joseph. "Ecosystems and Property Rights in Greater Yellowstone: The Legal System in Transition." In *The Greater Yellowstone Ecosystem: Redefining America's Wilderness Heritage*, edited by Robert B. Keiter and Mark S. Boyce, 77–84.New Haven, CT: Yale University Press, 1991.

———. *Mountains without Handrails: Reflections on the National Parks.* Ann Arbor: University of Michigan Press, 1980.

Schiappa, Edward. *Defining Reality: Definitions and the Politics of Meaning.* Carbondale: Southern Illinois University Press, 2003.

———. "Towards a Pragmatic Approach to Definition: 'Wetlands' and the Politics of Meaning." In *Environmental Pragmatism*, edited by Andrew Light and Eric Katz, 209–30. New York: Routledge, 1996.

Schneider, Daniel W. "Environmental History and Ecosystem Management." *Environmental History* 10, no. 1 (2005): 89–90.

Schullery, Paul. *Searching for Yellowstone: Ecology and Wonder in the Last Wilderness.* New York: Houghton Mifflin, 1997.

———, ed. *The Yellowstone Wolf: A Guide and Sourcebook.* Norman: University of Oklahoma Press, 2003.

Schulte, Steven C. *Wayne Aspinall and the Shaping of the American West.* Boulder: University Press of Colorado, 2002.

Sellars, Richard West. *Preserving Nature in the National Parks: A History.* New Haven, CT: Yale University Press, 1997.

Selznick, Philip. *TVA and the Grass Roots.* Berkeley: University of California Press, 1949.

Senate Committee on Energy and Natural Resources. *Federal Forest Management*, 1st Session, April 5, 1995.

Senate Committee on Energy and Natural Resources. *Grazing*, 103rd Cong., 2nd Session, April 20, May 14, 1994.

Senate Committee on Energy and Natural Resources and the House Committee on Resources. *Interior Columbia Basin Ecosystem Management Plan*, 1st Session, May 15, 1997.

Senate Subcommittee on the Department of the Interior and Related Agencies of the Committee on Appropriations and the Subcommittee on Forests and Public Lands Management of the Committee on Energy and Natural Resources. *Columbia Basin Ecosystem Management Plan*, 2nd Session, April 29; May 1, 4, and 28; October 15; December 7, 1998.

Shaffer, Mark L. "Minimum Population Sizes for Species Conservation." *BioScience* 31, no. 2 (February 1981): 131–34.

Shands, William E., Anne Black, and Jim Giltmier. "From New Perspectives to Ecosystem Management." *George Wright Forum* 11, no. 1 (1994): 35–47.

Shapiro, Martin. "Of Interests and Values: The New Politics and the New Political Science." In *The New Politics of Public Policy*, edited by Marc K. Landy and Martin A. Levin. Baltimore, MD: Johns Hopkins University Press, 1995.

Sher, Victor M. "Travels with Strix: The Spotted Owl's Journey through the Federal Courts." *Public Land & Resources Law Review* 14 (1993): 41–79.

Shindler, Bruce, Peter List, and Brent Steel. "Managing Federal Forests: Public Attitudes in Oregon and Nationwide." *Journal of Forestry* 91, no. 7 (July 1993): 36–42.

Shindler, Bruce, Brent Steel, and Peter List. "Public Judgments of Adaptive Management: A Response from Forest Communities." *Journal of Forestry* 94, no. 6 (June 1996): 4–12.

Shindler, Bruce A., Mark Brunson, and George H. Stankey. "Social Acceptability of Forest Conditions and Management Practices: A Problem Analysis." Portland, OR: USDA Forest Service, 2002.

Shipman, George A., and David F. Paulsen. "The Bureau of Land Management: United States Department of the Interior: The Organization of the Washington Office." Seattle: University of Washington, DC: Institute for Administrative Research, 1963.

———. "The Bureau of Land Management: United States Department of the Interior: The Program Action System." Seattle: University of Washington, DC: Institute for Administrative Research, 1964.

Simmons, Randy T. "Fixing the Endangered Species Act." In *Breaking the Environmental Policy Gridlock*, edited by Terry L. Anderson, 82–114. Stanford, CA: Hoover Institution Press, 1997.

Simon, Herbert A. *Administrative Behavior: A Study of Decision-Making Processes in Administrative Organization.* New York: Macmillan, 1947.

———. "The Architecture of Complexity." *Proceedings of the American Philosophical Society* 106, no. 6 (December 1962): 467–82.

———. "The Organization of Complex Systems." In *Hierarchy Theory: The Challenge of Complex Systems*, edited by Howard H. Pattee, 3–27. New York: George Braziller, 1973.

———. "Proverbs of Administration." *Public Administration Review* 6, no. 1 (Winter 1946): 53–67.

Skillen, James R. *The Nation's Largest Landlord: The Bureau of Land Management in the American West.* Lawrence: University Press of Kansas, 2009.

Sonner, Scott. "Scientists Favored Owl Plan Bush Administration Resists." *Seattle Times*, August 29, 1990.

Soulé, Michael E. "Conservation Biology and the 'Real World.'" In *Conservation*

Biology: The Science of Scarcity and Diversity, edited by Michael E. Soulé. Sunderland, MA: Sinauer Associates, 1986.

Soulé, Michael E., and Bruce A. Wilcox. "Conservation Biology: Its Scope and Its Challenge." In *Conservation Biology: An Evolutionary-Ecological Perspective*, edited by Michael E. Soulé and Bruce A. Wilcox, 1–12. Sunderland, MA: Sinauer Associates, Inc., 1980.

Stankey, George H., Roger N. Clark, and Bernard T. Bormann. "Adaptive Management of Natural Resources: Theory, Concepts, and Management Institutions." Portland, OR: USDA Forest Service, Pacific Northwest Research Station, 2005.

Stankey, George H., David N. Cole, Robert C. Lucas, Margaret E. Peterson, and Sidney Frissell. "The Limits of Acceptable Change (LAC) System for Wilderness Planning." Ogden, UT: Forest Service Intermountain Forest and Range Experiment Station, 1985.

Stankey, George H., and Bruce Shindler. "Adaptive Management Areas: Achieving the Promise, Avoiding the Peril." Pacific Northwest Research Station: US Forest Service, 1997.

Stankey, G. H., B. T. Bormann, C. Ryan, B. Shindler, V. Sturtevant, R. N. Clark, and C. Philpot. "Adaptive Management and the Northwest Forest Plan: Rhetoric and Reality." *Journal of Forestry* 101 (2003): 40–46.

"Status of the Interior Columbia Basin: Summary of Scientific Findings." Portland, OR: USDA Forest Service, 1996.

Steen, Harold K. *The US Forest Service: A History*. Seattle: University of Washington Press, 1976.

Stelle, Will. "Executive Branch Policies and Activities in Ecosystem Management." In *Ecosystem Management: Status and Potential*, 269–72. Washington, DC: GPO, 1994.

Subcommittee on Public Lands, National Parks and Forests. *Northern Rocky Mountain Gray Wolf Recovery Act of 1990*, 2nd Session, September 19, 1990.

Subcommittee on Public Lands and the Subcommittee on National Parks and Recreation. *Greater Yellowstone Ecosystem*, 99th Congress, 1st Session, October 24, 1985.

Swanson, F. J., and Jerry F. Franklin. "New Forestry Principles from Ecosystem Analysis of Pacific Northwest Forests." *Ecological Applications* 2, no. 3 (August 1992): 262–74.

Swedlow, Brendon. "Scientists, Judges, and Spotted Owls: Policymakers in the Pacific Northwest." *Duke Environmental Law and Policy Forum* 13 (Spring 2003): 187–278.

Tansley, A. G. "The Use and Abuse of Vegetational Concepts and Terms." *Ecology* 16, no. 3 (July 1935): 284–307.

Taylor, Andrew. "Biological Survey Bill Halted after Heated House Debate." *CQ Weekly*, October 9, 1993, 1.

Taylor, Frederick. *Principles of Scientific Management.* New York: Harper and Sons Publishers, 1911.

Thomas, Craig W. *Bureaucratic Landscapes: Interagency Cooperation and the Preservation of Biodiversity*. Cambridge, MA: The MIT Press, 2003.

Thomas, Jack Ward. Interview by Harold K. Steen, May 10–12, 2001. Durham, NC: Forest History Society.

———. *The Journals of a Forest Service Chief*, edited by Harold K. Steen. Seattle, WA: Forest History Society and University of Washington Press, 2004.

———. "Sustainability of the Northwest Forest Plan: Still to Be Tested." In *Forest Futures: Science, Politics, and Policy for the Next Century*, edited by Karen Arabas and Joe Bowersox, 3–22. Boulder, CO: Rowman & Littlefield, 2004.

Thomas, Jack Ward, and Jim Baca. "Interior Columbia Basin Ecosystem Management Project Charter." Portland, OR: US Forest Service and Bureau of Land Management, 1994.

Thomas, Jack Ward, and Michael P. Dombeck. "Decision Notice/Decision Record for the Interim Strategies for Managing Anadromous Fish-Producing Watersheds in Eastern Oregon and Washington, Idaho, and Portions of California." Washington, DC: Bureau of Land Management and US Forest Service, 1995.

Thomas, Jack Ward, Eric D. Forsman, Joseph B. Linit, E. Charles Meslow, Barry B. Noon, and Jared Verner. "A Conservation Strategy for the Northern Spotted Owl: Report of the Interagency Scientific Committee to Address the Conservation of the Northern Spotted Owl." Portland, OR: USDA Forest Service, USDI Bureau of Land Management, USDI Fish and Wildlife Service, USDI National Park Service, 1990.

Thomas, Jack Ward, Jerry F. Franklin, John Gordon, and K. Norman Johnson. "The Northwest Forest Plan: Origins, Components, Implementation Experience, and Suggestions for Change." *Conservation Biology* 20, no. 2 (April 2006): 277–87.

Thomas, Jack Ward, and Hal Salwasser. "Bringing Conservation Biology into a Position of Influence in Natural Resource Management." *Conservation Biology* 3, no. 2 (June 1989): 123–27.

Thompson, James R. "Reinvention as Reform: Assessing the National Performance Review." *Public Administration Review* 60, no. 6 (November-December 2000): 508–21.

Thornton, Robert D. "Habitat Conservation Plans: Frayed Safety Nets or Creative Partnerships?" *Natural Resources and Environment* 16, no. 2 (Fall 2001): 94–101.

———. "The Search for a Conservation Planning Paradigm: Section 10 of the ESA." *Natural Resources and Environment* 8, no. 1 (Summer 1993): 21–23, 65–67.

Tipple, Terence J., and J. Douglas Wellman. "Herbert Kaufman's Forest Ranger Thirty Years Later: From Simplicity and Homogeneity to Complexity and Diversity." *Public Administration Review* 51, no. 5 (September-October 1991): 421–28.

Tuchmann, E. Thomas, Kent P. Connaughton, Lisa E. Freedman, and Clarence B. Moriwaki. "The Northwest Forest Plan: A Report to the President and Congress." Washington, DC: USDA Office of Forestry and Economic Assistance, 1996.

"Upper Columbia River Basin Draft Environmental Impact Statement." Portland, OR: Bureau of Land Management and US Forest Service, 1997.

Urban, Dean L. "Landscape Ecology and Ecosystem Management." In *Sustainable Ecological Systems: Implementing an Ecological Approach to Land Management*, edited by W. Wallace Covington and Leonard F. Debano, 127–36. Flagstaff, AZ: USDA Forest Service Rocky Mountain Forest and Range Experiment Station, 1993.

US Fish and Wildlife Service. "Service Declares Wyoming Gray Wolf Recovered under the Endangered Species Act and Returns Management Authority to the State." Press release, August 31, 2012.

US Fish and Wildlife Service. "Wolves for Yellowstone? A Report to the United States Congress, Volume I and II." Billings, MT: US Fish and Wildlife Service, 1990.

Vaughn, Jacqueline, and Hanna J. Cortner. *George W. Bush's Healthy Forests: Reframing the Environmental Debate.* Boulder: University Press of Colorado, 2005.

Vig, Norman J. "Presidential Leadership and the Environment: From Reagan to Clinton." In *Environmental Policy in the 1990s: Reform or Reaction?*, edited by Norman J. Vig and Michael E. Kraft, 95–118. Washington, DC: CQ Press, 1997.

Vig, Norman J., and Michael E. Kraft, eds. *Environmental Policy: New Directions for the Twenty-First Century.* 6th ed. Washington, DC: CQ Press, 2006.

Walker, W. R., and S. C. Richardson. "The Federal Wetlands Manual: Swamped by Controversy." Blacksburg: Virginia Water Resources Research Center, 1991.

Walters, C., and C. S. Holling. "Large-Scale Management Experiments and Learning by Doing." *Ecology* 71, no. 6 (December 1990): 2060–68.

Walters, Carl. *Adaptive Management of Renewable Resources.* New York: McGraw Hill, 1986.

Walters, Carl J., and Ray Hilborn. "Ecological Optimization and Adaptive Management." *Annual Review of Ecology and Systematics* 9 (1978): 157–88.

Wellock, Thomas R. "The Dickey Bird Scientists Take Charge: Science, Policy, and the Spotted Owl." *Environmental History* 15 (July 2010): 381–414.

Westoby, Mark, Brian Walker, and Imanuel Noy-Meir. "Opportunistic Management for Rangelands Not at Equilibrium." *Journal of Range Management* 42, no. 4 (July 1989): 266–74.

White, Leonard D. *The Republican Era: 1869–1901: A Study in Administrative History.* New York: MacMillan, 1958.

White, P. J., Rick L. Wallen, Chris Geremia, John J. Treanor, and Douglas W. Blanton. "Management of Yellowstone Bison and Brucellosis Transmission Risk: Implications for Conservation and Restoration." *Biological Conservation* 144, no. 5 (May 2011): 1322–34.

Whittaker, Robert H. "A Criticism of the Plant Association and Climatic Climax Concept." *Northwest Science* 25, no. 1 (1951): 17–31.

Williams, Gerald W. "FEMAT to SEIS: Transforming the Social Assessment." Paper presented at Integrating Social Science in Ecosystem Management: A National Challenge, Helen, GA, December 12–14, 1994.

———. "New Perspectives, New Forestry, or Biological Diversity: How Did We

Get Here?" Paper presented at the Biological Diversity/Silviculture Workshop, Gleneden Beach, OR, May 6, 1991.

Wilson, Edward O., and Frances M. Peter, eds. *Biodiversity*. Washington, DC: National Academies Press, 1988.

Wilson, Matthew A. "The Wolf in Yellowstone: Science, Symbol, or Politics?" *Society and Natural Resources* 10, no. 5 (2008): 453–68.

Wilson, Woodrow. "The Study of Administration." *Political Science Quarterly* 2, no. 2 (June 1887): 197–222.

Winter, Charles E. *Four Hundred Million Acres: The Public Lands and Resources*. New York: Arno Press, 1979.

Worster, Donald. *Nature's Economy: A History of Ecological Ideas*. 2nd ed. New York: Cambridge University Press, 1994.

Yaffee, Steven L., Ali F. Phillips, Irene C. Frentz, Paul W. Hardy, Sussanne M. Maleki, and Barara E. Thorpe, eds. *Ecosystem Management in the United States: An Assessment of Current Experience*. Washington, DC: Island Press, 1996.

Yaffee, Steven Lewis. *The Wisdom of the Spotted Owl: Policy Lessons for a New Century*. Washington, DC: Island Press, 1994.

Yandle, Bruce, ed. *Land Rights: The 1990s Property Rights Rebellion*. Lanham, MD: Rowman and Littlefield, 1995.

"A Year of Change: To Restore America's Greatness: Department of the Interior, January 20, 1981–January 20, 1982." Washington, DC: Department of the Interior, 1982.

Zimmerer, Karl S. "Human Geography and the 'New Ecology': Prospect and Promise of Integration." *Annals of the Association of American Geographers* 84, no. 1 (March 1994): 108–25.

Zinn, Jeffrey, and M. Lynne Corn. "Executive Summary." In *Ecosystem Management: Status and Potential*, 1–14. Washington, DC: GPO, 1994.

Index